Student Solutions Manual for

Chemistry

PRINCIPLES & PRACTICE

by

Reger • Goode • Mercer

John M. DeKorte

Northern Arizona University

SAUNDERS COLLEGE PUBLISHING

Harcourt Brace Jovanovich College Publishers

Fort Worth • Philadelphia • Boston • New York • Chicago • Orlando
San Francisco • Atlanta • Dallas • London • Toronto • Austin • San Antonio

Printed in the United States of America.

DeKorte: Student Solutions Manual for CHEMISTRY: PRINCIPLES AND PRACTICE
by Reger, Goode, and Mercer

ISBN 0-03-096755-4

345 021 98765432

TO THE STUDENT

Chemistry is a subject that is not easy to learn through casual reading. Rather, you will find that chemistry is best learned by learning the thought processes behind the various lines of reasoning. This can only occur when you begin to interact with the material to the degree that is necessary to clarify your understanding. In addition, you will find that chemistry is a science with many principles that lend themselves to quantitative expression and the only way to master the intricacies of logical approaches to quantitative problem solving is to practice solving many such problems.

The end-of-chapter exercises in Chemistry: Principles and Practice by Reger, Goode and Mercer have been carefully chosen to test your understanding of chemistry and will require you to continually apply your knowledge to new situations. This Student Solutions Manual contains detailed explanations and solutions for all of the even-numbered exercises having answers in the appendix (all of the colored coded even-numbered exercises). However, it should only be used to assist you after you have first attempted to answer the exercises on your own or to confirm your answers. This will give you the greatest opportunity to refine your approaches to problem solving and develop confidence in your ability to answer the exercises successfully.

The exercises have been solved using the same methods that are used in the text. In many cases, you will find that helpful comments have been added to enhance your ability to understand the material. In addition, you will find references to pertinent Sections and Examples in the text to go to find help while working the exercises. We have included these references to point out the usefulness of the text in terms of working the exercises and to encourage you to use the text to work the exercises before consulting this manual.

The word processing for this manual was a task which our daughter Carrie graciously accepted. Her interpretation of the handwritten solutions and presentation of the material in excellent format is very much appreciated.

We think you will find this manual to be very helpful in your study of general chemistry and wish you the best as you discover the power of chemistry to interpret the world in which we live. We have tried to keep this manual as error-free as possible but probably have not succeeded completely. Please let us know about any errors or problems.

We thank the authors, Reger, Goode, and Mercer, for their appreciation and cooperation. A special word of thanks goes to my wife Carolyn for her support and understanding during the preparation of this manual.

John M. DeKorte
Chemistry Department
Northern Arizona University
Flagstaff, AZ 86011

Contents

Chapter 1: Introduction to Chemistry

1.8 ***See Section 1.2.***

(a) intensive, physical (b) intensive, physical (c) intensive, chemical (d) extensive, physical

(e) intensive, physical

1.10 ***See Section 1.2.***

Physical: b, c, e, f, h Chemical: a, d, g

1.12 ***See Section 1.2.***

Chemical: corrosive, reacts with practically all substances, burns some substances giving a bright flame, prevents cavities.

Physical: pale yellow color, melting point of -219.6°C, boiling point of -188.1°C, commonly exists in gaseous physical state.

1.16 ***See Section 1.2.***

(a) homogenous mixture (b) heterogeneous mixture (c) element (d) heterogeneous mixture

(e) homogenous mixture

1.18 ***See Table 1.3 and the list of elements inside the front cover of the text.***

(a) silver, Ag (b) carbon, C (c) sulfur, S (d) bromine, Br (e) helium, He

1.20 ***See Section 1.3.***

group: vertical column in periodic table.
period: horizontal row in periodic table.

Number the columns 1 through 18, starting with Group IA and moving left-to-right across the table to obtain the appropriate Arabic numerals. This gives: Roman = Arabic when Arabic is 7 or less, Roman = VIII when Arabic is 8, 9 or 10 and Roman = (Arabic - 10) when Arabic is 11 or greater.

1.22 ***See Section 1.3.***

(a) Zr, Hf (b) S, Se, Te, or Po (c) Cl, Br, I, or At (d) Be, Mg, Ca or Sr

(e) He, Ne, Kr, Xe, or Rn (f) P, As, Sb, or Bi

1.24 ***See Section 1.3.***

(a) S, representative (b) Fe, transition (c) K, representative (d) Eu, inner transition

(e) Br, representative

1.26 *See Section 1.3.*

(a) Na, sodium (b) Cl, chlorine (c) Ra, radium (d) Ne, neon

1.28 *See Section 1.3.*

Sr and Ra should have similar properties, since both are members of Group IIA.

1.32 *See Table 1.5.*

(a) mass, kilogram (b) distance, meter (c) temperature, kelvin (d) time, second

1.34 *See Sections 1.5 and 6, including table 1.7.*

$$? \text{ distance of 1500-meter race in mi} = 1,500 \text{ m} \times \frac{1 \text{ km}}{10^3 \text{ m}} \times \frac{1 \text{ mi}}{1.609 \text{ km}} = \mathbf{0.9323 \text{ mi}}$$

1.36 *See Sections 1.5 and 6, including Table 1.7.*

$$? \text{ volume of wine (in qt)} = 12 \text{ bottles} \times \frac{750 \text{ mL}}{1 \text{ bottle}} \times \frac{1 \text{ L}}{10^3 \text{ mL}} \times \frac{1.057 \text{ qt}}{1 \text{ L}} = \mathbf{9.51 \text{ qt}}$$

Note: 12 bottles is taken to be an exact quantity.

1.38 *See Section 1.5.*

(a) $T_F = T_C\left[\frac{1.8^\circ \text{ F}}{1.0^\circ \text{ C}}\right] + 32^\circ \text{ F}$ yields $\mathbf{T_C = \left[\frac{1.0^\circ \text{ C}}{1.8^\circ \text{ F}}\right](T_F - 32^\circ \text{ F})}$

(b) $T_K = T_C + 273.15$ and $T_F = T_C\left[\frac{1.8^\circ \text{ F}}{1.0^\circ \text{ C}}\right] + 32^\circ \text{ F}$ yield

$$T_F = (T_K - 273.15)\left[\frac{1.8^\circ \text{F}}{1.0^\circ \text{C}}\right] + 32^\circ \text{F} = \mathbf{T_K\left[\frac{1.8^\circ \text{ F}}{1.0^\circ \text{ C}}\right] - 459.67^\circ \text{ F}}$$

(c) $T_K = T_C + 273.15$ and $T_C = \left[\frac{1.0^\circ \text{C}}{1.8^\circ \text{F}}\right](T_F - 32^\circ)$ yield $\mathbf{T_K = \left[\frac{1.0^\circ \text{C}}{1.8^\circ \text{F}}\right](T_F - 32^\circ \text{ F}) + 273.15}$

1.40 *See Section 1.5.*

$T_K = T_C + 273.15$ yields $T_C = T_K - 273.15$, so $T_C = 4.21 - 273.15 = \mathbf{-268.94^\circ C}$

$T_F = (T_K - 273.15)\left[\frac{1.8^\circ \text{F}}{1.0^\circ \text{C}}\right] + 32^\circ \text{F}$ yields $T_F = (4.21 - 273.15)\left[\frac{1.8^\circ \text{F}}{1.0^\circ \text{C}}\right] + 32^\circ \text{F} = \mathbf{-452.09^\circ F}$

1.42 *See Sections 1.2 and 5.*

Volume of metal = volume of water displaced = 48.5 mL - 30.0 mL = 18.5 mL

$$\text{Density of metal} = \frac{\text{mass}}{\text{volume}} = \frac{47.8\ \text{g}}{18.5\ \text{mL}} = \mathbf{2.58\ g/mL}\ \textbf{or}\ \mathbf{2.58\ g/cm^3}$$

1.44 *See Section 1.5.*

$$?\ \text{volume of benzene (in L)} = 2.50\ \text{kg} \times \frac{10^3\ \text{g}}{\text{kg}} \times \frac{1\ \text{cm}^3}{0.879\ \text{g}} \times \frac{1\ \text{L}}{10^3\ \text{cm}^3} = \mathbf{2.84\ L}$$

1.46 *See Sections 1.2 and 5.*

$$\text{Volume of lead brick (in cm}^3) = (8.50\ \text{in} \times 5.10\ \text{in} \times 3.20\ \text{in}) \times \left(\frac{2.54\ \text{cm}}{\text{in}}\right)^3 = 2.27 \times 10^3\ \text{cm}^3$$

$$?\ \text{mass of lead brick (in g)} = 2.27 \times 10^3\ \text{cm}^3 \times \frac{11.4\ \text{g}}{\text{cm}^3} = \mathbf{2.59 \times 10^4\ g}$$

1.50 *See Section 1.6.*

(a) 5 (b) 2; zeros preceding 7 indicate decimal place (c) 2 (d) 3 (e) 6

(f) 5; zeros preceding 1 indicate decimal place

1.52 *See Section 1.6.*

(a) $17.2 \times 12.55 = 216$ — Three significant digits in answer, since there are just three significant digits in the quantity 17.2.

(b) $1.4 \times \frac{1.11}{42.33} = 3.7 \times 10^{-2}$ — Two significant digits in answer, since there are just two significant digits in the quantity 1.4.

(c) $18.33 \times 0.0122 = 0.224$ — Three significant digits in answer, since there are just three significant digits in the quantity 0.0122.

(d) $25.7 - 25.25 = 0.4$ — One decimal place in answer, since there is just one decimal place in the quantity 25.7.

(e) $19.5 + 2.35 + 0.037 = 21.9$ — One decimal place in answer, since there is just one decimal place in the quantity 19.5.

(f) $2.00 \times 10^3 - 1.7 \times 10^1 = 1.98 \times 10^3$ — Two decimal places in answer, since there is just two decimal places in the quantity 2.00×10^3.

1.54 *See Section 1.6.*

(a) $\frac{(25.12 - 1.75) \times 0.01920}{(24.339 - 23.15)} = 0.377$ — Subtraction of quantities in denominator gives 1.19 to two decimal places, and 1.19 leads to just three significant digits for answer to multiplication and division.

(b) $\dfrac{55.4}{(26.3-18.904)}=7.5$

Subtraction of quantities in denominator gives 7.4 to one decimal place, and 7.4 leads to just two significant digits for answer to division.

(c) $(0.9221\times 27.977)+(0.0470\times 28.976)+(0.309\times 29.974)=36.42$

The answer for the last two multiplications are restricted to three significant digits and thus two decimal places, since there are just three significant digits in the quantities 0.0470 and 0.309. This limits the answer for the addition to just two decimal places.

1.58 ***See Section 1.5 and 1.6.***

$$\text{? speed of sound (in mi/hr)} = \frac{340\text{ m}}{\text{s}}\times\frac{1\text{ km}}{10^3\text{ m}}\times\frac{1\text{ mi}}{1.609\text{ km}}\times\frac{60\text{ s}}{\text{min}}\times\frac{60\text{ min}}{\text{hr}} = \mathbf{761\ mi/hr}$$

1.60 ***See Sections 1.5 and 1.6.***

$$\text{? distance in one light year (in mi)} = 1\text{ year}\times\frac{365.25\text{ days}}{\text{yr}}\times\frac{24\text{ hr}}{\text{day}}\times\frac{60\text{ min}}{\text{hr}}\times\frac{60\text{ s}}{\text{min}}\times\frac{3.00\text{ x }10^8\text{ m}}{\text{s}}$$

$$\times\frac{1\text{ km}}{10^3\text{ m}}\times\frac{1\text{ mi}}{1.609\text{ km}} = \mathbf{5.88\ x\ 10^{12}\ mi}$$

1.62 ***See Sections 1.5 and 1.6.***

Since area = volume/thickness,

$$\text{? area (in m}^2) = \frac{1.00\text{ gal}}{8.00\text{ x }10^{-2}\text{ mm}}\times\frac{4\text{ qt}}{\text{gal}}\times\frac{1\text{ L}}{1.057\text{ qt}}\times\frac{10^3\text{ mL}}{1\text{ L}}\times\frac{1\text{ cm}^3}{1\text{ mL}}\times\left(\frac{1\text{ m}}{10^2\text{ cm}}\right)^3\times\frac{10^3\text{ mm}}{1\text{ m}} = \mathbf{47.3\ m^2}$$

1.64 ***See Sections 1.5 and 1.6.***

Since grams can be converted to volume using density and area = volume/thickness,

$$\text{? volume (in cm}^3) = 28.5\text{ g}\times\frac{1\text{ cm}^3}{19.3\text{ g}} = 1.48\text{ cm}^3$$

and

$$\text{? area (in ft}^2) = \frac{1.48\text{ cm}^3}{1.27\text{ x }10^{-5}\text{ cm}}\times\left(\frac{1\text{ in}}{2.54\text{ cm}}\right)^2\times\left(\frac{1\text{ ft}}{12\text{ in}}\right)^2 = \mathbf{125\ ft^2}$$

1.66 ***See Sections 1.2 and 1.5.***

Volume of metal = volume of water displaced = $19.40\text{ mL}-12.35\text{ mL}=7.05\text{ mL}$

$$\text{Density of metal} = \frac{\text{mass}}{\text{volume}} = \frac{134.412\text{ g}}{7.05\text{ mL}} = \mathbf{19.1\ g/mL\ \ or\ \ 19.1 g/cm^3}$$

Chapter 2: Atoms, Molecules, and Ions

2.2 *See Section 2.1.*

The ratio of the mass of S in combination with the fixed mass of 1.0 g O is $\frac{1.0}{0.67}$ or $\frac{3.0}{2.0}$ in accord with the Law of Multiple Proportions.

2.12 *See Section 2.2 and Example 2.1.*

(a) $^{15}_{7}N$ (b) $^{70}_{31}Ga$ (c) $^{40}_{18}Ar$

2.14 *See Section 2.2 and Example 2.3.*

(a) $^{79}_{33}As$: 33 protons, 46 neutrons, 33 electrons (b) $^{51}_{23}V$: 23 protons, 28 neutrons, 23 electrons

(c) $^{128}_{52}Te$: 52 protons, 76 neutrons, 52 electrons

2.16 *See Section 2.2 and Example 2.3.*

(a) $^{16}_{8}O^{2-}$ (b) $^{79}_{34}Se^{2-}$ (c) $^{59}_{28}Ni^{2+}$

2.18 *See Section 2.2 and Example 2.4.*

symbol	$^{23}Na^{+}$	$^{40}Ca^{2+}$	$^{81}Br^{-}$	$^{128}Te^{2-}$
atomic number	11	20	35	52
mass number	23	40	81	128
charge	1+	2+	1-	2-
number protons	11	20	35	52
number electrons	10	18	36	54
number neutrons	12	20	46	76

2.24 *See Section 2.3 and Table 2.1.*

(a) I^- (b) Mg^{2+} (c) O^{2-} (d) Na^+

2.26 *See Section 2.3 and Example 2.5.*

(a) CaS (b) Mg_3N_2 (c) FeF_2

2.28 *See Section 2.3, Table 2.1 and Example 2.5.*

(a) $CaCl_2$ (b) Rb_2S (c) Li_3N (d) Y_2Se_3

2.30 *See Table 2.2.*

(a) hydroxide ion, OH^- (b) chlorate ion, ClO_3^- (c) permanganate ion, MnO_4^-

2.32 *See Section 2.3, Tables 2.1 and 2.2 and Example 2.6.*

(a) magnesium nitrite, $Mg(NO_2)_2$ (b) lithium phosphate, Li_3PO_4 (c) barium cyanide, $Ba(CN)_2$

2.34 *See Section 2.4 and Tables 2.2 and 2.4.*

(a) LiI, lithium iodide (b) Mg_3N_2, magnesium nitride (c) Na_3PO_4, sodium phosphate

(d) $Ba(ClO_4)_2$, barium perchlorate

2.36 *See Section 2.4, Tables 2.2, 3 and 4 and Example 2.9.*

(a) $CoCl_3$, cobalt(III) chloride (b) $FeSO_4$, iron(II) sulfate (c) CuO, copper(II) oxide

2.38 *See Sections 2.3 and 4, Tables 2.1, 2 , 3 and 4 and Examples 2.5, 6, 7 and 8.*

(a) maganese(III) sulfide, Mn_2S_3 (b) iron(II) cyanide, $Fe(CN)_2$ (c) potassium sulfide, K_2S

(d) mercury(II) chloride, $HgCl_2$

2.40 *See Section 2.4, Table 2.5 and Example 2.10.*

(a) sulfur tetrafluoride, SF_4 (b) nitrogen trichloride, NCl_3 (c) dinitrogen pentoxide, N_2O_5

(d) chlorine trifluoride, ClF_3

2.42 *See Section 2.4, Table 2.5 and Example 2.11.*

(a) PBr_5, phosphorous pentabromide (b) SeO_2, selenium dioxide (c) B_2Cl_4, diboron tetrachloride

(d) S_2Cl_2, disulfur dichloride

2.44 *See Sections 2.3 and 4, Tables 2.1 and 3 and Example 2.5.*

potassium chloride, KCl

2.46 *See Section 2.4, Table 2.2 and Example 2.7.*

$Mg(OH)_2$, magnesium hydroxide

2.48 *See Section 2.2 and Example 2.3.*

(a) ${}^{35}_{17}Cl^-$ (b) ${}^{39}_{19}K^+$

2.50	*See Sections 2.3 and 4 and Tables 2.1 through 5.*

(a) NO, nitrogen monoxide, molecular

(b) $Y_2(SO_4)_3$, yttrium sulfate, ionic

(c) Na_2O, sodium oxide, ionic

(d) NBr_3, nitrogen tribromide, molecular

Note: Ionic compounds generally contain combinations of metals with nonmetals whereas molecular compounds generally contain just nonmetals.

2.52	*See Section 2.2 and Example 2.4.*

symbol	$^{70}Ga^{3+}$	$^{103}Rh^{3+}$	$^{114}In^{+}$	$^{28}Si^{2-}$
atomic number	31	45	49	14
mass number	70	103	114	28
charge	3+	3+	1+	2-
number protons	31	45	49	14
number electrons	28	42	48	16
number neutrons	39	58	65	14

2.54	*See Sections 1.3 and 2.3 and 4.*

(a) calcium chloride, $CaCl_2$

(b) carbon dioxide, CO_2

(c) iron(III) oxide, Fe_2O_3

2.56	*See Section 2.2 and Example 3.5.*

(a) $^{23}_{11}Na^{+}$

(b) $^{121}_{51}Sb^{3+}$

(c) $^{84}_{36}Kr$

Chapter 3: Stoichiometry I

3.2 ***See Section 3.1 and Examples 3.1 and 2.***

(a)	Unbalanced:	$C_5H_{12} + O_2$	→	$CO_2 + H_2O$	Start with C_5H_{12}.
	Step 1:	$C_5H_{12} + O_2$	→	$\underline{5}CO_2 + \underline{6}H_2O$	Balances C, H.
	Step 2 :	$C_5H_{12} + \underline{8}O_2$	→	$5CO_2 + 6H_2O$	Balances O.
(b)	Unbalanced:	$NH_3 + O_2$	→	$N_2 + H_2O$	Start with NH_3.
	Step 1:	$\underline{2}NH_3 + O_2$	→	$N_2 + H_2O$	Balances N.
	Step 2:	$2NH_3 + O_2$	→	$N_2 + \underline{3}H_2O$	Balances H.
	Step 3:	$2NH_3 + \underline{3/2}O_2$	→	$N_2 + 3H_2O$	Balances O.
	Step 4:	$\underline{4}NH_3 + \underline{3}O_2$	→	$\underline{2}N_2 + \underline{6}H_2O$	Gives whole number coefficients.
(c)	Unbalanced:	$KOH + H_2SO_4$	→	$K_2SO_4 + H_2O$	Start with K_2SO_4.
	Step 1:	$\underline{2}KOH + H_2SO_4$	→	$K_2SO_4 + H_2O$	Balances K.
	Step 2:	$2KOH + H_2SO_4$	→	$K_2SO_4 + \underline{2}H_2O$	Balances O, H.

Note: S balance in starting equation is maintained through Steps 1 and 2.

3.4 ***See Section 3.1 and Examples 3.1 and 2.***

(a)	Unbalanced:	$N_2H_4 + N_2O_4$	→	$N_2 + H_2O$	Start with N_2O_4.
	Step 1:	$N_2H_4 + N_2O_4$	→	$N_2 + \underline{4}H_2O$	Balances O.
	Step 2:	$\underline{2}N_2H_4 + N_2O_4$	→	$N_2 + 4H_2O$	Balances H.
	Step 3:	$2N_2H_4 + N_2O_4$	→	$\underline{3}N_2 + 4H_2O$	Balances N.
(b)	Unbalanced:	$F_2 + H_2O$	→	$HF + O_2$	Start with H_2O.
	Step 1:	$F_2 + \underline{2}H_2O$	→	$HF + O_2$	Balances O.
	Step 2:	$F_2 + 2H_2O$	→	$\underline{4}HF + O_2$	Balances H.
	Step 3:	$\underline{2}F_2 + 2H_2O$	→	$4HF + O_2$	Balances F.
(c)	Unbalanced:	$Na_2O + H_2O$	→	$NaOH$	Start with Na_2O.
	Step 1:	$Na_2O + H_2O$	→	$\underline{2}NaOH$	Balances Na, O, H.

3.6 ***See Section 3.1 and Example 3.3.***

(a)	Unbalanced:	$C_6H_{12} + O_2$	→	$CO_2 + H_2O$	Start with C_6H_{12}.
	Step 1:	$C_6H_{12} + O_2$	→	$\underline{6}CO_2 + \underline{6}H_2O$	Balances C, H.
	Step 2:	$C_6H_{12} + \underline{9}O_2$	→	$6CO_2 + 6H_2O$	Balances O.
(b)	Unbalanced:	$C_4H_8 + O_2$	→	$CO_2 + H_2O$	Start with C_4H_8.
	Step 1:	$C_4H_8 + O_2$	→	$\underline{4}CO_2 + \underline{4}H_2O$	Balances C, H.
	Step 2:	$C_4H_8 + \underline{6}O_2$	→	$4CO_2 + 4H_2O$	Balances O.
(c)	Unbalanced:	$C_3H_6O + O_2$	→	$CO_2 + H_2O$	Start with C_3H_6O.
	Step 1:	$C_3H_6O + O_2$	→	$\underline{3}CO_2 + \underline{3}H_2O$	Balances C, H.
	Step 2:	$C_3H_6O + \underline{4}O_2$	→	$3CO_2 + 3H_2O$	Balances O.

(d)

Unbalanced:	$C_4H_6O_2 + O_2 \rightarrow CO_2 + H_2O$	Start with $C_4H_6O_2$.
Step 1:	$C_4H_6O_2 + O_2 \rightarrow \underline{4}CO_2 + \underline{3}H_2O$	Balances C, H.
Step 2:	$C_4H_6O_2 + \underline{9/2}O_2 \rightarrow 4CO_2 + 3H_2O$	Balances O.
Step 3:	$\underline{2}C_4H_6O_2 + \underline{9}O_2 \rightarrow \underline{8}CO_2 + \underline{6}H_2O$	Gives whole number coefficients.

3.8 ***See Section 3.1 and Example 3.4.***

(a)

Unbalanced:	$H_2SO_4 + KOH \rightarrow H_2O + K_2SO_4$	Start with K_2SO_4.
Step 1:	$H_2SO_4 + \underline{2}KOH \rightarrow H_2O + K_2SO_4$	Balances K.
Step 2:	$H_2SO_4 + 2KOH \rightarrow \underline{2}H_2O + K_2SO_4$	Balances O, H.

(b)

Unbalanced:	$HCl + Ca(OH)_2 \rightarrow H_2O + CaCl_2$	Start with $CaCl_2$.
Step 1:	$\underline{2}HCl + Ca(OH)_2 \rightarrow H_2O + CaCl_2$	Balances Cl.
Step 2:	$2HCl + Ca(OH)_2 \rightarrow \underline{2}H_2O + CaCl_2$	Balances O, H.

(c) Balanced: $HNO_3 + LiOH \rightarrow H_2O + LiNO_3$

3.10 ***See Section 3.1 and Examples 3.1 and 2.***

Unbalanced:	$N_2H_4 + O_2 \rightarrow NO_2 + H_2O$	Start with N_2H_4.
Step 1:	$N_2H_4 + O_2 \rightarrow \underline{2}NO_2 + \underline{2}H_2O$	Balances N, H.
Step 2:	$N_2H_4 + \underline{3}O_2 \rightarrow 2NO_2 + 2H_2O$	Balances O.

3.12 ***See Section 3.1 and Examples 3.1 and 2.***

Unbalanced:	$S_8 + Cl_2 \rightarrow S_2Cl_2$	Start with S_2Cl_2.
Step 1:	$S_8 + Cl_2 \rightarrow \underline{4}S_2Cl_2$	Balances S.
Step 2:	$S_8 + \underline{4}Cl_2 \rightarrow 4S_2Cl_2$	Balances Cl.

3.14 ***See Section 3.2 and Example 3.5.***

Multiply the mass of each isotope by its decimal fraction, and add these numbers together to obtain the atomic weight of the element.

Atomic weight $= (0.604)(68.93\mu) + (0.396)(70.925\mu) = \mathbf{69.7\mu}$

The element is gallium, **Ga**.

3.16 ***See Section 3.2 and Example 3.6.***

(a) $? \text{ mol S} = 9.40 \text{ g S} \times \dfrac{1 \text{ mol S}}{32.1 \text{ g S}} = \mathbf{0.293 \text{ mol S}}$

(b) $? \text{ g Al} = 3.3 \text{ mol Al} \times \dfrac{27.0 \text{ g Al}}{1 \text{ mol Al}} = \mathbf{89 \text{ g Al}}$

(c) $? \text{ g Cl} = 3.0 \times 10^{25} \text{ atoms Cl} \times \dfrac{1 \text{ mol Cl}}{6.022 \times 10^{23} \text{ atoms Cl}} \times \dfrac{35.5 \text{ g Cl}}{1 \text{ mol Cl}} = \mathbf{1.8 \times 10^{3} \text{ g Cl}}$

3.18 ***See Section 3.2 and Example 3.7.***

(a) Formula weight for NaOH:

1[Na] x 23.0	=	23.0	
1[O] x 16.0	=	16.0	
1[H] x 1.0	=	1.0	
		40.0	Hence, molar mass for NaOH is **40.0 g/mol**.

(b) Molecular weight for C_2H_4:

2[C] x 12.0	=	24.0	
4[H] x 1.0	=	4.0	
		28.0	Hence, molar mass for C_2H_4 is **28.0 g/mol**.

(c) Formula weight for $Mg(OH)_2$:

1[Mg] x 24.3	=	24.3	
2[O] x 16.0	=	32.0	
2[H] x 1.0	=	2.0	
		58.3	Hence, molar mass for $Mg(OH)_2$ is **58.3 g/mol**.

3.20 ***See Section 3.2 and Examples 3.8 and 9.***

(a) $? \text{ mol } C_6H_6 = 14.3 \text{ g } C_6H_6 \times \dfrac{1 \text{ mol } C_6H_6}{78.0 \text{ g } C_6H_6} = \mathbf{0.183 \text{ mol } C_6H_6}$

(b) $? \text{ g SiH}_4 = 3.22 \times 10^{22} \text{ molecules SiH}_4 \times \dfrac{1 \text{ mol SiH}_4}{6.022 \times 10^{23} \text{ molecules SiH}_4} \times \dfrac{32.1 \text{ g SiH}_4}{1 \text{ mol SiH}_4} = \mathbf{1.72 \text{ g SiH}_4}$

3.22 ***See Section 3.2 and Examples 3.8, 9 and 10.***

(a) Molecular weight for $C_{22}H_{25}NO_6$:

22[C] x 12.0	=	264.0
25[H] x 1.0	=	25.0
1[N] x 14.0	=	14.0
6[O] x 16.0	=	96.0
		399.0

Hence, molar mass for $C_{22}H_{25}NO_6$ is **399.0 g/mol**.

(b) $? \text{ g } C_{22}H_{25}NO_6 = 3.2 \times 10^{22} \text{ molecules } C_{22}H_{25}NO_6 \times \dfrac{1 \text{ mol } C_{22}H_{25}NO_6}{6.022 \times 10^{23} \text{ molecules } C_{22}H_{25}NO_6}$

$\times \dfrac{399.0 \text{ g } C_{22}H_{25}NO_6}{1 \text{ mol } C_{22}H_{25}NO_6} = \mathbf{21 \text{ g } C_{22}H_{25}NO_6}$

(c) $? \text{ mol } C_{22}H_{25}NO_6 = 326 \text{ g } C_{22}H_{25}NO_6 \times \dfrac{1 \text{ mol } C_{22}H_{25}NO_6}{399.0 \text{ g } C_{22}H_{25}NO_6} = \mathbf{0.817 \text{ mol } C_{22}H_{25}NO_6}$

(d) $? \text{ atoms C} = 50 \text{ molecules } C_{22}H_{25}NO_6 \times \dfrac{22 \text{ atoms C}}{1 \text{ molecule } C_{22}H_{25}NO_6} = \mathbf{1.1 \times 10^3 \text{ atoms C}}$

3.24 *See Section 3.2 and Example 3.8.*

(a) $? \text{ mol } K_2SO_4 = 2.2 \text{ g } K_2SO_4 \times \dfrac{1 \text{ mol } K_2SO_4}{174.3 \text{ g } K_2SO_4} = \mathbf{0.013 \text{ mol } K_2SO_4}$

(b) $? \text{ mol } C_8H_{12}N_4 = 6.4 \text{ g } C_8H_{12}N_4 \times \dfrac{1 \text{ mol } C_8H_{12}N_4}{164.0 \text{ g } C_8H_{12}N_4} = \mathbf{0.039 \text{ mol } C_8H_{12}N_4}$

(c) $? \text{ mol } Fe(C_5H_5)_2 = 7.13 \text{ g } Fe(C_5H_5)_2 \times \dfrac{1 \text{ mol } Fe(C_5H_5)_2}{185.8 \text{ g } Fe(C_5H_5)_2} = \mathbf{0.0384 \text{ mol } Fe(C_5H_5)_2}$

3.26 *See Section 3.2 and Example 3.10.*

(a) $? \text{ g } N_2O_4 = 7.55 \text{ mol } N_2O_4 \times \dfrac{92.0 \text{ g } N_2O_4}{1 \text{ mol } N_2O_4} = \mathbf{695 \text{ g } N_2O_4}$

(b) $? \text{ g } CaCl_2 = 9.2 \text{ mol } CaCl_2 \times \dfrac{111.0 \text{ g } CaCl_2}{1 \text{ mol } CaCl_2} = \mathbf{1.0 \times 10^3 \text{ g } CaCl_2}$

(c) $? \text{ g } CO = 0.44 \text{ mol } CO \times \dfrac{28.0 \text{ g } CO}{1 \text{ mol } CO} = \mathbf{12 \text{ g } CO}$

3.28 *See Section 3.2 and Examples 3.9 and 10.*

(a) $? \text{ g } NO_2 = 3.50 \text{ mol } NO_2 \times \dfrac{46.0 \text{ g } NO_2}{1 \text{ mol } NO_2} = \mathbf{1.61 \times 10^2 \text{ g } NO_2}$

(b) $? \text{ molecules } NO_2 = 3.5 \text{ mol } NO_2 \times \dfrac{6.022 \times 10^{23} \text{ molecules } NO_2}{1 \text{ mol } NO_2} = \mathbf{2.1 \times 10^{24} \text{ molecules } NO_2}$

(c) $? \text{ atoms N} = 2.1 \times 10^{24} \text{ molecules } NO_2 \times \dfrac{1 \text{ atom N}}{1 \text{ molecule } NO_2} = \mathbf{2.1 \times 10^{24} \text{ atoms N}}$

$? \text{ atoms O} = 2.1 \times 10^{24} \text{ molecules } NO_2 \times \dfrac{2 \text{ atoms O}}{1 \text{ molecule } NO_2} = \mathbf{4.2 \times 10^{24} \text{ atoms O}}$

3.30 *See Section 3.3 and Example 3.11.*

(a) Molecular weight for C_4H_8:

4[C] x 12.0	=	48.0
8[H] x 1.0	=	8.0
molecular weight	=	56.0

Molar mass for C_4H_8 is 56.0 g/mol.

Percent C by mass:

$\% \text{ C} = \dfrac{48.0 \text{ g C}}{56.0 \text{ g } C_4H_8} \times 100\% = \mathbf{85.7\% \text{ C}}$

Percent H by mass:

$\% \text{ H} = \dfrac{8.0 \text{ g H}}{56.0 \text{ g } C_4H_8} \times 100\% = \mathbf{14.3\% \text{ H}}$

(b) Molecular weight for $C_3H_4N_2$:

3[C] x 12.0	=	36.0
4[H] x 1.0	=	4.0
2[N] x 14.0	=	28.0
molecular weight	=	68.0

Molar mass for $C_3H_4N_2$ is 68.0 g/mol.

Percent C by mass:
$$\%\,C = \frac{36.0\ g\ C}{68.0\ g\ C_3H_4N_2} \times 100\% = \mathbf{52.9\%\ C}$$
Percent H by mass:
$$\%\,H = \frac{4.0\ g\ H}{68.0\ g\ C_3H_4N_2} \times 100\% = \mathbf{5.9\%\ H}$$
Percent N by mass:
$$\%\,N = \frac{28.0\ g\ N}{68.0\ g\ C_3H_4N_2} \times 100\% = \mathbf{41.2\%\ N}$$

(c) Formula weight for Fe_2O_3:

2[Fe] x 55.8	=	111.6
3[O] x 16.0	=	48.0
formula weight	=	159.6

Molar mass for Fe_2O_3 is 159.6 g/mol.

Percent Fe by mass:
$$\%\,Fe = \frac{111.6\ g\ Fe}{159.6\ g\ Fe_2O_3} \times 100\% = \mathbf{69.9\%\ Fe}$$
Percent O by mass:
$$\%\,O = \frac{48.0\ g\ O}{159.6\ g\ Fe_2O_3} \times 100\% = \mathbf{30.1\%\ O}$$

3.32 ***See Section 3.3 and Example 3.11.***

Molecular weight for CO:

1[C] x 12.0 amu	=	12.0
1[O] x 16.0 amu	=	16.0
molecular weight	=	28.0

Molar mass for CO is 28.0 g/mol.

Percent C by mass:
$$\%\,C = \frac{12.0\ g\ C}{28.0\ g\ CO} \times 100\% = \mathbf{42.9\%\ C}$$

$$?\ g\ C = 4.9\ g\ CO \times \frac{42.9\ g\ C}{100.0\ g\ CO} = \mathbf{2.1\ g\ C}$$

Note: Percent by mass means g/100 g.

3.34 ***See Section 3.3 and Example 3.11.***

$$(a)\ ?\ g\ C = 4.32\ g\ CO_2 \times \frac{12.0\ g\ C}{44.0\ g\ CO_2} = \mathbf{1.18\ g\ C}$$

$$(b)\ ?g\ C = 2.21\ g\ C_2H_4 \times \frac{24.0\ g\ C}{28.0\ g\ C_2H_4} = \mathbf{1.89\ g\ C}$$

$$(c)\ ?\ g\ C = 0.0443\ g\ CS_2 \times \frac{12.0\ g\ C}{76.2\ g\ CS_2} = \mathbf{0.00698\ g\ C}$$

3.36 ***See Section 3.3 and Example 3.14.***

$$?\ mol\ C = 0.831\ g\ C \times \frac{1\ mol\ C}{12.011\ g\ C} = 0.0692\ mol\ C$$

$$\text{relative mol C} = \frac{0.0692\ mol\ C}{0.0692} = 1.00\ mol\ C$$

$$?\ mol\ H = 0.139\ g\ H \times \frac{1\ mol\ H}{1.008\ g\ H} = 0.138\ mol\ H$$

$$\text{relative mol H} = \frac{0.138\ mol\ H}{0.0692} = 1.99\ mol\ H$$

The empirical formula is $\mathbf{CH_2}$.

3.38 ***See Section 3.3 and Example 3.14.***

$? \text{ mol C} = 0.571 \text{ g C} \times \frac{1 \text{ mol C}}{12.011 \text{ g C}} = 0.0475 \text{ mol C}$ relative mol C $= \frac{0.0475 \text{ mol C}}{0.0238} = 2.00 \text{ mol C}$

$? \text{ mol H} = 0.072 \text{ g H} \times \frac{1 \text{ mol H}}{1.008 \text{ g H}} = 0.071 \text{ mol H}$ relative mol H $= \frac{0.071 \text{ mol H}}{0.0238} = 3.0 \text{ mol H}$

$? \text{ mol N} = 0.333 \text{ g N} \times \frac{1 \text{ mol N}}{14.007 \text{ g N}} = 0.0238 \text{ mol N}$ relative mol N $= \frac{0.0238 \text{ mol N}}{0.0238} = 1.00 \text{ mol N}$

The empirical formula is $\mathbf{C_2H_3N}$.

3.40 ***See Section 3.3 and Example 3.14.***

$? \text{ mol C} = 1.11 \text{ g C} \times \frac{1 \text{ mol C}}{12.011 \text{ g C}} = 0.0924 \text{ mol C}$ relative mol C $= \frac{0.0924 \text{ mol C}}{0.0231} = 4.00 \text{ mol C}$

$? \text{ mol H} = 0.187 \text{ g H} \times \frac{1 \text{ mol H}}{1.008 \text{ gH}} = 0.186 \text{ mol H}$ relative mol H $= \frac{0.186 \text{ mol H}}{0.0231} = 8.05 \text{ mol H}$

$? \text{ mol O} = 0.370 \text{ g O} \times \frac{1 \text{ mol O}}{15.999 \text{ g O}} = 0.0231 \text{ mol O}$ relative mol O $= \frac{0.0231 \text{ mol O}}{0.0231} = 1.00 \text{ mol O}$

The empirical formula is $\mathbf{C_4H_8O}$.

3.42 ***See Section 3.3 and Example 3.12.***

Assume the sample has a mass of 100.00 g and therefore contains 44.06 g Fe and 55.93 g Cl.

$? \text{ mol Fe} = 44.06 \text{ g Fe} \times \frac{1 \text{ mol Fe}}{55.847 \text{ g Fe}} = 0.7889 \text{ mol Fe}$ relative mol Fe $= \frac{0.7889 \text{ mol Fe}}{0.7889} = 1.00 \text{ mol Fe}$

$? \text{ mol Cl} = 55.94 \text{ g Cl} \times \frac{1 \text{ mol Cl}}{35.453 \text{ g Cl}} = 1.578 \text{ mol Cl}$ relative mol Cl $= \frac{1.578 \text{ mol Cl}}{0.7889} = 2.00 \text{ mol Cl}$

The empirical formula is $\mathbf{FeCl_2}$.

3.44 ***See Section 3.3 and Example 3.12.***

Assume the sample has a mass of 100.00 g and therefore contains 26.89 g Ti, 67.44 g C and 5.67 g H.

$? \text{ mol Ti} = 26.89 \text{ g Ti} \times \frac{1 \text{ mol Ti}}{47.88 \text{ g Ti}} = 0.5616 \text{ mol Ti}$ relative mol Ti $= \frac{0.5616 \text{ mol Ti}}{0.5616} = 1.000 \text{ mol Ti}$

$? \text{ mol C} = 67.44 \text{ g C} \times \frac{1 \text{ mol C}}{12.011 \text{ g C}} = 5.615 \text{ mol C}$ relative mol C $= \frac{5.615 \text{ mol C}}{0.5616} = 9.998 \text{ mol C}$

$? \text{ mol H} = 5.67 \text{ g H} \times \frac{1 \text{ mol H}}{1.008 \text{ g H}} = 5.625 \text{ mol H}$ relative mol H $= \frac{5.625 \text{ mol H}}{0.5616} = 10.02 \text{ mol H}$

The empirical formula is **$TiC_{10}H_{10}$**. The compound is $Ti(C_5H_5)_2$.

3.46 ***See Section 3.3 and Example 3.13.***

$$? \text{ mol C} = 3.80 \text{ g } CO_2 \times \frac{1 \text{ mol } CO_2}{44.01 \text{ g } CO_2} \times \frac{1 \text{ mol C}}{1 \text{ mol } CO_2} = 0.0863 \text{ mol C}$$

$$? \text{ mol H} = 1.04 \text{ g } H_2O \times \frac{1 \text{ mol } H_2O}{18.02 \text{ g } H_2O} \times \frac{2 \text{ mol H}}{1 \text{ mol } H_2O} = 0.115 \text{ mol H}$$

? g O = total mass of sample - mass C - mass H

$? \text{ g C} = 0.0863 \text{ mol C} \times \frac{12.011 \text{ g C}}{1 \text{ mol C}} = 1.037 \text{ g C}$ $? \text{ g H} = 0.115 \text{ mol H} \times \frac{1.008 \text{ g H}}{1 \text{ mol H}} = 0.116 \text{ g H}$

$? \text{ g O} = 2.074 \text{ g} - 1.037 \text{ g C} - 0.116 \text{ g H} = 0.921 \text{ g O}$ $? \text{ mol O} = 0.921 \text{ g O} \times \frac{1 \text{ mol O}}{15.999 \text{ g O}} = 0.0576 \text{ mol O}$

relative mol C $= \frac{0.0863 \text{ mol C}}{0.0576} = 1.50 \text{ mol C}$ relative mol H $= \frac{0.115 \text{ mol H}}{0.0576} = 2.00 \text{ mol H}$

relative mol O $= \frac{0.0576 \text{ mol O}}{0.0576} = 1.00 \text{ mol O}$

Multiplying each of these by the same smallest integer that gives whole numbers of relative moles of atoms for each yields:

relative mol C = 1.50 mol C x 2 = 3.00 mol C
relative mol H = 1.00 mol H x 2 = 4.00 mol H
relative mol O = 1.00 mol O x 2 = 2.00 mol O

The empirical formula is **$C_3H_4O_2$**.

3.48 ***See Section 3.3 and Example 3.13.***

$$? \text{ mol C} = 2.60 \text{ g } CO_2 \times \frac{1 \text{ mol } CO_2}{44.01 \text{ g } CO_2} \times \frac{1 \text{ mol C}}{1 \text{ mol } CO_2} = 0.0591 \text{ mol C}$$

$$? \text{ mol H} = 0.799 \text{ g } H_2O \times \frac{1 \text{ mol } H_2O}{18.02 \text{ g } H_2O} \times \frac{2 \text{ mol H}}{1 \text{ mol } H_2O} = 0.0887 \text{ mol H}$$

? g O = total mass of sample – mass C – mass H – mass N

$$? \text{ g C} = 0.0591 \text{ mol C} \times \frac{12.011 \text{ g C}}{1 \text{ mol C}} = 0.710 \text{ g C}$$

$$? \text{ g H} = 0.0887 \text{ mol H} \times \frac{1.008 \text{ g H}}{1 \text{ mol H}} = 0.0894 \text{ g H}$$

$$? \text{ g N} = 1.48 \text{ g sample} \times \frac{0.340 \text{ g N}}{2.43 \text{ g sample}} = 0.207 \text{ g N}$$

$$? \text{ g O} = 1.48 \text{ g} - 0.710 \text{ g C} - 0.0894 \text{ g H} - 0.207 \text{ g N} = 0.47 \text{g O}$$

$$? \text{ mol N} = 0.207 \text{ g N} \times \frac{1 \text{ mol N}}{14.01 \text{ g N}} = 0.0148 \text{ mol N}$$

$$? \text{ mol O} = 0.47 \text{ g O} \times \frac{1 \text{ mol O}}{15.999 \text{ g O}} = 0.029 \text{ mol O}$$

$$\text{relative mol C} = \frac{0.0591 \text{ mol C}}{0.029} = 2.0 \text{ mol C}$$

$$\text{relative mol H} = \frac{0.0887 \text{ mol H}}{0.029} = 3.0 \text{ mol H}$$

$$\text{relative mol N} = \frac{0.0148 \text{ mol N}}{0.029} = 0.51 \text{ mol N}$$

$$\text{relative mol O} = \frac{0.029 \text{ mol O}}{0.029} = 1.0 \text{ mol O}$$

Multiplying each of these by the same smallest integer that gives whole numbers of relative moles of atoms for each yields:

relative mol C = 2.0 mol C x 2 = 4.0 mol C
relative mol H = 3.0 mol H x 2 = 6.0 mol H
relative mol N = 0.51 mol N x 2 = 1.0 mol N
relative mol O = 1.0 mol O x 2 = 2.0 mol O

The empirical formula is $\mathbf{C_4H_6NO_2}$.

3.50 ***See Section 3.3 and Example 3.12.***

(a) Empirical formula weight for C_2H_4O:

2[C] x 12.0	=	24.0
4[H] x 1.0	=	4.0
1[O] x 16.0	=	16.0
formula weight	=	44.0

$$n = \frac{\text{molecular weight}}{\text{empirical formula weight}} = \frac{132}{44.0} = 3$$

The molecular formula is $\mathbf{C_6H_{12}O_3}$.

(b) Empirical formula weight for $C_3H_4NO_3$:

3[C] x 12.0	=	36.0
4[H] x 1.0	=	4.0
1[N] x 14.0	=	14.0
3[O] x 16.0	=	48.0
formula weight	=	102.0

$$n = \frac{\text{molecular weight}}{\text{empirical formula weight}} = \frac{408}{102.0} = 4$$

The molecular formula is $\mathbf{C_{12}H_{16}N_4O_{12}}$.

3.52 ***See Section 3.3 and Example 3.12.***

Assume the sample has a mass of 100.0 g and therefore contains 62.0 g C, 10.4 g H and 27.5 g O.

$$? \text{ mol C} = 62.0 \text{ g C} \times \frac{1 \text{ mol C}}{12.011 \text{ g C}} = 5.16 \text{ mol C}$$

$$\text{relative mol C} = \frac{5.16 \text{ mol C}}{1.72} = 3.00 \text{ mol C}$$

$? \text{ mol H} = 10.4 \text{ g H} \times \frac{1 \text{ mol H}}{1.008 \text{ g H}} = 10.3 \text{ mol H}$ relative mol H $= \frac{10.3 \text{ mol H}}{1.72} = 5.99$ mol H

$? \text{ mol O} = 27.5 \text{ g O} \times \frac{1 \text{ mol O}}{15.999 \text{ g O}} = 1.72 \text{ mol O}$ relative mol O $= \frac{1.72 \text{ mol O}}{1.72} = 1.00$ mol O

The empirical formula is C_3H_6O.

Empirical formula weight for C_3H_6O:

3[C] x 12.0	=	36.0
6[H] x 1.0	=	6.0
1[O] x 16.0	=	16.0
formula weight	=	58.0

$$n = \frac{\text{molecular weight}}{\text{empirical formula weight}} = \frac{174}{58.0} = 3$$

The molecular formula is $\mathbf{C_9H_{18}O_3}$.

3.54 ***See Section 3.3 and Example 3.12.***

Assume the sample has a mass of 100.0 g and therefore contains 40.0 g C, 6.71 g H and 53.3 g O.

$? \text{ mol C} = 40.0 \text{ g C} \times \frac{1 \text{ mol C}}{12.011 \text{ g C}} = 3.33 \text{ mol C}$ relative mol C $= \frac{3.33 \text{ mol C}}{3.33} = 1.00$ mol C

$? \text{ mol H} = 6.71 \text{ g H} \times \frac{1 \text{ mol H}}{1.008 \text{ g H}} = 6.66 \text{ mol H}$ relative mol H $= \frac{6.66 \text{ mol H}}{3.33} = 2.00$ mol H

$? \text{ mol O} = 53.3 \text{ g O} \times \frac{1 \text{ mol O}}{15.999 \text{ g O}} = 3.33 \text{ mol O}$ relative mol O $= \frac{3.33 \text{ mol O}}{3.33} = 1.00$ mol O

The empirical formula is CH_2O.

Empirical formula weight for CH_2O:

1[C] x 12.0	=	12.0
2[H] x 1.0	=	2.0
1[O] x 16.0	=	16.0
formula weight	=	30.0

$$n = \frac{\text{molecular weight}}{\text{empirical formula weight}} = \frac{60}{30.0} = 2$$

The molecular formula for acetic acid is $\mathbf{C_2H_4O_2}$. This is usually written as CH_3CO_2H.

3.56 ***See Sections 3.1 and 4 and Examples 3.3 and 15.***

(a)

Unbalanced:	$C_3H_6 + O_2$	→	$CO_2 + H_2O$	Start with C_3H_6.
Step 1:	$C_3H_6 + O_2$	→	$\underline{3}CO_2 + \underline{3}H_2O$	Balances C, H.
Step 2:	$C_3H_6 + \underline{9/2}O_2$	→	$3CO_2 + 3H_2O$	Balances O.
Step 3:	$\underline{2}C_3H_6 + \underline{9}O_2$	→	$\underline{6}CO_2 + \underline{6}H_2O$	Gives whole number coefficients.

(b) *Strategy:* $g\ C_3H_6 \rightarrow mol\ C_3H_6 \rightarrow mol\ CO_2 \rightarrow g\ CO_2$

$$? \text{ g } CO_2 = 2.45 \text{ g } C_3H_6 \times \frac{1 \text{ mol } C_3H_6}{42.0 \text{ g } C_3H_6} \times \frac{6 \text{ mol } CO_2}{2 \text{ mol } C_3H_6} \times \frac{44.0 \text{ g } CO_2}{1 \text{ mol } CO_2} = \mathbf{7.70 \text{ g } CO_2}$$

3.58 ***See Sections 3.1 and 4 and Examples 3.1, 2 and 15.***

Unbalanced:	$P_4 + Cl_2$	→	PCl_5	Start with P_4.
Step 1:	$P_4 + Cl_2$	→	$\underline{4}PCl_5$	Balances P.
Step 2:	$P_4 + \underline{10}Cl_2$	→	$4PCl_5$	Balances Cl.

Strategy: $g\ P_4 \rightarrow mol\ P_4 \rightarrow mol\ Cl_2 \rightarrow g\ Cl_2$

$$?\ g\ Cl_2 = 0.567\ g\ P_4 \times \frac{1\ mol\ P_4}{123.9\ g\ P_4} \times \frac{10\ mol\ Cl_2}{1\ mol\ P_4} \times \frac{70.9\ g\ Cl_2}{1\ mol\ Cl_2} = \mathbf{3.24\ g\ Cl_4}$$

3.60 ***See Sections 3.1 and 4 and Examples 3.1, 2 and 15.***

Unbalanced:	$Al + H_2SO_4$	→	$Al_2(SO_4)_3 + H_2$	Start with $Al_2(SO_4)_3$.
Step 1:	$Al + \underline{3}H_2SO_4$	→	$Al_2(SO_4)_3 + H_2$	Balances SO_4 units.
Step 2:	$\underline{2}Al + 3H_2SO_4$	→	$Al_2(SO_4)_3 + H_2$	Balances Al.
Step 3:	$2Al + 3H_2SO_4$	→	$Al_2(SO_4)_3 + \underline{3}H_2$	Balances H.

Strategy: $g\ H_2 \rightarrow mol\ H_2 \rightarrow mol\ Al \rightarrow g\ Al$

$$?\ g\ Al = 13.2\ g\ H_2 \times \frac{1\ mol\ H_2}{2.02\ g\ H_2} \times \frac{2\ mol\ Al}{3\ mol\ H_2} \times \frac{27.0\ g\ Al}{1\ mol\ Al} = \mathbf{118\ g\ Al}$$

3.62 ***See Sections 3.1 and 4 and Examples 3.1, 2, 16 and 17.***

Unbalanced:	$Al + HCl$	→	$AlCl_3 + H_2$	Star with $AlCl_3$.
Step 1:	$Al + \underline{3}HCl$	→	$AlCl_3 + H_2$	Balances Cl.
Step 2:	$Al + 3HCl$	→	$AlCl_3 + \underline{3/2}H_2$	Balances H.
Step 3:	$\underline{2}Al + \underline{6}HCl$	→	$\underline{2}AlCl_3 + \underline{3}H_2$	Gives whole number coefficients.

Strategy: $g\ Al \rightarrow mol\ Al \rightarrow mol\ AlCl_3 \rightarrow g\ AlCl_3$

$$?\ g\ AlCl_3 = 3.3\ g\ Al \times \frac{1\ mol\ Al}{27.0\ g\ Al} \times \frac{2\ mol\ AlCl_3}{2\ mol\ Al} \times \frac{133.4\ g\ AlCl_3}{1\ mol\ AlCl_3} = 16.3\ AlCl_3$$

$$\text{percent yield} = \frac{\text{actual yield}}{\text{theoretical yield}} \times 100\% = \frac{3.5\ g\ AlCl_3}{16.3\ g\ AlCl_3} \times 100\% = \mathbf{21\%}$$

3.64 ***See Section 3.1 and 4 and Examples 3.1, 2, and 16.***

Unbalanced:	$Li + O_2$	→	Li_2O	Start with Li_2O.
Step 1:	$Li + O_2$	→	$\underline{2}Li_2O$	Balances O.
Step 2:	$\underline{4}Li + O_2$	→	$2Li_2O$	Balances Li.

Strategy: $g\ Li \rightarrow mol\ Li \rightarrow mol\ Li_2O \rightarrow g\ Li_2O$

$$?\ g\ LiO = 0.45\ g\ Li \times \frac{1\ mol\ Li}{6.9\ g\ Li} \times \frac{2\ mol\ Li_2O}{4\ mol\ Li} \times \frac{29.8\ Li_2O}{1\ mol\ Li_2O} = \mathbf{0.97\ g\ Li_2O}$$

3.66 *See Sections 3.1 and 5 and Examples 3.1, 2 and 18.*

Unbalanced:	$N_2 + H_2$	→	NH_3	Start with NH_3.
Step 1:	$N_2 + H_2$	→	$\underline{2}NH_3$	Balances N.
Step 2:	$N_2 + \underline{3}H_2$	→	$2NH_3$	Balances H.

Strategy: $g\ N_2 \rightarrow mol\ N_2 \rightarrow mol\ NH_3 \rightarrow g\ NH_3$

$$?\ \text{g NH}_3 \text{ based on N}_2 = 14\ \text{g N}_2 \times \frac{1\ \text{mol N}_2}{28.0\ \text{g N}_2} \times \frac{2\ \text{mol NH}_3}{1\ \text{mol N}_2} \times \frac{17.0\ \text{g NH}_3}{1\ \text{mol NH}_3} = 17.0\ \text{g NH}_3$$

Strategy: $g\ H_2 \rightarrow mol\ H_2 \rightarrow mol\ NH_3 \rightarrow g\ NH_3$

$$?\ \text{g NH}_3 \text{ based on H}_2 = 1.0\ \text{g H}_2 \times \frac{1\ \text{mol H}_2}{2.0\ \text{g H}_2} \times \frac{2\ \text{mol NH}_3}{3\ \text{mol H}_2} \times \frac{17.0\ \text{g NH}_3}{1\ \text{mol NH}_3} = 5.7\ \text{g NH}_3$$

H_2 is the limiting reactant because it produces less NH_3. The maximum amount of NH_3 which can be produced from 14 g N_2 and 1.0 g H_2 is **5.7 g**.

3.68 *See Section 3.5 and Example 3.18.*

$$Zn(s) + 2AgNO_3(aq) \rightarrow 2\ Ag(s) + Zn(NO_3)_2(aq)$$

Strategy: $g\ Zn \rightarrow mol\ Zn \rightarrow mol\ Ag \rightarrow g\ Ag$

$$?\ \text{g Ag based on Zn} = 3.22\ \text{g Zn} \times \frac{1\ \text{mol Zn}}{65.4\ \text{g Zn}} \times \frac{2\ \text{mol Ag}}{1\ \text{mol Zn}} \times \frac{107.9\ \text{g Ag}}{1\ \text{mol Ag}} = 10.6\ \text{g Ag}$$

Strategy: $g\ AgNO_3 \rightarrow mol\ AgNO_3 \rightarrow mol\ Ag \rightarrow g\ Ag$

$$?\ \text{g Ag based on AgNO}_3 = 4.35\ \text{g AgNO}_3 \times \frac{1\ \text{mol AgNO}_3}{169.9\ \text{g AgNO}_3} \times \frac{2\ \text{mol Ag}}{2\ \text{mol AgNO}_3} \times \frac{107.9\ \text{g Ag}}{1\ \text{mol Ag}} = 2.76\ \text{g Ag}$$

$AgNO_3$ is the limiting reactant because it produces less Ag. The maximum amount of Ag which can be produced from 3.22 g Zn and 4.35 g $AgNO_3$ is **2.76 g**.

3.70 *See Section 3.4 and Examples 3.16 and 17.*

Use the balanced equation that was obtained in solving 3.56: $2C_3H_6 + 9O_2 \rightarrow 6CO_2 + 6H_2O$.

Strategy: $g\ C_3H_6 \rightarrow mol\ C_3H_6 \rightarrow mol\ H_2O \rightarrow g\ H_2O$

$$?\ \text{g H}_2\text{O} = 33.5\ \text{g C}_3\text{H}_6 \times \frac{1\ \text{mol C}_3\text{H}_6}{42.0\ \text{g C}_3\text{H}_6} \times \frac{6\ \text{mol H}_2\text{O}}{2\ \text{mol C}_3\text{H}_6} \times \frac{18.0\ \text{g H}_2\text{O}}{1\ \text{mol H}_2\text{O}} = 43.1\ \text{g H}_2\text{O}$$

$$\text{percent yield} = \frac{\text{actual yield}}{\text{theorectical yield}} \times 100\% = \frac{16.1\ \text{g H}_2\text{O}}{43.1\ \text{g H}_2\text{O}} \times 100\% = \mathbf{37.4\%}$$

3.72 *See Section 3.4 and 5 and Example 3.18 and 19.*

(a) Balanced: $HNO_3 + NaOH \rightarrow H_2O + NaNO_3$

Strategy: $g\ HNO_3 \rightarrow mol\ HNO_3 \rightarrow mol\ NaNO_3 \rightarrow g\ NaNO_3$

$$?\ g\ NaNO_3 \text{ based on } HNO_3 = 21.2\ g\ HNO_3 \times \frac{1\ mol\ HNO_3}{63.0\ g\ HNO_3} \times \frac{1\ mol\ NaNO_3}{1\ mol\ HNO_3} \times \frac{85.0\ g\ NaNO_3}{1\ mol\ NaNO_3} = 28.6\ NaNO_3$$

Strategy: $g\ NaOH \rightarrow mol\ NaOH \rightarrow mol\ NaNO_3 \rightarrow g\ NaNO_3$

$$?\ g \text{ based on } NaOH = 23.1\ g\ NaOH \times \frac{1\ mol\ NaOH}{40.0\ g\ NaOH} \times \frac{1\ mol\ NaNO_3}{1\ mol\ NaOH} \times \frac{85.0\ g\ NaNO_3}{1\ mol\ NaNO_3} = 49.1\ g\ NaNO_3$$

HNO_3 is the limiting reactant because it produces less $NaNO_3$. The maximum amount of $NaNO_3$ which can be produced from 21.2 g HNO_3 and 23.1 NaOH is 28.6 g $NaNO_3$. This is the theorectical yield of the reaction.

$$\text{percent yield} = \frac{\text{actual yield}}{\text{theorectical yield}} \times 100\% = \frac{12.9\ g\ NaNO_3}{28.6\ g\ NaNO_3} \times 100\% = \mathbf{45.1\%}$$

(b) Since there is an excess of NaOH, some will remain at the end of the reaction.
Assuming complete reaction and conservation of mass,

$$\text{initial g } HNO_3 + \text{initial g NaOH} = \text{g } H_2O \text{ expected} + \text{g } NaNO_3 \text{ expected} + \text{g excess NaOH}$$

$$?\ g\ H_2O \text{ based on } HNO_3 = 21.2\ g\ HNO_3 \times \frac{1\ mol\ HNO_3}{63.0\ g\ HNO_3} \times \frac{1\ mol\ H_2O}{1\ mol\ HNO_3} \times \frac{18.0\ g\ H_2O}{1\ mol\ H_2O} = 6.1\ g\ H_2O$$

$$\begin{aligned}\text{g excess NaOH} &= \text{initial g } HNO_3 + \text{initial g NaOH} - \text{g } H_2O \text{ expected} - \text{g } NaNO_3 \text{ expected} \\ &= 21.2\ g + 23.1\ g - 6.1\ g - 28.6\ g = \mathbf{9.6\ g\ NaOH}\end{aligned}$$

3.74 *See Sections 3.2 and 3.*

Strategy: $tons\ Cu \rightarrow tons\ CuFeS_2 \rightarrow tons\ ore$

$$?\ ton\ ore = (0.990 \times 20.0\ tons)Cu \times \frac{183.6\ tons\ CuFeS_2}{63.5\ tons\ Cu} \times \frac{100\ tons\ ore}{10\ tons\ CuFeS_2} = \mathbf{572\ tons\ ore}$$

3.76 *See Sections 3.2 and 3.3.*

Assume the sample contains 100.0 g and therefore contains 25.5 g Cu, 12.8 g S, 57.7 g O and 4.0 g H.

Strategy: $g\ Cu \rightarrow mol\ Cu \rightarrow mol\ CuSO_4$

$$?\ mol\ CuSO_4 = 25.5\ g\ Cu \times \frac{1\ mol\ Cu}{63.5\ g\ Cu} \times \frac{1\ mol\ CuSO_4}{1 mol\ Cu} = 0.402\ mol\ CuSO_4$$

Strategy: $g\ H \rightarrow mol\ H \rightarrow mol\ H_2O$

$$? \text{ mol } H_2O = 4.0 \text{ g H} \times \frac{1 \text{ mol H}}{1.008 \text{ g H}} \times \frac{1 \text{ mol } H_2O}{2 \text{ mol H}} = 2.0 \text{ mol } H_2O$$

$$\frac{\text{mol } H_2O}{\text{mol } CuSO_4} = \frac{2.0 \text{ mol}}{0.402 \text{ mol}} = 5.0$$

The value of x in $CuSO_4 \cdot x\ H_2O$ is **5.0**.

3.78 ***See Section 3.3 and Examples 3.12.***

Assume the sample contains 100.00 g and therefore contains 71.56 g C, 6.71 g H, 4.91 g N and 16.82 g O.

$$? \text{ mol C} = 71.56 \text{ g C} \times \frac{1 \text{ mol C}}{12.011 \text{ g C}} = 5.96 \text{ mol C} \qquad \text{relative mol C} = \frac{5.96 \text{ mol C}}{0.350} = 17.0 \text{ mol C}$$

$$? \text{ mol H} = 6.71 \text{ g H} \times \frac{1 \text{ mol H}}{1.008 \text{ g H}} = 6.66 \text{ mol H} \qquad \text{relative mol H} = \frac{6.66 \text{ mol H}}{0.350} = 19.0 \text{ mol H}$$

$$? \text{ mol N} = 4.91 \text{ g N} \times \frac{1 \text{ mol N}}{14.007 \text{ g N}} = 0.350 \text{ mol N} \qquad \text{relative mol N} = \frac{0.350 \text{ mol N}}{0.350} = 1.00 \text{ mol N}$$

$$? \text{ mol O} = 16.82 \text{ g O} \times \frac{1 \text{ mol O}}{15.999 \text{ g O}} = 1.05 \text{ mol O} \qquad \text{relative mol O} = \frac{1.05 \text{ mol O}}{0.350} = 3.00 \text{ mol O}$$

The empirical formula is $C_{17}H_{19}NO_3$.

Empirical formula weight for $C_{17}H_{19}NO_3$:

17[C] x 12.0	=	204.0
19[H] x 1.0	=	19.0
1[N] x 14.0	=	14.0
3[O] x 16.0	=	48.0
formula weight	=	285.0

The empirical formula wieght of morphine is numerically equal to the molar mass of morphine. Hence, the molecular formula for morphine is $\mathbf{C_{17}H_{19}NO_3}$.

3.80 ***See Sections 3.1 and 4 and Examples 3.1, 2, 16 and 17.***

Unbalanced:	$NaWCl_6$	→	$Na_2WCl_6 + WCl_6$	Start with Na_2WCl_6.
Step 1:	$\underline{2}NaWCl_6$	→	$Na_2WCl_6 + WCl_6$	Balances Na, W, Cl.

Strategy: $g\ NaWCl_6 \rightarrow mol\ NaWCl_6 \rightarrow mol\ WCl_6 \rightarrow g\ WCl_6$

$$? \text{ g } WCl_6 = 5.64 \text{ g } NaWCl_6 \times \frac{1 \text{ mol } NaWCl_6}{419.6 \text{ g } NaWCl_6} \times \frac{1 \text{ mol } WCl_6}{2 \text{ mol } NaWCl_6} \times \frac{396.6 \text{ g } WCl_6}{1 \text{ mol } WCl_6} = 2.67 \text{ g } WCl_6$$

$$\text{percent yield} = \frac{\text{actual yield}}{\text{theorectical yield}} \times 100\% = \frac{1.52 \text{ g } WCl_6}{2.67 \text{ g } WCl_6} \times 100\% = \mathbf{56.9\%}$$

3.82 ***See Section 3.4 and Example 3.16.***

The amount of hydrogen gas produced depends on the amount of excess zinc remaining from the reaction with $CuSO_4$. Hence, this needs to be determined first.

$$? \text{ g Zn excess} = \text{g Zn initial} - \text{g Zn reacting with CuSO}_4$$

Strategy: $g\ CuSO_4 \rightarrow mol\ CuSO_4 \rightarrow mol\ Zn \rightarrow g\ Zn$

$$? \text{ g Zn reacting with CuSO}_4 = 1.20 \text{ g CuSO}_4 \times \frac{1 \text{ mol CuSO}_4}{159.6 \text{ g CuSO}_4} \times \frac{1 \text{ mol Zn}}{1 \text{ mol CuSO}_4} \times \frac{65.4 \text{ g Zn}}{1 \text{ mol Zn}} = 0.492 \text{ g Zn}$$

$$? \text{ g Zn remaining and reacting with H}_2\text{SO}_4 = 1.20 \text{ g} - 0.492 \text{ g Zn reacting with CuSO}_4 = 0.71 \text{ g}$$

Strategy: $g\ Zn \rightarrow mol\ Zn \rightarrow mol\ H_2 \rightarrow g\ H$

$$? \text{ g H}_2 = 0.71 \text{ g Zn} \times \frac{1 \text{ mol Zn}}{65.4 \text{ g Zn}} \times \frac{1 \text{ mol H}_2}{1 \text{ mol Zn}} \times \frac{2.0 \text{ g H}_2}{1 \text{ mol H}_2} = \mathbf{0.022 \text{ g H}_2}$$

Chapter 4: Stoichiometry II

4.2 *See Section 4.1 and Example 4.1.*

$$\text{molarity} = \frac{\text{moles of solute}}{\text{liters of solution}}$$

$$?\ \text{mol KOH} = 8.23\ \text{g KOH} \times \frac{1\ \text{mol KOH}}{56.1\ \text{g KOH}} = 0.147\ \text{mol KOH}$$

$$?\ \text{L soln} = 250\ \text{mL} \times \frac{1\ \text{L}}{10^3\ \text{L}} = 0.250\ \text{L soln}$$

$$?\ \text{molarity} = \frac{0.147\ \text{mol KOH}}{.250\ \text{L soln}} = \mathbf{0.588}\ \boldsymbol{M}\ \mathbf{KOH}$$

4.4 *See Section 4.1 and Example 4.4.*

$$?\ \text{g AgNO}_3 = 300\ \text{mL} \times \frac{1\ \text{L}}{10^3\ \text{mL}} \times \frac{1.00\ \text{mol AgNO}_3}{1\ \text{L}} \times \frac{169.9\ \text{g AgNO}_3}{1\ \text{mol AgNO}_3} = \mathbf{51.0\ g\ AgNO_3}$$

4.6 *See Section 4.1 and Example 4.4.*

Solving M(con) x V(con) = M(dil) x V(dil) for V(con) gives

$$V(\text{con}) = \frac{M(\text{dil}) \times V(\text{dil})}{M(\text{con})} \qquad V(\text{con}) = \frac{0.45M \times 2.5\ \text{L}}{2.3M} = \mathbf{0.49\ L\ of\ 2.3}\ \boldsymbol{M}\ \mathbf{HCl}$$

4.8 *See Section 4.1 and Example 4.5.*

Solving M(con) x V(con) = M(dil) x V(dil) for M(dil) gives

$$M(\text{dil}) = \frac{M(\text{con}) \times V(\text{con})}{V(\text{dil})} \qquad M(\text{dil}) = \frac{1.0\ M \times 0.050\ \text{L}}{2.0\ \text{L}} = \mathbf{0.025}\ \boldsymbol{M}\ \mathbf{C_6H_{12}O_6}$$

4.10 *See Section 4.1 and Examples 4.1 and 5.*

(a) $$\text{molarity} = \frac{\text{moles of solute}}{\text{liters of solution}}$$

$$?\ \text{L soln} = 2.0\ \text{L soln}$$

$$?\ \text{mol NaOH} = 3.56\ \text{g NaOH} \times \frac{1\ \text{mol NaOH}}{40.0\ \text{g NaOH}} = 0.089\ \text{mol NaOH}$$

$$?\ \text{molarity} = \frac{0.089\ \text{mol NaOH}}{2.0\ \text{L soln}} = \mathbf{0.044}\ \boldsymbol{M}\ \mathbf{NaOH}$$

(b) Solving M(con) x V(con) = M(dil) x V(dil) for M(dil) gives

$$M(\text{dil}) = \frac{M(\text{con}) \times V(\text{con})}{V(\text{dil})} \qquad M(\text{dil}) = \frac{1.4\ M \times 0.0250\ \text{L}}{2.0\ \text{L}} = \mathbf{0.018}\ \boldsymbol{M}\ \mathbf{NaOH}$$

4.12 ***See Section 4.1 and Example 4.2.***

$$\text{(a) ? mol HNO}_3 = 0.033\ \text{L HNO}_3\ \text{soln} \times \frac{3.11\ \text{mol HNO}_3}{1\ \text{L HNO}_3\ \text{soln}} = \mathbf{0.10\ mol\ HNO_3}$$

$$\text{(b) ? mol HNO}_3 = 1.0\ \text{L HNO}_3\ \text{soln} \times \frac{3.2\ \text{mol HNO}_3}{1\ \text{L HNO}_3\ \text{soln}} = \mathbf{3.2\ mol\ HNO_3}$$

4.14 ***See Section 4.1 and Example 4.3.***

$$\text{(a) ? g HCl} = 3.13\ \text{L HCl soln} \times \frac{2.21\ \text{mol HCl}}{1\ \text{L HCl soln}} \times \frac{36.5\ \text{g HCl}}{1\ \text{mol HCl}} = \mathbf{252\ g\ HCl}$$

$$\text{(b) ? g KCl} = 1.5\ \text{L KCl soln} \times \frac{1.2\ \text{mol KCl}}{1\ \text{L KCl soln}} \times \frac{74.6\ \text{g KCl}}{1\ \text{mol KCl}} = \mathbf{1.3 \times 10^2\ g\ KCl}$$

4.16 ***See Section 4.1 and Examples 4.1 and 5.***

$$\text{? g KSCN} = 1.00\ \text{L KSCN soln} \times \frac{0.20\ \text{mol KSCN}}{1\ \text{L KSCN soln}} \times \frac{97.2\ \text{g KSCN}}{1\ \text{mol KSCN}} = \mathbf{19\ g\ KSCN}$$

4.18 ***See Section 4.1 and Examples 4.1 and 5.***

$$\text{molarity} = \frac{\text{moles of solute}}{\text{liters of solution}} \qquad \text{? L soln} = 1.00\ \text{L soln}$$

$$\text{? mol NaOH} = 6.00\ \text{g NaOH} \times \frac{1\ \text{mol NaOH}}{40.0\ \text{g NaOH}} = 0.150\ \text{mol NaOH} \qquad \text{molarity} = \frac{0.150\ \text{mol NaOH}}{1.00\ \text{L}} = 0.150\ M$$

Solving M(con) x V(con) = M(dil) x V(dil) for M(dil) gives

$$M(\text{dil}) = \frac{M(\text{con}) \times V(\text{con})}{V(\text{dil})} \qquad M(\text{dil}) = \frac{1.50\ M \times 0.100\ \text{L}}{5.00\ \text{L}} = \mathbf{0.00300}\ \boldsymbol{M}\ \mathbf{NaOH}$$

4.20 ***See Sections 3.1 and 4.2 and Examples 3.4 and 4.6.***

Balanced: $HCl + NaOH \rightarrow H_2O + NaCl$

Strategy: L HCl soln → mol HCl → mol NaOH → g NaOH

$$\text{? g NaOH} = 0.100\ \text{L HCl soln} \times \frac{1.3\ \text{mol HCl}}{1\ \text{L HCl soln}} \times \frac{1\ \text{mol NaOH}}{1\ \text{mol HCl}} \times \frac{40.0\ \text{g NaOH}}{1\ \text{mol NaOH}} = \mathbf{5.2\ g\ NaOH}$$

4.22 *See Sections 3.1 and 4.2 and Examples 3.4 and 4.7.*

Unbalanced:	$HNO_3 + Ca(OH)_2 \rightarrow H_2O + Ca(NO_3)_2$	Start with $Ca(NO_3)_2$.
Step 1:	$\underline{2}HNO_3 + Ca(OH)_2 \rightarrow H_2O + Ca(NO_3)_2$	Balances NO_3 units.
Step 2:	$2HNO_3 + Ca(OH)_2 \rightarrow \underline{2}H_2O + Ca(NO_3)_2$	Balances H, O.

Strategy: g $Ca(OH)_2$ → mol $Ca(OH)_2$ → mol HNO_3 → L HNO_3

$$? \text{ L HNO}_3 \text{ soln} = 22 \text{ g Ca(OH)}_2 \times \frac{1 \text{ mol Ca(OH)}_2}{74.1 \text{ g Ca(OH)}_2} \times \frac{2 \text{ mol HNO}_3}{1 \text{ mol Ca(OH)}_2} \times \frac{1 \text{ L HNO}_3 \text{ soln}}{0.66 \text{ mol HNO}_3} = \mathbf{0.90 \text{ L HNO}_3 \text{ soln}}$$

4.24 *See Sections 3.1 and 4.2 and Examples 3.4 and 4.7.*

Balanced: $HCl + NaOH \rightarrow H_2O + NaCl$

Strategy: g NaOH → mol NaOH → mol HCl → M HCl soln

$$? \text{ mol HCl} = 2.3 \text{ g NaOH} \times \frac{1 \text{ mol NaOH}}{40.0 \text{ g NaOH}} \times \frac{1 \text{ mol HCl}}{1 \text{ mol NaOH}} = 0.058 \text{ mol HCl}$$

$$? \text{ } M \text{ HCl soln} = \frac{0.058 \text{ mol HCl}}{0.135 \text{ L HCl soln}} = \mathbf{0.43 \text{ } \mathit{M} \text{ HCl}}$$

4.26 *See Sections 3.1, 3.5 and 4.2 and Examples 3.4, 3.18 and 4.6.*

Unbalanced:	$H_2SO_4 + Ba(OH)_2 \rightarrow H_2O + BaSO_4$	Start with $Ba(OH)_2$.
Step 1:	$H_2SO_4 + Ba(OH)_2 \rightarrow \underline{2}H_2O + BaSO_4$	Balances H and O.

Strategy: L H_2SO_4 soln → mol H_2SO_4 → mol $BaSO_4$ → g $BaSO_4$

$$? \text{ g BaSO}_4 \text{ based on H}_2\text{SO}_4 = 0.355 \text{ L H}_2\text{SO}_4 \text{ soln} \times \frac{0.032 \text{ mol H}_2\text{SO}_4}{1 \text{ L H}_2\text{SO}_4 \text{ soln}}$$

$$\times \frac{1 \text{ mol BaSO}_4}{1 \text{ mol H}_2\text{SO}_4} \times \frac{233.4 \text{ g BaSO}_4}{1 \text{ mol BaSO}_4} = 2.6 \text{ g BaSO}_4$$

Strategy: L $Ba(OH)_2$ soln → mol $Ba(OH)_2$ → mol $BaSO_4$ → g $BaSO_4$

$$? \text{ g BaSO}_4 \text{ based on Ba(OH)}_2 = 0.266 \text{ L Ba(OH)}_2 \text{ soln} \times \frac{0.015 \text{ mol Ba(OH)}_2}{1 \text{ L Ba(OH)}_2 \text{ soln}}$$

$$\times \frac{1 \text{ mol BaSO}_4}{1 \text{ mol Ba(OH)}_2} \times \frac{233.4 \text{ g BaSO}_4}{1 \text{ mol BaSO}_4} = 0.93 \text{ g BaSO}_4$$

$Ba(OH)_2$ is the limiting reactant because it produces less $BaSO_4$. The maximum amount of $BaSO_4$ which can be produced from 355 mL of 0.032 *M* H_2SO_4 and 266 mL of 0.015 *M* $Ba(OH)_2$ is **0.93 g**.

4.28 ***See Sections 3.1 and 4.2 and Examples 3.2 and 4.7.***

Unbalanced:	$HCl + Ba(OH)_2 \rightarrow H_2O + BaCl_2$	Start with $BaCl_2$.
Step 1:	$\underline{2}HCl + Ba(OH)_2 \rightarrow H_2O + BaCl_2$	Balances Cl.
Step 2:	$2HCl + Ba(OH)_2 \rightarrow \underline{2}H_2O + BaCl_2$	Balances H and O.

Strategy: $L\ HCl \rightarrow mol\ HCl \rightarrow mol\ Ba(OH)_2 \rightarrow M\ Ba(OH)_2\ soln$

$$? \text{ mol Ba(OH)}_2 = \left[\left(0.075 \text{ L HCl soln} \times \frac{0.10 \text{ mol HCl}}{1 \text{ L HCl soln}}\right) + \left(0.035 \text{ L HCl soln} \times \frac{0.012 \text{ mol HCl}}{1 \text{ L HCl soln}}\right)\right]$$

$$\times \frac{1 \text{ mol Ba(OH)}_2}{2 \text{ mol HCl}} = 4.0 \text{ x } 10^{-3} \text{ mol Ba(OH)}_2$$

$$? \, M \text{ Ba(OH)}_2 \text{ soln} = \frac{4.0 \text{ x } 10^{-3} \text{ mol Ba(OH)}_2}{0.125 \text{ L Ba(OH)}_2 \text{ soln}} = \mathbf{3.2 \text{ x } 10^{-2}} \; \boldsymbol{M} \textbf{ Ba(OH)}_\mathbf{2}$$

4.30 ***See Section 4.2 and Table 4.1.***

(a) BaI_2, (c) Na_2CO_3 and (d) $(NH_4)_2SO_4$ are classified as soluble salts. In Table 4.1, see Rules 3, 1, and 1, respectively.

4.32 ***See Section 4.2, Table 4.1 and Example 4.9.***

(a) The ions present from the reactants are Na^+(aq), OH^-(aq), Mg^{2+}(aq) and Cl^-(aq). The two possible products are NaCl and $Mg(OH)_2$. Rule 5 in Table 4.1 indicates $Mg(OH)_2$ is insoluble. The net ionic equation for the reaction is:

$$Mg^{2+}(aq) + 2\ OH^- \rightarrow Mg(OH)_2(s)$$

(b) The ions present from the reactants are Na^+(aq), NO_3^-(aq), Mg^{2+}(aq) and Br^-(aq). The two possible products are NaBr and $MgNO_3$. Rules 1 and 2 in Table 1.4 indicate both of these compounds are soluble. Hence, no precipitate is formed.

(c) The ions present from the reactants are Ba^{2+} (aq), ClO_4^-(aq), Na^+(aq) and CO_3^{2-}(aq). The two possible products are $BaCO_3$ and $NaClO_4$. Rule 6 in Table 4.1 indicates $BaCO_3$ is insoluble. The net ionic equation for the reaction is:

$$Ba^{2+}(aq) + CO_3^{2-}(aq) \rightarrow BaCO_3(s)$$

4.34 ***See Section 4.2 and Table 4.1.***

Rule 3 in Table 4.1 indicates only the chlorides of Ag^+, Pb^{2+} and Hg_2^{2+} are insoluble. Hence, the solution must contain Pb^{2+}.

4.36 ***See Section 4.2 and Table 4.1.***

Rule 6 indicates all carbonates, except those of the Group IA metals and NH_4^+, are insoluble. Hence, both Ca^{2+} and Ba^{2+} would be expected to form carbonate precipitates. However, Rule 4 in Table 4.1 indicates only the sulfates of Ba^{2+}, Pb^{2+}, Hg^{2+} and Hg_2^{2+} are insoluble. Since no sulfate precipitate forms, the solution cannot contain Ba^{2+}. It must therefore contain Ca^{2+} to give the carbonate precipitate.

4.38 See Section 4.2 and Table 4.1.

Rule 4 in Table 4.1 indicates Hg^{2+} forms a sulfate precipitate whereas Ca^{2+} does not. Hence, one can test for Hg^{2+} by adding Na_2SO_4. If a precipitate forms, Hg^{2+} is present. If no precipitate forms, Hg^{2+} is absent, and Ca^{2+} is present. Rule 6 in Table 4.1 indicates Na_2CO_3 or Na_3PO_4 can be added to test for Ca^{2+} in the absence of Hg^{2+}. If Hg^{2+} is absent and a carbonate or phosphate precipitate is obtained, Ca^{2+} is present.

4.40 See Sections 3.5 and 4.2 and Examples 3.18, 4.6 and 4.9.

The ions present from the reactants are $Ag^+(aq)$, $NO_3^-(aq)$, $Na^+(aq)$ and $Br^-(aq)$. The two possible products are AgBr and $NaNO_3$. Rule 3 in Table 4.1 indicates AgBr is insoluble.

Overall: $AgNO_3(aq) + NaBr(aq) \rightarrow AgBr(s) + NaNO_3(aq)$

Complete ionic: $Ag^+(aq) + NO_3^-(aq) + Na^+(aq) + Br^-(aq) \rightarrow AgBr(s) + Na^+(aq) + NO_3^-(aq)$

Net ionic: $Ag^+(aq) + Br^-(aq) \rightarrow AgBr(s)$

Strategy: $L\ AgNO_3\ soln \rightarrow mol\ AgNO_3 \rightarrow mol\ AgBr \rightarrow g\ AgBr$

$$? \text{ g AgBr based on AgNO}_3 = 0.345 \text{ L AgNO}_3 \text{ soln} \times \frac{0.33 \text{ mol AgNO}_3}{1 \text{ L AgNO}_3 \text{ soln}} \times \frac{1 \text{ mol AgBr}}{1 \text{ mol AgNO}_3}$$

$$\times \frac{187.8 \text{ g AgBr}}{1 \text{ mol AgBr}} = 21\text{g AgBr}$$

Stategy: $LNaBr\ soln \rightarrow mol\ NaBr \rightarrow mol\ AgBr \rightarrow g\ AgBr$

$$? \text{ g AgBr based on NaBr} = 0.100 \text{ L NaBr soln} \times \frac{1.3 \text{ mol NaBr}}{1 \text{ L NaBr soln}} \times \frac{1 \text{ mol AgBr}}{1 \text{ mol NaBr}} \times \frac{187.8 \text{ g AgBr}}{1 \text{ mol AgBr}} = 24 \text{ g AgBr}$$

$AgNO_3$ is the limiting reactant because it produces less AgBr. The maximum amount of AgBr which can be produced from 345 mL of 0.33 *M* $AgNO_3$ and 100 mL of 1.3 *M* NaBr is **21 g**.

4.42 See Section 4.2, Table 4.1 and Example 4.10.

The ions present from the reactants are $K^+(aq)$, $CO_3^{2-}(aq)$, $Ca^{2+}(aq)$ and $Cl^-(aq)$. The two possible products are KCl and $CaCO_3$. Rule 6 in Table 4.1 indicates $CaCO_3$ is insoluble. The overall and net ionic equations are:

$$CaCl_2(aq) + K_2CO_3(aq) \rightarrow CaCO_3(s) + 2KCl(aq)$$

$$Ca^{2+}(aq) + CO_3^{2-}(aq) \rightarrow CaCO_3(s)$$

Strategy: $g\ CaCO_3 \rightarrow mol\ CaCO_3 \rightarrow mol\ CaCl_2 \rightarrow M\ CaCl_2\ soln$

$$? \text{ mol CaCl}_2 = 4.5 \text{ g CaCO}_3 \times \frac{1 \text{ mol CaCO}_3}{100.1 \text{ g CaCO}_3} \times \frac{1 \text{ mol CaCl}_2}{1 \text{ mol CaCO}_3} = 0.045 \text{ mol CaCl}_2$$

$$? \, M \text{ CaCl}_2 \text{ soln} = \frac{0.045 \text{ mol CaCl}_2}{0.300 \text{ L CaCl}_2 \text{ soln}} = \mathbf{0.15}\ \boldsymbol{M}\ \mathbf{CaCl_2}$$

4.44 See Sections 3.1 and 4.2 and Examples 3.2 and 4.6.

Unbalanced: $Pb(CH_3COO)_2(aq) + Na_2SO_4(aq) \rightarrow PbSO_4(s) + NaCH_3COO(aq)$ Start with $Pb(CH_3COO)_2$.
Step 1: $Pb(CH_3COO)_2(aq) + Na_2SO_4(aq) \rightarrow PbSO_4(s) + \underline{2}NaCH_3COO(aq)$ Balances CH_3COO units and Na.

Strategy: L $Pb(CH_3COO)_2$ soln → mol $Pb(CH_3COO)_2$ → mol $PbSO_4$ → g $PbSO_4$

$$? \text{ g } PbSO_4 = 0.0200 \text{ L } Pb(CH_3COO)_2 \text{ soln} \times \frac{2.55 \text{ mol } Pb(CH_3COO)_2}{1 \text{ L } Pb(CH_3COO)_2 \text{ soln}} \times \frac{1 \text{ mol } PbSO_4}{1 \text{ mol } Pb(CH_3COO)_2}$$

$$\times \frac{303.3 \text{ g } PbSO_4}{1 \text{ mol } PbSO_4} = \mathbf{15.5 \text{ g } PbSO_4}$$

4.46 See Section 4.2, Table 4.1 and Example 4.6.

The ions present from the reactants are $NH_4^+(aq)$, $SO_4^-(aq)$, $Ba^{2+}(aq)$ and $Cl^-(aq)$. The two possible products are NH_4Cl and $BaSO_4$. Rule 4 in Table 4.1 indicates $BaSO_4$ is insoluble.

Unbalanced: $(NH_4)_2SO_4(aq) + BaCl_2(aq) \rightarrow BaSO_4(s) + NH_4Cl(aq)$ Start with $(NH_4)_2SO_4$.
Step 1: $(NH_4)_2SO_4(aq) + BaCl_2(aq) \rightarrow BaSO_4(s) + \underline{2}NH_4Cl(aq)$ Balances NH_4 units and Cl.

Strategy: L $(NH_4)_2SO_4$ soln → mol $(NH_4)_2SO_4$ → mol $BaSO_4$ → g $BaSO_4$

$$? \text{ g } BaSO_4 = 0.021 \text{ L } (NH_4)_2SO_4 \text{ soln} \times \frac{3.5 \text{ mol } (NH_4)_2SO_4}{1 \text{ L } (NH_4)_2SO_4 \text{ soln}} \times \frac{1 \text{ mol } BaSO_4}{1 \text{ mol } (NH_4)_2SO_4}$$

$$\times \frac{233.4 \text{ g } BaSO_4}{1 \text{ mol } BaSO_4} = \mathbf{17 \text{ g } BaSO_4}$$

4.48 See Section 4.3 and Example 4.11.

Balanced Overall Equation: $HCl(aq) + KOH(aq) \rightarrow H_2O(l) + KCl(aq)$

Strategy: L KOH soln → mol KOH → mol HCl → M HCl soln

$$? \text{ mol HCl} = 0.0334 \text{ L KOH soln} \times \frac{2.2 \text{ mol KOH}}{1 \text{ L KOH soln}} \times \frac{1 \text{ mol HCl}}{1 \text{ mol KOH}} = 0.073 \text{ mol HCl}$$

$$? M \text{ HCl soln} = \frac{0.073 \text{ mol HCl}}{0.100 \text{ L HCl soln}} = \mathbf{0.73 \ \mathit{M} \text{ HCl}}$$

4.50 See Section 4.4 and Example 4.14.

$$\text{percent Ca} = \frac{\text{g Ca}}{\text{g sample}} \times 100\%$$

$$? \text{ g Ca} = 1.22 \text{ g } CaCO_3 \times \frac{40.1 \text{ g Ca}}{100.1 \text{ g } CaCO_3} = 0.489 \text{ g Ca}$$

$$\text{percent Ca} = \frac{0.489 \text{ g Ca}}{2.11 \text{ g sample}} \times 100\% = \mathbf{23.2\% \text{ Ca}}$$

4.52 *See Section 4.4 and Example 4.14.*

(a) The negative sign for the enthalpy change means heat is given off by the system.

(b) $? \text{ enthalpy change} = 3.00 \text{ g C} \times \frac{1 \text{ mol C}}{12.0 \text{ g C}} \times \frac{-393.5 \text{ kJ}}{1 \text{ mol C}} = \mathbf{-98.4 \text{ kJ}}$

4.54 *See Section 4.4 and Example 4.14.*

$$? \text{ enthalpy change} = 2.20 \text{ g } N_2 \times \frac{1 \text{mol } N_2}{28.0 \text{ g } N_2} \times \frac{+180 \text{ kJ}}{1 \text{ mol } N_2} = \mathbf{+14.1 \text{ kJ}}$$

4.56 *See Section 4.4 and Example 4.14.*

$$? \text{ enthalpy change} = 12.2 \text{ g Al} \times \frac{1 \text{ mol Al}}{27.0 \text{ g Al}} \times \frac{-852 \text{ kJ}}{2 \text{ mol Al}} = \mathbf{-192 \text{ kJ}}$$

4.58 *See Section 4.4 and Example 4.14.*

$$? \text{ enthalpy change} = 10.0 \text{ g } C_8H_{18} \times \frac{1 \text{ mol } C_8H_{18}}{114.1 \text{ g } C_8H_{18}} \times \frac{-5.45 \text{ x } 10^3 \text{ kJ}}{1 \text{ mol } C_8H_{18}} = \mathbf{-479 \text{ kJ}}$$

4.60 *See Section 4.5 and Example 4.16.*

Temperature change $= 39.1^\circ\text{C} - 22.5^\circ\text{C} = 16.6^\circ\text{C}$ or 16.6 K

$$q = 300 \text{ g} \times 4.184 \text{ x } 10^{-3} \frac{\text{kJ}}{\text{g} \cdot \text{K}} \times 16.6 \text{ K} = \mathbf{20.8 \text{ kJ}}$$

4.62 *See Section 4.5 and Example 4.16.*

Temperature change $= 34.2^\circ\text{C} - 22.5^\circ\text{C} = 11.7^\circ\text{C}$ or 11.7 K

$$q = 20.0 \text{ g} \times 0.900 \frac{\text{J}}{\text{g} \cdot \text{K}} \times 11.7 \text{ K} = \mathbf{211 \text{ J}}$$

4.64 *See Section 4.5 and Example 4.16.*

Temperature change $= 17.1^\circ\text{C} - 22.1^\circ\text{C} = -5.0^\circ\text{C}$ or -5.0 K

$$q = 107 \text{ g} \times 4.184 \frac{\text{J}}{\text{g} \cdot \text{K}} \times -5.0 \text{ K} = -2.2 \text{ x } 10^3 \text{ J or } \mathbf{-2.2 \text{ kJ}}$$

The decrease in the temperature of the solution indicates the process is endothermic.

4.66 *See Section 4.5 and Example 4.18.*

Temperature change $= 23.1^\circ\text{C} - 21.2^\circ\text{C} = 1.9^\circ\text{C}$ or 1.9 K

$$q = 300 \text{ g} \times 4.184 \frac{\text{J}}{\text{g} \cdot \text{K}} \times 1.9 \text{ K} = 2.4 \text{ x } 10^3 \text{ J or } \mathbf{2.4 \text{ kJ}}$$

The increase in the temperature of the mixture indicates the process is **exothermic**.

4.68 ***See Section 4.5 and Example 4.18.***

Temperature change $= 19.7^\circ C - 22.8^\circ C = -3.1^\circ C$ or $-3.1\ K$

$$q \text{ for } 0.100 \text{ mol } NH_4SCN = 523\ g \times 4.184 \frac{J}{g \cdot K} \times -3.1\ K = -6.8 \times 10^3\ J \text{ or } -6.8\ kJ$$

$$q \text{ for } 2.00 \text{ mol } NH_4SCN = \frac{-6.8\ kJ}{0.100 \text{ mol } NH_4SCN} \times 2.00 \text{ mol } NH_4SCN = -1.4 \times 10^2\ kJ$$

The decrease in the temperature of the mixture indicates the process is **endothermic**. The enthalpy change for the reaction is **$+1.4 \times 10^2$ kJ.**

4.70 ***See Section 4.4 and Example 4.14.***

$$?\ \Delta H = \frac{-95.44\ kJ}{2.00\ g\ C_8H_{18}} \times 228.3\ g\ C_8H_{18} \text{ for 2 moles } C_8H_{18} = \mathbf{-1.09 \times 10^4\ kJ}$$

4.72 ***See Section 4.4 and Example 4.14.***

$$?\ \Delta H = \frac{-5.65\ kJ}{3.00\ g\ NO} \times 60.0\ g\ NO \text{ for 2 moles NO} = \mathbf{-1.13 \times 10^2\ kJ}$$

4.74 ***See Section 4.4 and Example 4.15.***

$$?\ g\ CH_4 = 1.00\ g\ H_2O \times \frac{1\ mol\ H_2O}{18.0\ g\ H_2O} \times \frac{44.0\ kJ}{1\ mol\ H_2O} \times \frac{1\ mol\ CH_4}{890\ kJ} \times \frac{16.0\ g\ CH_4}{1\ mol\ CH_4} = \mathbf{0.0439\ g\ CH_4}$$

4.76 ***See Section 4.6 and Example 4.19.***

The best approach for working Hess's Law problems involves finding one substance that appears in the desired final equation and in just one of the equations to be combined. Then use the latter equation to begin finding the algebraic combination of equations needed to obtain the desired final equation. In this case, C_2H_2, H_2 and C_2H_6 each appear separately in the given equations. Hence, any one of the given equations can be used to begin with. The desired combination is:

$2CO_2(g) + 3H_2O(l) \rightarrow C_2H_6(g) + 7/2O_2(g)$		$\Delta H = +1560$ kJ
$C_2H_2(g) + 5/2O_2(g) \rightarrow 2CO_2(g) + H_2O(l)$		$\Delta H = -1300$ kJ
$2H_2(g) + O_2(g) \rightarrow 2H_2O(l)$		$\Delta H = -572$ kJ
$C_2H_2(g) + 2H_2(g) \rightarrow C_2H_6(g)$		**$\Delta H = -312$ kJ**

The first equation was reversed to provide $C_2H_6(g)$ as a product in the appropriate numbers in the final equation, and the sign of the ΔH was changed. The second equation was multiplied by one-half to provide $C_2H_2(g)$ in appropriate numbers in the final equation, and the ΔH for the second equation was also multiplied by one-half. The third equation was multiplied by two to provide H_2 as a reactant in appropriate numbers in the final equation, and the ΔH was also multiplied by two. $H_2O(l)$ and $O_2(g)$ appear on both sides of the equation in equal numbers and are not shown in the final equation.

4.78 *See Section 4.6, Example 4.19. and the solution for 4.76.*

$Zn(s) + 2HCl(aq)$	$\rightarrow ZnCl_2(aq) + H_2(g)$	$\Delta H =$	-152.4 kJ
$ZnCl_2(aq) + H_2O(l)$	$\rightarrow ZnO(s) + 2HCl(aq)$	$\Delta H =$	+90.2 kJ
$H_2(g) + 1/2O_2(g)$	$\rightarrow H_2O(l)$	$\Delta H =$	-285.8 kJ
$Zn(s) + 1/2O_2(g)$	$\rightarrow ZnO(s)$	**ΔH =**	**-348.0 kJ**

4.80 *See Sections 4.1 and 2 and Examples 4.1, 4 and 7.*

$$?\,M \text{ original NaOH} = \frac{5.30\text{ g NaOH}}{1.00\text{ L NaOH soln}} \times \frac{1\text{ mol NaOH}}{40.0\text{ g NaOH}} = \frac{0.132\text{ mol NaOH}}{\text{L NaOH soln}} = 0.132\,M\text{ NaOH}$$

Solving M(con) x V(con) = M(dil) x V(dil) for M(dil) gives

$$M(\text{dil}) = \frac{M(\text{con}) \times V(\text{con})}{V(\text{dil})} \qquad M(\text{dil}) = \frac{0.132\,M \times 0.100\text{ L}}{.500\text{ L}} = 0.00264\,M\text{ NaOH(dilute)}$$

Strategy: L H_2SO_4 soln $\rightarrow$ mol H_2SO_4 $\rightarrow$ mol NaOH $\rightarrow$ L NaOH soln

$$?\text{ L dil NaOH} = 0.033\text{ L H}_2\text{SO}_4\text{ soln} \times \frac{0.022\text{ mol H}_2\text{SO}_4}{1\text{ L H}_2\text{SO}_4\text{ soln}} \times \frac{2\text{ mol NaOH}}{1\text{ mol H}_2\text{SO}_4} \times \frac{1\text{ L dilute NaOH soln}}{0.0264\text{ mol NaOH}}$$

=.055 L dilute NaOH soln

4.82 *See Section 4.1 and Example 4.3.*

$$?\text{ g NaOH} = 0.080\text{ L NaOH soln} \times \frac{2.33\text{ mol NaOH}}{1\text{ L NaOH soln}} \times \frac{40.0\text{ g NaOH}}{1\text{ mol NaOH}} = 7.5\text{ g NaOH}$$

$$?\text{ g KOH} = 0.080\text{ L KOH soln} \times \frac{1.22\text{ mol KOH}}{1\text{ L KOH soln}} \times \frac{56.1\text{ g KOH}}{1\text{ mol KOH}} = 5.5\text{ g KOH}$$

Total mass of solid = 7.5 g NaOH + 5.5 g KOH = **13.0 g**.

4.84 *See Section 4.1.*

$$?\text{ mol Cl}^- \text{ from NaCl} = 0.150\text{ L NaCl soln} \times \frac{1.5\text{ mol NaCl}}{1\text{ L NaCl soln}} \times \frac{1\text{ mol Cl}^-}{1\text{ mol NaCl}} = 0.22\text{ mol Cl}^-$$

$$?\text{ mol Cl}^- \text{ from CaCl}_2 = 0.200\text{ L CaCl}_2\text{ soln} \times \frac{2.0\text{ mol CaCl}_2}{1\text{ L CaCl}_2\text{ soln}} \times \frac{2\text{ mol Cl}^-}{1\text{ mol CaCl}_2} = 0.80\text{ mol Cl}^-$$

Total mol Cl^- = 0.22 mol Cl^- + 0.80 mol Cl^- = 1.02 mol Cl^-.

$$?\text{ M Cl}^- = \frac{1.02\text{ mol Cl}^-}{.350\text{ L soln}} = \frac{2.91\text{ mol Cl}^-}{\text{L soln}} = \mathbf{2.91\,M\,Cl^-}$$

4.86 ***See Section 4.3 and Example 4.13.***

Balanced: $SnF_2(aq) + 2Pb^{2+}(aq) + 2Cl^-(aq) \rightarrow 2PbClF(s) + Sn^{2+}(aq)$

$$\text{percent } SnF_2 = \frac{\text{g } SnF_2}{\text{g sample}} \times 100\%$$

Strategy: $g\ PbClF \rightarrow mol\ PbClF \rightarrow mol\ SnF_2 \rightarrow g\ SnF_2$

$$?\ \text{g } SnF_2 = 0.105 \text{ g } PbClF \times \frac{1 \text{ mol } PbClF}{261.6 \text{ g } PbClF} \times \frac{1 \text{ mol } SnF_2}{2 \text{ mol } PbClF} \times \frac{156.7 \text{ g } SnF_2}{1 \text{ mol } SnF_2} = 0.0314 \text{ g } SnF_2$$

$$\text{percent } SnF_2 = \frac{0.0314 \text{ g } SnF_2}{10.50 \text{ g sample}} \times 100\% = \mathbf{0.299\%\ SnF_2}$$

4.88 ***See Section 4.2 and Example 4.7.***

Balanced: $3H_2SO_4(aq) + 3NaNO_2(aq) \rightarrow 2NO(g) + HNO_3(aq) + 3\ NaHSO_4(aq) + H_2O(l)$

Strategy: $g\ NO \rightarrow mol\ NO \rightarrow mol\ H_2SO_4 \rightarrow L\ H_2SO_4$

$$?\ \text{L } H_2SO_4 \text{ soln} = 2.44 \text{ g NO} \times \frac{1 \text{ mol NO}}{30.0 \text{ g NO}} \times \frac{3 \text{ mol } H_2SO_4}{2 \text{ mol NO}} \times \frac{1 \text{ L } H_2SO_4 \text{ soln}}{1.22 \text{ mol } H_2SO_4} = \mathbf{0.100\ L\ H_2SO_4\ soln}$$

4.90 ***See Section 4.6 and Example 4.19.***

$C_2H_2(g)$	$\rightarrow$	$2C(s) + H_2(g)$	$\Delta H = -227$ kJ
$2C(s) + H_2(g) + 2Cl_2(g)$	$\rightarrow$	$C_2H_2Cl_4(l)$	$\Delta H = +130$ kJ
$C_2H_2(g) + 2Cl_2(g)$	$\rightarrow$	$C_2H_2Cl_4(l)$	$\mathbf{\Delta H = -97\ kJ}$

Chapter 5: The Gaseous State

5.6 *See Section 5.1 and Example 5.1.*

(a) $? \text{ atm} = 334 \text{ torr} \times \dfrac{1 \text{ atm}}{760 \text{ torr}} = \mathbf{0.439\ atm}$

(b) $? \text{ atm} = 3944 \text{ Pa} \times \dfrac{1 \text{ kPa}}{10^3 \text{ Pa}} \times \dfrac{1 \text{ atm}}{101.325 \text{ kPa}} = \mathbf{0.03892\ atm}$

(c) $? \text{ torr} = 2.4 \text{ atm} \times \dfrac{760 \text{ torr}}{1 \text{ atm}} = \mathbf{1.8 \times 10^3\ torr}$

5.8 *See Section 5.1 and recall $T_K = T_C + 273$.*

(a) $T_K = 45 + 273 = \mathbf{318\ K}$ (b) $T_K = -28 + 273 = \mathbf{245\ K}$ (c) $T_K = 230 + 273 = \mathbf{503\ K}$

5.10 *See Section 5.3.*

State One:	State Two:
$P_1 = 1.02$ atm	$P_2 = ?$
$T_1 = 39 + 273 = 312$ K	$T_2 = 499 + 273 = 772$ K

Solving $\dfrac{P_1}{T_1} = \dfrac{P_2}{T_2}$ for P_2 gives $P_2 = P_1 \times \dfrac{T_2}{T_1}$ $\qquad P_2 = 1.02 \text{ atm} \times \dfrac{772 \text{ K}}{312 \text{ K}} = \mathbf{2.52\ atm}$

Since the temperature increase occurred at constant volume, an increase in pressure is expected.

5.12 *See Section 5.3 and Example 5.3.*

State One:	State Two:
$V_1 = 39.6$ mL	$V_2 = ?$
$T_1 = 27 + 273 = 300$ K	$T_2 = 127 + 273 = 400$ K

Solving $\dfrac{V_1}{T_1} = \dfrac{V_2}{T_2}$ for V_2 gives $V_2 = V_1 \times \dfrac{T_2}{T_1}$ $\qquad V_2 = 39.6 \text{ mL} \times \dfrac{400 \text{ K}}{300 \text{ K}} = \mathbf{52.8\ mL}$

Since the temperature increase occurred at constant pressure, an increase in volume is expected.

Note: It is not necessary to convert volumes to liters when using Boyle's Law. Since V_1 and V_2 appear on opposite sides of the equations, conversion factors for converting mL to L would appear on opposite sides of the equation in equal numbers and therefore cancel each other out.

5.14 ***See Section 5.3, Example 5.2 and note for Solution 5.12 concerning units.***

State One:
$P_1 = 399$ torr
$V_1 = 166$ mL

State Two:
$P_2 = 1$ atm or 760 torr
$V_2 = ?$

Solving $P_1 V_1 = P_2 V_2$ for V_2 gives $V_2 = V_1 \times \frac{P_1}{P_2}$

$$V_2 = 166\text{ mL} \times \frac{399\text{ torr}}{760\text{ torr}} = \mathbf{87.2\ mL}$$

Since the pressure increase occurred at constant temperature, a decrease in volume is expected.

5.16 ***See Section 5.3.***

State One:
$n_1 = 77.4 \times 10^6$ mol
$V_1 = 4.50 \times 10^4\ ft^3$
$T_1 = -5 + 273 = 268$ K

State Two:
$n_2 = 72.1 \times 10^6$ mol
$V_2 = ?$
$T_2 = +7 + 273 = 280$ K

Solving $P_1 V_1 = n_1RT_1$ for R and $P_2 V_2 = n_2RT_2$ for R gives $\frac{P_1V_1}{n_1T_1} = R = \frac{P_2V_2}{n_2T_2}$

At constant pressure, this yields $\frac{V_1}{n_1T_1} = \frac{V_2}{n_2T_2}$ which gives $V_2 = V_1 \times \frac{n_2}{n_1} \times \frac{T_2}{T_1}$

$$V_2 = 4.50 \times 10^4\ ft^3 \times \frac{72.1 \times 10^6\ \text{mol}}{77.4 \times 10^6\ \text{mol}} \times \frac{280\text{ K}}{268\text{ K}} = \mathbf{4.38 \times 10^4\ ft^3}$$

5.18 ***See Section 5.3 and Example 5.4.***

State One:
$P_1 = 2.11$ atm
$V_1 = 900$ mL
$T_1 = 0 + 273 = 273$ K

State Two:
$P_2 = 4.33$ atm
$V_2 = ?$
$T_2 = 22 + 273 = 295$ K

Solving $\frac{P_1V_1}{T_1} = \frac{P_2V_2}{T_2}$ for V_2 gives $V_2 = V_1 \times \frac{P_1}{P_2} \times \frac{T_2}{T_1}$

$$V_2 = 900\text{ mL} \times \frac{2.11\text{ atm}}{4.33\text{ atm}} \times \frac{295\text{ K}}{273\text{ K}} = \mathbf{474\ mL}$$

5.20 ***See Section 5.4.***

Known Quantities: $P = 230\text{ torr} \times \frac{1\text{ atm}}{760\text{ torr}} = 0.303$ atm $\quad V = 3.22$ L $\quad T = 33 + 273 = 306$ K

Solving PV = nRT for n gives $n = \frac{PV}{RT}$

$$n = \frac{(0.303\text{ atm})(3.22\text{ L})}{\left(0.0821\frac{\text{L}\cdot\text{atm}}{\text{mol}\cdot\text{K}}\right)(306\text{ K})} = \mathbf{0.0388\ mol\ H_2}$$

5.22 ***See Section 5.4.***

Known Quantities: $P = 100$ atm $V = 3.00$ L $T = 27 + 273 = 300$ K

Solving PV = nRT for n gives $n = \frac{PV}{RT}$

$$n = \frac{(100\ \text{atm})(3.00\ \text{L})}{\left(0.0821\frac{\text{L}\cdot\text{atm}}{\text{mol}\cdot\text{K}}\right)(300\ \text{K})} = \mathbf{12.2\ mol}$$

5.24 ***See Section 5.4 and Table 5.1.***

Known Quantities: $n = 82.3$ mol $P = 1.01 \times 10^5\ \text{Pa} \times \frac{1\ \text{mmHg}}{133.3\ \text{Pa}} \times \frac{1\ \text{atm}}{760\ \text{mmHg}} = 0.997\ \text{atm}$

$T = 25 + 273 = 298$ K

Solving PV = nRT for V gives $V = \frac{nRT}{P}$

$$V = \frac{(82.3\ \text{mol})\left(0.0821\frac{\text{L}\cdot\text{atm}}{\text{mol}\cdot\text{K}}\right)(298\ \text{K})}{(0.997\ \text{atm})} = \mathbf{2.02 \times 10^3\ L\ H_2}$$

5.26 ***See Section 5.4 and Example 5.6.***

Known Quantities: $m = 0.550$ g $P = 744\ \text{torr} \times \frac{1\ \text{atm}}{760\ \text{torr}} = 0.979\ \text{atm}$

$V = 258\ \text{mL} \times \frac{1\ \text{L}}{10^3\ \text{mL}} = 0.258\ \text{L}$ $T = 22 + 273 = 295$ K

Solving PV = nRT for n gives $n = \frac{PV}{RT}$

$$n = \frac{(0.979\ \text{atm})(0.258\ \text{L})}{\left(0.0821\frac{\text{L}\cdot\text{atm}}{\text{mol}\cdot\text{K}}\right)(295\ \text{K})} = 0.0104\ \text{mol}$$

and using $\mathcal{M} = \frac{m}{n}$ yields $\mathcal{M} = \frac{0.550\ \text{g}}{0.0104\ \text{mol}} = \mathbf{52.9\ g/mol}$

5.28 ***See Sections 5.2 and 4 and Example 5.7.***

Known Quantities: $m = 2.41$ g $P = 1.00$ atm $V = 1.00$ L $T = 0 + 273 = 273$ K

Solving PV = nRT for n gives $n = \frac{PV}{RT}$

$$n = \frac{(1.00\ \text{atm})(1.00\ \text{L})}{\left(0.0821\frac{\text{L}\cdot\text{atm}}{\text{mol}\cdot\text{K}}\right)(273\ \text{K})} = 0.0446\ \text{mol}$$

and using $\mathcal{M} = \frac{m}{n}$ yields $\mathcal{M} = \frac{2.41\ g}{0.0446\ mol} = \mathbf{54.0\ g/mol}$

Note: An alternative solution which is based on the value of the molar volume of a gas at STP is given by:

$$\mathcal{M} = 2.41\ \frac{g}{L} \times 22.4\ \frac{L}{mol} = \mathbf{54.0\ g/mol}$$

5.30	***See Section 5.5.***

Unbalanced:	$Li + H_2O$	→	$LiOH + H_2$	Start with H_2, since Li and O are balanced..
Step 1:	$Li + H_2O$	→	$LiOH + \underline{1/2}H_2$	Balances H.
Step 2:	$\underline{2}Li + \underline{2}H_2O$	→	$\underline{2}LiOH + H_2$	Gives whole number coefficients.

Strategy: $g\ Li \rightarrow mol\ Li \rightarrow mol\ H_2$

$$?\ mol\ H_2 = 0.0223\ g\ Li \times \frac{1\ mol\ Li}{6.94\ g\ Li} \times \frac{1\ mol\ H_2}{2\ mol\ Li} = 0.00161\ mol\ H_2$$

Known Quantities: $n = 0.00161\ mol\ H_2$ $P = 1.33\ atm$ $T = 33 + 273 = 306\ K$

Solving PV = nRT for V gives $V = \frac{nRT}{P}$ $V = \frac{(0.00161\ mol)\left(0.0821 \frac{L \cdot atm}{mol \cdot K}\right)(306\ K)}{(1.33\ atm)} = \mathbf{0.0304\ L\ H_2}$

5.32	***See Section 5.5.***

Unbalanced:	$KClO_3$	→	$KCl + O_2$	Start with $KClO_3$.
Step 1:	$\underline{2}KClO_3$	→	$KCl + \underline{3}O_2$	Balances O.
Step 2:	$2KClO_3$	→	$\underline{2}KCl + 3O_2$	Balances K and Cl.

Strategy: $g\ KClO_3 \rightarrow mol\ KClO_3 \rightarrow mol\ O_2$

$$?\ mol\ O = 4.42\ g\ KClO_3 \times \frac{1\ mol\ KClO_3}{122.6\ g\ KClO_3} \times \frac{3\ mol\ O_2}{2\ mol\ KClO_3} = 0.0541\ mol\ O_2$$

Known Quantities: $n = 0.0541\ mol\ O_2$ $P = 760\ torr = 1.00\ atm$ $T = 23 + 273 = 296\ K$

Solving PV = nRT for V gives $V = \frac{nRT}{P}$ $V = \frac{(0.0541\ mol)\left(0.0821 \frac{L \cdot atm}{mol \cdot K}\right)(296\ K)}{(1.00\ atm)} = \mathbf{1.31\ L\ O_2}$

5.34	***See Section 5.5 and Example 5.8.***

Balanced: $Zn + H_2SO_4 \rightarrow ZnSO_4 + H_2$

Strategy: g Zn → mol Zn → mol H_2

$$? \text{ mol } H_2 \text{ based on Zn} = 1.33 \text{ g Zn} \times \frac{1 \text{ mol Zn}}{65.4 \text{ g Zn}} \times \frac{1 \text{ mol } H_2}{1 \text{ mol Zn}} = 0.0203 \text{ mol } H_2$$

Strategy: L H_2SO_4 *soln → mol* H_2SO_4 *→ mol* H_2

$$? \text{ mol } H_2 \text{ based on } H_2SO_4 = 0.300 \text{ L } H_2SO_4 \text{ soln} \times \frac{2.33 \text{ mol } H_2SO_4}{1 \text{ L } H_2SO_4 \text{ soln}} \times \frac{1 \text{ mol } H_2}{1 \text{ mol } H_2SO_4} = 0.699 \text{ mol } H_2$$

Zn is the limiting reactant, since it yields less H_2.

Known Quantities: $n = 0.0203 \text{ mol } H_2$ $P = 1.12 \text{ atm}$ $T = 25 + 273 = 298 \text{ K}$

Solving PV = nRT for V gives $V = \frac{nRT}{P}$ $$V = \frac{(0.0203 \text{ mol})\left(0.0821 \frac{\text{L}\cdot\text{atm}}{\text{mol}\cdot\text{K}}\right)(298 \text{ K})}{(1.12 \text{ atm})} = \mathbf{0.443 \text{ L } H_2}$$

5.36 ***See Section 5.5.***

Unbalanced:	NaN_3	→	$Na + N_2$	Start with NaN_3.
Step 1:	$\underline{2}NaN_3$	→	$Na + \underline{3}N_2$	Balances N.
Step 2:	$2NaN_3$	→	$\underline{2}Na + 3N_2$	Balances Na.

Strategy: g NaN_3 *→ mol* NaN_3 *→ mol* N_2

$$? \text{ mol } N_2 = 1.88 \text{ g } NaN_3 \times \frac{1 \text{ mol } NaN_3}{65.0 \text{ g } NaN_3} \times \frac{3 \text{ mol } N_2}{2 \text{ mol } NaN_3} = 0.0434 \text{ mol } N_2$$

Known Quantities: $n = 0.0434 \text{ mol } N_2$ $P = 755 \text{ torr} \times \frac{1 \text{ atm}}{760 \text{ torr}} = 0.993 \text{ atm}$ $T = 24 + 273 = 297 \text{ K}$

Solving PV = nRT for V gives $V = \frac{nRT}{P}$ $$V = \frac{(0.0434 \text{ mol})\left(0.0821 \frac{\text{L}\cdot\text{atm}}{\text{mol}\cdot\text{K}}\right)(297 \text{ K})}{0.993 \text{ atm}} = \mathbf{1.07 \text{ L } N_2}$$

5.38 ***See Section 5.5 and Example 5.9.***

Unbalanced:	$H_2S + O_2$	→	$SO_2 + H_2O$	Start with O_2, since H and S are balanced.
Step 1:	$H_2S + \underline{3/2}O_2$	→	$SO_2 + H_2O$	Balances O.
Step 2:	$\underline{2}H_2S + \underline{3}O_2$	→	$\underline{2}SO_2 + \underline{2}H_2O$	Gives whole number coefficients.

$$? \text{ L } SO_2 = 2.44 \text{ L } H_2S \times \frac{2 \text{ L } SO_2}{2 \text{ L } H_2S} = \mathbf{2.44 \text{ L } SO_2}$$ $$? \text{ L } O_2 = 2.44 \text{ L } H_2S \times \frac{3 \text{ L } O_2}{2 \text{ L } H_2S} = \mathbf{3.66 \text{ L } O_2}$$

5.40 ***See Section 5.6.***

$$P_T = P_{H_2} + P_{Ar} = 1.22 \text{ atm} + 4.33 \text{ atm} = \mathbf{5.55 \text{ atm}}$$

5.42 ***See Sections 5.3, 4 and 6.***

The partial pressure of the gas already in the 3.11 L container, gas 1, will remain the same. The partial pressure of the gas that is added to the 3.11 L container, gas 2, can be calculated using Boyle's Law or the Ideal Gas Law. The approach of Boyle's Law will be used here.

State One:	State Two:
$P_1 = 2.55$ atm	$P_2 = ?$
$V_1 = 2.11$ L	$V_2 = 3.11$ L
$T_1 = 27 + 273 = 300$ K	$T_2 = 27 + 273 = 300$ K

Solving $\frac{P_1 V_1}{T_1} = \frac{P_2 V_2}{T_2}$ for P_2 at a constant temperature gives $P_2 = P_1 \times \frac{V_1}{V_2}$ $\quad P_2 = 2.55 \text{ atm} \times \frac{2.11 \text{ L}}{3.11 \text{ L}} = 1.73 \text{ atm}$

Hence, the total pressure of the mixture of gases 1 and 2 in the 3.11 L container is given by $P_T = P_{gas\ 1} + P_{gas\ 2}$ and is $P_T = 4.33 \text{ atm} + 1.73 \text{ atm} = \mathbf{6.06\ atm}$.

Note: The alternative approach involves using PV = nRT to calculate the moles of gas in the 2.11 L container. The pressure that would be exerted by this same number of moles of gas in the 3.11 L container is then calculated by using PV = nRT a second time. The pressure exerted by this second gas is added to the pressure exerted by the first gas to obtain the total pressure.

5.44 ***See Section 5.6 and Example 5.10.***

Known Quantities: $n_{O_2} = 0.322$ mol $\quad V_{O_2} = 3.22$ L $\quad T_{O_2} = 100 + 273 = 373$ K

$n_{N_2} = 1.53$ mol $\quad V_{N_2} = 3.22$ L $\quad T_{N_2} = 100 + 273 = 373$ K

Solving PV = nRT for P gives $P_{O_2} = \frac{n_{O_2} RT}{V_{O_2}}$ $\quad P_{O_2} = \frac{(0.322 \text{ mol})\left(0.0821 \frac{\text{L}\cdot\text{atm}}{\text{mol}\cdot\text{K}}\right)(373 \text{ K})}{3.22 \text{ L}} = 3.06 \text{ atm}$

and $\quad P_{N_2} = \frac{n_{N_2} RT}{V_{N_2}}$ $\quad P_{N_2} = \frac{(1.53 \text{ mol})\left(0.0821 \frac{\text{L}\cdot\text{atm}}{\text{mol}\cdot\text{K}}\right)(373 \text{ K})}{3.22 \text{ L}} = 14.6 \text{ atm}$

Hence, $P_T = P_{O_2} + P_{N_2} = 3.06 \text{ atm} + 14.6 \text{ atm} = \mathbf{17.7\ atm}$

5.46 ***See Sections 5.4 and 6 and Example 5.5.***

Known Quantities: $n_{H_2} = 10.5 \text{ g } H_2 \times \frac{1 \text{ mol } H_2}{2.02 \text{ g } H_2} = 5.20 \text{ mol } H_2$ $\quad V = 30.0$ L $\quad T = 120 + 273 = 393$ K

$P_{Ar} = 1.53$ atm

Solving PV = nRT for P gives $P_{H_2} = \frac{nRT}{V}$ $\quad P_{H_2} = \frac{(5.20 \text{ mol})\left(0.0821 \frac{\text{L}\cdot\text{atm}}{\text{mol}\cdot\text{K}}\right)(393 \text{ K})}{30.0 \text{ L}} = \mathbf{5.59\ atm}$

Hence, $P_T = P_{H_2} + P_{Ar} = 5.59 \text{ atm} + 1.53 \text{ atm} = \mathbf{7.12\ atm}$

5.48 *See Section 5.6, Table 5.3 and Example 5.11.*

Solving $P_T = P_{O_2} + P_{H_2O}$ for P_{O_2} gives $P_{O_2} = P_T - P_{H_2O}$ At 26° C, $P_{O_2} = 755 \text{ torr} - 25.21 \text{ torr} = \mathbf{730\ torr}$

5.50 *See Section 5.6 and Example 5.12.*

Known Quantities: $n_{H_2} = 0.220$ mol $n_{N_2} = 0.432$ mol $P_T = 5.22$ atm

Hence, $\chi_{H_2} = \dfrac{0.220 \text{ mol}}{0.652 \text{ mol}} = 0.337$ and $P_{H_2} = \chi_{H_2} P_T = (0.337)(5.22 \text{ atm}) = \mathbf{1.76\ atm}$

5.54 *See Section 5.7.*

Since μ_{rms} increases with decreasing molar mass ($\mathcal{M}$), we obtain the following order of increasing rms speeds at a given temperature: $\mathbf{O_2 < N_2 < Ne}$.

5.56 *See Section 5.7.*

Since μ_{rms} increases with decreasing molar mass ($\mathcal{M}$) and increases with increasing absolute temperature, we obtain the following order of increasing rms speeds: **Ar at 25°C < Ne at 25°C < Ne at 100°C.**

5.58 *See Section 5.7 and Example 5.10.*

For Ne atoms at 100°C,

$$\mu_{rms} = \sqrt{\frac{3RT}{\mathcal{M}}} = \sqrt{\frac{(3)(8.314 \text{ kg}\cdot\text{m}^2\cdot\text{s}^{-2}\cdot\text{mol}^{-1}\cdot\text{K}^{-1})(373 \text{ K})}{0.02018 \text{ kg}\cdot\text{mol}^{-1}}} = \mathbf{679\ m/s}$$

5.60 *See Section 5.7 and Example 5.10.*

Solving Equation 3 for $\mathcal{M}$ gives

$$\mathcal{M} = \frac{3RT}{(\mu_{rms})^2} = \frac{(3)(8.314 \text{ kg}\cdot\text{m}^2\cdot\text{s}^{-2}\cdot\text{mol}^{-1}\cdot\text{K}^{-1})(301 \text{ K})}{(518 \text{ m}\cdot\text{s}^{-1})^2} = 0.0280 \text{ kg}\cdot\text{mol}^{-1} = \mathbf{28.0\ g/mol}$$

5.64 *See Section 5.8.*

$$\frac{\text{rate of effusion of He}}{\text{rate of effusion of Ne}} = \sqrt{\frac{\mathcal{M}_{Ne}}{\mathcal{M}_{He}}} = \sqrt{\frac{20.18}{4.003}} = \sqrt{5.041} = \mathbf{2.245}$$

5.66 *See Section 5.8.*

The calculation in Section 5.8 indicates He diffuses 3.16 times faster than Ar. Hence, the volume loss for the argon balloon should be approximately one-third that for the helium balloon. It is 50 mL/3.16 = **16 mL.**

5.68 ***See Section 5.8 and Example 5.14.***

$$\frac{t_x}{t_{H_2}} = \sqrt{\frac{\mathcal{M}_x}{\mathcal{M}_{H_2}}} \text{ yields } \mathcal{M}_x = \mathcal{M}_{H_2} \times \frac{(t_x)^2}{(t_{H_2})^2} = 2.02 \text{ g/mol} \times \frac{(9.12 \text{ min})^2}{(1.20 \text{ min})^2} = \mathbf{117 \text{ g/mol}}$$

5.72 ***See Section 5.9.***

(a) Oxygen gas at 100°C is more likely to follow the Ideal Gas Law than oxygen gas at -100°C. Deviations caused by the actual size of the gas particles and the attractive forces become less important in gas samples at higher temperatures, since the gas particles are further apart and have higher rms speeds.

(b) Nitrogen gas is more likely to follow the Ideal Gas Law at -100°C than xenon at -100°C, since -100°C is closer to the boiling point of xenon (-107°C) than that of nitrogen (-196°C). Deviations caused by the actual size of the gas particles and the attractive forces between the particles are likely to be high near the boiling point/condensation point of a substance.

(c) Argon gas at 1 atm pressure is more likely to follow the Ideal Gas Law than argon gas at 50 atm pressure. At high pressures, the gas particles are forced closer together causing the volumes of the gas particles and attractive forces between particles to become significant.

5.74 ***See Section 5.9, Table 5.4 and Example 5.16.***

Solving the Ideal Gas Law equation for pressure gives $P = \frac{nRT}{V}$

$$P = \frac{(10.2 \text{ mol})\left(0.0821 \frac{\text{L}\cdot\text{atm}}{\text{mol}\cdot\text{K}}\right)(803 \text{ K})}{3.23 \text{ L}} = \mathbf{208 \text{ atm}}$$

Solving the van der Waals equation for pressure gives $P = \frac{nRT}{V\text{-}bn} - \frac{an^2}{V^2}$

$$P = \frac{(10.2 \text{ mol})\left(0.0821 \frac{\text{L}\cdot\text{atm}}{\text{mol}\cdot\text{K}}\right)(803 \text{ K})}{3.23 \text{ L} - \left(0.0322 \text{ L}\cdot\text{mol}^{-1}\right)(10.2 \text{ mol})} - \frac{\left(1.34 \text{ L}^2\cdot\text{atm}\cdot\text{mol}^{-2}\right)(10.2 \text{ mol})^2}{(3.23 \text{ L})^2} = \mathbf{218 \text{ atm}}$$

The pressure calculated using the Ideal Gas Law is less than that calculated using the van der Waals equation. Under these conditions, the correction for the actual size of the Ar atoms (bn) dominates the correction for attractive forces between Ar atoms ($-an^2/V^2$), since the former leads to an increase in pressure and the latter to a decrease in pressure.

5.76 ***See Section 5.3.***

A tire gauge measures the difference in pressure between the pressure of the gas in the tire and atmospheric pressure: $P_{gauge} = P_{gas} - P_{atm}$. Hence, we have $P_{gas} = P_{gauge} + P_{atm}$ and

State One:

$P_1 = 32 \text{ psi} + 15 \text{ psi} = 47 \text{ psi}$

State Two:

$P_2 = ?$

$$T_1 = \left(\frac{1.0^\circ C}{1.8^\circ F}\right)(90-32)^\circ F + 273 = 305 \text{ K}$$

$$T_2 = \left(\frac{1.0^\circ C}{1.8^\circ F}\right)(32-32)^\circ F + 273 = 273 \text{ K}$$

Solving $\frac{P_1}{T_1} = \frac{P_2}{T_2}$ for P_2 gives $P_2 = P_1 \times \frac{T_2}{T_1}$

$$P_2 = 47 \text{ psi} \times \frac{273 \text{ K}}{305 \text{ K}} = \mathbf{42 \text{ psi}}$$

This means the pressure of the gas in the tire decreases by 5 psi and $P_{gauge} = P_{gas} - P_{atm} = 42 \text{ psi} - 15 \text{ psi} = \mathbf{27 \text{ psi}}$.

5.78 ***See Section 5.4 and Example 5.5.***

Known Quantities: $n_{O_2} = 0.24 \text{ kg } O_2 \times \frac{10^3 \text{ g } O_2}{\text{kg } O_2} \times \frac{1 \text{ mol } O_2}{32.0 \text{ g } O_2} = 7.5 \text{ mol } O_2$ $V = 2.8 \text{ L}$

$T = 20 + 273 = 293 \text{ K}$

Solving PV = nRT for P gives $P = \frac{nRT}{V}$

$$P = \frac{(7.5 \text{ mol})\left(0.0821 \frac{\text{L}\cdot\text{atm}}{\text{mol}\cdot\text{K}}\right)(293 \text{ K})}{2.8 \text{ L}} = \mathbf{64 \text{ atm}}$$

5.80 See Sections 3.1 and 4 and 5.4 and Examples 3.3 and 5.5.

Unbalanced:	$C_{56}H_{108}O_6 + O_2 \rightarrow CO_2 + H_2O$	Start with $C_{56}H_{108}O_6$.
Step 1:	$C_{56}H_{108}O_6 + O_2 \rightarrow \underline{56}CO_2 + \underline{54}H_2O$	Balances C and H.
Step 2:	$C_{56}H_{108}O_6 + \underline{80}O_2 \rightarrow 56CO_2 + 54H_2O$	Balances O.

Strategy: $lb\ C_{56}H_{108}O_6 \rightarrow g\ C_{56}H_{108}O_6 \rightarrow mol\ C_{56}H_{108}O_6 \rightarrow mol\ O_2$

$$? \text{ mol } O_2 = 5.0 \text{ lb } C_{56}H_{108}O_6 \times \frac{453.6 \text{ g } C_{56}H_{108}O_6}{1 \text{ lb } C_{56}H_{108}O_6} \times \frac{1 \text{ mol } C_{56}H_{108}O_6}{876.9 \text{ g } C_{56}H_{108}O_6}$$

$$\times \frac{80 \text{ mol } O_2}{1 \text{ mol } C_{56}H_{108}O_6} = 2.1 \text{ x } 10^2 \text{ mol } O_2$$

Known Quantities: $n_{O_2} = 2.1 \text{ x } 10^2 \text{ mol}$ $P = 1.00 \text{ atm}$ $T = 22 + 273 = 295 \text{ K}$

Solving PV = nRT for V gives $V = \frac{nRT}{P}$

$$V = \frac{(2.1 \text{ x } 10^2 \text{ mol})\left(0.0821 \frac{\text{L}\cdot\text{atm}}{\text{mol}\cdot\text{K}}\right)(295 \text{ K})}{1.00 \text{ atm}} = \mathbf{5.1 \text{ x } 10^3 \text{ L } O_2}$$

5.82 See Sections 3.1 and 5 and 5.4 and 5 and Examples 3.18.

Unbalanced:	$H_2 + O_2 \rightarrow H_2O$	Start with H_2O.
Step 1:	$H_2 + O_2 \rightarrow \underline{2}H_2O$	Balances O.
Step 2:	$\underline{2}H_2 + O_2 \rightarrow 2H_2O$	Balances H.

Known Quantities: $P = 1.22$ atm $V_{O_2} = 4.33$ L $V_{H_2} = 6.77$ L $T = 27 + 273 = 300$ K

Solving PV = nRT for n gives $n_{O_2} = \frac{PV_{O_2}}{RT}$ $$n_{O_2} = \frac{(1.22 \text{ atm})(4.33 \text{ L})}{\left(0.0821 \frac{\text{L}\cdot\text{atm}}{\text{mol}\cdot\text{K}}\right)(300 \text{ K})} = 0.214 \text{ mol } O_2$$

and $n_{H_2} = \frac{PV_{H_2}}{RT}$ $$n_{H_2} = \frac{(1.22 \text{ atm})(6.77 \text{ L})}{\left(0.0821 \frac{\text{L}\cdot\text{atm}}{\text{mol}\cdot\text{K}}\right)(300 \text{ K})} = 0.335 \text{ mol } H_2$$

$$? \text{ mol } H_2O \text{ based on } O_2 = 0.214 \text{ mol } O_2 \times \frac{2 \text{ mol } H_2O}{1 \text{ mol } O_2} = 0.428 \text{ mol } H_2O$$

$$? \text{ mol } H_2O \text{ based on } H_2 = 0.335 \text{ mol } H_2 \times \frac{2 \text{ mol } H_2O}{2 \text{ mol } H_2} = 0.335 \text{ mol } H_2O$$

H_2 is the limiting reactant, since it gives less moles of

H_2O. $? \text{ g } H_2O = 0.335 \text{ mol } H_2O \times \frac{18.02 \text{ g } H_2O}{1 \text{ mol } H_2O} = \mathbf{6.04 \text{ g } H_2O}$

5.84 See Sections 3.4 and 5.4 and 5.

Balanced: $2LiOH(s) + CO_2(g) \rightarrow Li_2CO_3(s) + H_2O(l)$

Known Quantities for CO_2: $P = 1.00$ atm $V = 400$ L $T = 24 + 273 = 297$ K

Solving PV = nRT for n gives $n_{CO_2} = \frac{PV_{CO_2}}{RT}$ $$n_{CO_2} = \frac{(1.00 \text{ atm})(400 \text{ L})}{\left(0.0821 \frac{\text{L}\cdot\text{atm}}{\text{mol}\cdot\text{K}}\right)(297 \text{ K})} = 16.4 \text{ mol } CO_2$$

Strategy: mol CO_2 → mol LiOH → g LiOH

$$? \text{ g LiOH} = 16.4 \text{ mol } CO_2 \times \frac{2 \text{ mol LiOH}}{1 \text{ mol } CO_2} \times \frac{23.9 \text{ g LiOH}}{1 \text{ mol LiOH}} = \mathbf{784 \text{ g LiOH}}$$

Chapter 6: Electrons in Atoms

6.2	*See Section 6.1.*

In a vacuum, all electromagnetic radiation travels at the same speed, $c = 3.00 \times 10^8 \text{ m·s}^{-1}$. The product of wavelength times frequency is equal to this speed. This means the shorter the wavelength of the light, the higher its frequency. Hence, light waves having wavelengths of 560 nm and 720 nm travel at equal speed, but the 560 nm light waves have a higher frequency.

6.4	*See Section 6.1 and Figure 6.3.*

Solving $\lambda\nu = c$ for λ and recognizing that $1 \text{ Hz} = 1 \text{ s}^{-1}$ gives $\lambda = \dfrac{c}{\nu}$

$$\lambda = \frac{3.00 \times 10^8 \text{ m·s}^{-1}}{6.00 \times 10^{13} \text{ s}^{-1}} = \mathbf{5.00 \times 10^{-6} \text{ m}}$$

This wavelength is equivalent to 5.00×10^3 nm.
The radiation is in the **infrared** region of the electromagnetic spectrum.

6.6	*See Section 6.1.*

Solving $\lambda\nu = c$ for λ and recognizing that $1 \text{ Hz} = 1 \text{ s}^{-1}$ gives $\lambda = \dfrac{c}{\nu}$

$$\lambda = \frac{3.00 \times 10^8 \text{ m·s}^{-1}}{2.47 \times 10^{20} \text{ s}^{-1}} \times \frac{10^{12} \text{ pm}}{\text{m}} = \mathbf{1.21 \text{ pm}}$$

6.8	*See Section 6.1 and Examples 6.1 and 2.*

(a) Solving $\lambda\nu = c$ for ν gives $\nu = \dfrac{c}{\lambda}$

$$\nu = \frac{3.00 \times 10^8 \text{ m·s}^{-1}}{488 \text{ nm}} \times \frac{10^9 \text{ nm}}{1 \text{ m}} = 6.15 \times 10^{14} \text{ s}^{-1}$$

Substituting ν in $\Delta E = h\nu$ yields $\Delta E = \left(6.63 \times 10^{-34} \text{ J·s}\right) \times \left(6.15 \times 10^{14} \text{ s}^{-1}\right) = \mathbf{4.08 \times 10^{-19} \text{ J}}$

$$\text{(b) ? photons} = \frac{1 \text{ photon}}{4.08 \times 10^{-19} \text{ J}} \times 1.00 \frac{\text{J}}{\text{s}} = \mathbf{2.45 \times 10^{18} \frac{photons}{s}}$$

6.10 *See Section 6.1 and Examples 6.1 and 2.*

Solving $\lambda\nu = c$ for ν gives $\nu = \dfrac{c}{\lambda}$

$$\nu = \frac{3.00 \text{ x } 10^{8} \text{ m}\cdot\text{s}^{-1}}{589 \text{ nm}} \times \frac{10^{9} \text{ nm}}{1 \text{ m}} = 5.09 \text{ x } 10^{14} \text{ s}^{-1}$$

Substituting ν in $\Delta E = h\nu$ yields $\Delta E = \left(6.63 \text{ x } 10^{-34} \text{ J}\cdot\text{s}\right)\times\left(5.09 \text{ x } 10^{14} \text{ s}^{-1}\right) = \mathbf{3.37 \text{ x } 10^{-19} \text{ J}}$

6.12 *See Section 6.1 and Figures 6.13.*

(a) Solving $\lambda\nu = c$ for λ gives $\lambda = \dfrac{c}{\nu}$

$$\lambda = \frac{3.00 \text{ x } 10^{8} \text{ m}\cdot\text{s}^{-1}}{1.16 \text{ x } 10^{15} \text{ s}^{-1}} = \mathbf{2.59 \text{ x } 10^{-7} \text{ m}}$$

This wavelength is equivalent to 259 nm. The radiation is in the **ultraviolet** region of the electromagnetic spectrum.

(b) Light in the visible region of the electromagnetic spectrum (400-700 nm) has a longer wavelength. Hence, it does not have sufficient energy to dislodge electrons from carbon.

6.14 *See Section 6.1.*

$$? \text{ electrons} = 1.0 \text{ x } 10^{-6} \frac{\text{coul}}{\text{s}} \times \frac{1 \text{ electron}}{1.602 \text{ x } 10^{-19} \text{ coul}} = \mathbf{6.2 \text{ x } 10^{12} \frac{electrons}{s}}$$

An electron is dislodged from a metal by collision with a single photon, having an energy of $h\nu$, provided that energy exceeds the threshold energy. Hence, 6.2 x 10^{12} photons must be absorbed per second to produce 6.2 x 10^{12} electrons per second.

6.16 *See Section 6.2 and Example 6.7.*

Substituting in $\dfrac{1}{\lambda} = Z^2 R_h \left(\dfrac{1}{n_1^2} - \dfrac{1}{n_2^2}\right)$ and noting that Z = 3 for Li^{2+} gives

$$\frac{1}{\lambda} = (3)^2\left(1.097 \text{ x } 10^{7} \text{ m}^{-1}\right)\left(\frac{1}{4^2} - \frac{1}{6^2}\right) = (3)^2\left(1.097 \text{ x } 10^{7} \text{ m}^{-1}\right)\left(\frac{6^2 - 4^2}{4^2 \times 6^2}\right) = 3.428 \text{ x } 10^{6} \text{ m}^{-1}$$

Solving for λ yields $\lambda = \dfrac{1}{3.428 \text{ x } 10^{6} \text{ m}^{-1}} = \mathbf{2.917 \text{ x } 10^{-7} \text{ m}}$

This wavelength is equivalent to 292 nm. The radiation is in the **ultraviolet** region of the electromagnetic spectrum.

6.18 *See Section 6.2.*

At normal temperatures nearly all of the atoms of hydrogen are present in the ground state, so nearly all of the absorption lines arise from transitions from the ground state (n = 1) to the excited states (n > 1). On the other hand, electrons in excited states can return to more than one lower energy state (n = 1, 2, 3, etc.) giving more lines in the emission spectrum of hydrogen atoms.

6.20 *See Section 6.3 and Example 6.5.*

Solving $p = mv = \dfrac{h}{\lambda}$ for λ gives $\lambda = \dfrac{h}{mv}$

$$\lambda = \frac{6.63 \text{ x } 10^{-34} \text{ kg}\cdot\text{m}^2\cdot\text{s}^{-1}}{\left(9.11 \text{ x } 10^{-31} \text{ kg}\right)\left(2.19 \text{ x } 10^{6} \text{ m}\cdot\text{s}^{-1}\right)} = 3.32 \text{ x } 10^{-10} \text{ m}$$

This wavelength is equivalent to **332 pm**.

The circumference of the first Bohr orbit is given by $c = 2\pi r = (2)(3.143)(52.9 \text{ pm}) = \mathbf{332 \text{ pm}}$.

Hence, the wavelength of an electron traveling in the first Bohr orbit is exactly equal to the circumference of the first Bohr orbit.

6.22 *See Section 6.3 and Example 6.5.*

Solving $p = mv = \dfrac{h}{\lambda}$ for λ gives $\lambda = \dfrac{h}{mv}$

$$\text{(a) } \lambda = \frac{6.63 \text{ x } 10^{-34} \text{ kg}\cdot\text{m}^2\cdot\text{s}^{-1}}{(0.100 \text{ kg})\left(40.0 \text{ m}\cdot\text{s}^{-1}\right)} = \mathbf{1.66 \text{ x } 10^{-34} \text{ m}}$$

$$\text{(b) } \lambda = \frac{6.63 \text{ x } 10^{-34} \text{ kg}\cdot\text{m}^2\cdot\text{s}^{-1}}{(753 \text{ kg})\left(24.6 \text{ m}\cdot\text{s}^{-1}\right)} = \mathbf{3.58 \text{ x } 10^{-38} \text{ m}}$$

$$\text{(c) } \lambda = \frac{6.63 \text{ x } 10^{-34} \text{ kg}\cdot\text{m}^2\cdot\text{s}^{-1}}{\left(1.67 \text{ x } 10^{-27} \text{ kg}\right)\left(2.70 \text{ x } 10^{3} \text{ m}\cdot\text{s}^{-1}\right)} = \mathbf{1.47 \text{ x } 10^{-10} \text{ m}}$$

6.24 *See Section 6.3.*

Solving $p = mv = \dfrac{h}{\lambda}$ for v gives $v = \dfrac{h}{m\lambda}$

$$v = \frac{6.63 \text{ x } 10^{-34} \text{ kg}\cdot\text{m}^2\cdot\text{s}^{-1}}{\left(1.67 \text{ x } 10^{-27} \text{ kg}\right)(0.150 \text{ nm})} \times \frac{10^9 \text{ nm}}{\text{m}} = \mathbf{2.65 \text{ x } 10^{3} \text{ m/s}}$$

6.26 See Section 6.4 and Example 6.6.

(a) $n = 6$ and $\ell = 1$ are associated with the 6p subshell.
(b) $n = 3$ and $\ell = 0$ are associated with the 3s subshell.
(c) $n = 5$ and $\ell = 2$ are associated with the 5d subshell.
(d) $n = 4$ and $\ell = 0$ are associated with the 4s subshell.
(e) Since ℓ must be at least one less than n, a value of $\ell = 3$ is not possible when $n = 2$.

6.28 See Section 6.4.

(a) 3p corresponds to $n = 3$ and $\ell = 1$.
(b) 5d corresponds to $n = 5$ and $\ell = 2$.
(c) 7s corresponds to $n = 7$ and $\ell = 0$.
(d) 4f corresponds to $n = 4$ and $\ell = 3$.
(e) 2s corresponds to $n = 2$ and $\ell = 0$.

6.30 See Section 6.5 and Figure 6.14.

$2s < 3s = 3p_x = 3p_y < 4s = 4p_z = 4d_{xy}$

6.32 See Section 6.4.

(a) $n = 2$ and $\ell = 1$ are associated with the 2p subshell.
(b) Since ℓ must be at least one less than n, a value of $\ell = 2$ is not possible when $n = 2$.
(c) $n = 3$ and $\ell = 0$ are associated with the 3s subshell.
(d) Since allowed values for m_ℓ are all whole numbers from $-\ell$ to $+\ell$, a value $m_\ell = 1$ is not possible when $\ell = 0$.
(e) $n = 3$ and $\ell = 2$ are associated with the 3d subshell.
(f) $n = 5$ and $\ell = 0$ are associated with the 5s subshell.

6.44 See Section 6.5 and Figure 6.19.

(a) 2s < 2p < 3p < 3d < 5p
(b) 1s < 2s < 2p < 3s < 3d < 4d
(c) 1s < 2s < 2p < 3s < 3p < 3d < 4p

6.46 See Section 6.6.

(a) Be $1s^2 2s^2$ ↑↓ (1s) ↑↓ (2s)

Ne $1s^2 2s^2 2p^6$ ↑↓ (1s) ↑↓ (2s) ↑↓ ↑↓ ↑↓ (2p)

(b) N $1s^2 2s^2 2p^3$ ↑↓ (1s) ↑↓ (2s) ↑ ↑ ↑ (2p)

(c) H $1s^1$ ↑ (1s)

Li $1s^2 2s^1$ ↑↓ (1s) ↑ (2s)

B $1s^2 2s^2 2p^1$ ↑↓ (1s) ↑↓ (2s) ↑ _ _ (2p)

F $1s^2 2s^2 2p^5$ ↑↓ (1s) ↑↓ (2s) ↑↓ ↑↓ ↑ (2p)

6.48 ***See Section 6.6.***

Be and Ne; see solution for 6.46a.

6.50 ***See Section 6.6.***

(a) He, 1s (b) Be, 2s (c) C, 2p (d) F, 2p

6.52 ***See Sections 6.4 and 6 and Figure 6.19.***

Energy level diagram for C:

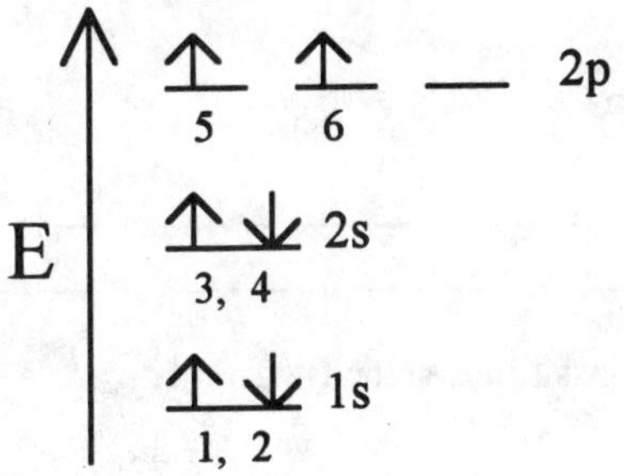

E ↑

2p: ↑ (5) ↑ (6) —

2s: ↑↓ (3, 4)

1s: ↑↓ (1, 2)

Quantum numbers for electrons in C:

electron	n	ℓ	m_ℓ	m_s
1	1	0	0	+1/2
2	1	0	0	−1/2
3	2	0	0	+1/2
4	2	0	0	−1/2
5	2	1	−1	+1/2
6	2	1	0	+1/2

The first electron placed in an orbital was arbitrarily assigned a m_s value of +1/2. The m_ℓ values chosen for the 2p electrons were arbitrarily chosen to be -1 and 0 in keeping with $m_\ell = -\ell$ to $+\ell$.

6.54 ***See Sections 6.4 and 6.***

Li $1s^2 2s^1$ The highest energy electron has $n = 2$, $\ell = 0$, $m_\ell = 0$ and $m_s = +1/2$, with the choice of assigning +1/2 for m_s being arbitrary.

6.56 *See Section 6.2 and Example 6.4.*

The Rydberg equation is $\frac{1}{\lambda} = R_h\left(\frac{1}{n_1^2} - \frac{1}{n_2^2}\right)$.

The two lowest energy lines in the Paschen series ($n_1 = 3$) involve transitions from $n_2 = 4$ and $n_2 = 5$ to $n_1 = 3$.

Substituting $n_1 = 3$ and $n_2 = 4$ into the Rydberg equation gives

$$\frac{1}{\lambda} = \left(1.097 \text{ x } 10^7 \text{ m}^{-1}\right)\left(\frac{1}{3^2} - \frac{1}{4^2}\right) = \left(1.097 \text{ x } 10^7 \text{ m}^{-1}\right)\left(\frac{4^2 - 3^2}{3^2 \times 4^2}\right) = 5.333 \text{ x } 10^5 \text{ m}^{-1}$$

Solving for λ in nm yields $\lambda = \frac{1}{5.333 \text{ x } 10^5 \text{ m}^{-1}} \times \frac{10^9 \text{ nm}}{1 \text{ m}} = \mathbf{1.875 \text{ x } 10^3 \text{ nm}}$

Substituting $n_1 = 3$ and $n_2 = 5$ into the Rydberg equation gives

$$\frac{1}{\lambda} = \left(1.097 \text{ x } 10^7 \text{ m}^{-1}\right)\left(\frac{1}{3^2} - \frac{1}{5^2}\right) = \left(1.097 \text{ x } 10^7 \text{ m}^{-1}\right)\left(\frac{5^2 - 3^2}{3^2 \times 5^2}\right) = 7.801 \text{ x } 10^5 \text{ m}^{-1}$$

Solving for λ in nm yields $\lambda = \frac{1}{7.801 \text{ x } 10^5 \text{ m}^{-1}} \times \frac{10^9 \text{ nm}}{1 \text{ m}} = \mathbf{1.282 \text{ x } 10^3 \text{ nm}}$

6.58 *See Insights into Chemistry: A Closer View.*

Recognizing that the momentum of the automobile is equal to the product of mass times velocity gives

$$\text{momentum, p} = 650 \text{ kg} \times 55\frac{\text{mi}}{\text{hr}} \times \frac{1.609 \text{ km}}{1 \text{ mi}} \times \frac{10^3 \text{ m}}{1 \text{ km}} \times \frac{1 \text{ hr}}{3600 \text{ s}} = 1.6 \text{ x } 10^4 \text{ kg}\cdot\text{m}\cdot\text{s}^{-1}$$

and $\Delta p = 0.1 \text{ x } 10^4 \text{ kg}\cdot\text{m}^2\cdot\text{s}^{-1}$

Solving $\Delta x \Delta p = 1.06 \text{ x } 10^{-34} \text{ kg}\cdot\text{m}^2\cdot\text{s}^{-1}$ for Δx gives $\Delta x = \frac{1.06 \text{ x } 10^{-34} \text{ kg}\cdot\text{m}^2\cdot\text{s}^{-1}}{\Delta p}$

and substituting for Δp gives $\Delta x = \frac{1.06 \text{ x } 10^{-34} \text{ kg}\cdot\text{m}^2\cdot\text{s}^{-1}}{0.1 \text{ x } 10^4 \text{ kg}\cdot\text{m}\cdot\text{s}^{-1}} = \mathbf{1 \text{ x } 10^{-37} \text{ m}}$

This amount of uncertainty in position is insignificant.

6.60 *See Section 6.6.*

(a) Li $1s^2 2s^1$

↑↓	↑
1s	2s

(b) F $1s^2\ 2s^2\ 2p^5$

↑↓	↑↓	↑↓ ↑↓ ↑
1s	2s	2p

(c) O $1s^2\ 2s^2\ 2p^4$

$$\underset{1s}{\uparrow\downarrow} \quad \underset{2s}{\uparrow\downarrow} \quad \underset{2p}{\uparrow\downarrow\ \uparrow\ \uparrow}$$

6.62 ***See Section 6.1 and 2.***

The Rydberg equation is $\frac{1}{\lambda} = R_h\left(\frac{1}{n_1^2} - \frac{1}{n_2^2}\right)$.

(a) Substituting $n_1 = 2$ and $n_2 = 3$ into the Rydberg equation gives

$$\frac{1}{\lambda} = \left(1.097 \text{ x } 10^7 \text{ m}^{-1}\right)\left(\frac{1}{2^2} - \frac{1}{3^2}\right) = \left(1.097 \text{ x } 10^7 \text{ m}^{-1}\right)\left(\frac{3^2 - 2^2}{2^2 \times 3^2}\right) = 1.524 \text{ x } 10^6 \text{ m}^{-1}$$

Substituting $\frac{1}{\lambda}$ in $\Delta E = h\nu = \frac{hc}{\lambda} = hc\frac{1}{\lambda}$ yields

$$\Delta E = \left(6.63 \text{ x } 10^{-34} \text{ J}\cdot\text{s}\right)\left(3.00 \text{ x } 10^8 \text{ m}\cdot\text{s}^{-1}\right)\left(1.524 \text{ x } 10^6 \text{ m}^{-1}\right) = \mathbf{3.02 \text{ x } 10^{-19} \text{ J}}$$

Substituting $n_1 = 32$ and $n_2 = 33$ into the Rydberg equation gives

$$\frac{1}{\lambda} = \left(1.097 \text{ x } 10^7 \text{ m}^{-1}\right)\left(\frac{1}{32^2} - \frac{1}{33^2}\right) = \left(1.097 \text{ x } 10^7 \text{ m}^{-1}\right)\left(\frac{33^2 - 32^2}{32^2 \times 33^2}\right) = 6.394 \text{ x } 10^2 \text{ m}^{-1}$$

Substituting $\frac{1}{\lambda}$ in $\Delta E = h\nu = \frac{hc}{\lambda} = hc\frac{1}{\lambda}$ yields

$$\Delta E = \left(6.63 \text{ x } 10^{-34} \text{ J}\cdot\text{s}\right)\left(3.00 \text{ x } 10^8 \text{ m}\cdot\text{s}^{-1}\right)\left(6.394 \text{ x } 10^2 \text{ m}^{-1}\right) = \mathbf{1.27 \text{ x } 10^{-22} \text{ J}}$$

The difference between the $n = 2$ and $n = 3$ levels is much larger than the energy difference between the $n = 32$ and $n = 33$ levels.

(b) The largest energy difference that can be observed for the hydrogen atom is the difference between the $n = 1$ level and the $n = \infty$ level. Sustituting $n_1 = 1$ and $n_2 = \infty$ into the Rydberg equation gives

$$\frac{1}{\lambda} = \left(1.097 \text{ x } 10^7 \text{ m}^{-1}\right)\left(\frac{1}{1^2} - \frac{1}{\infty^2}\right) = 1.097 \text{ x } 10^7 \text{ m}^{-1}$$

Substituting $\frac{1}{\lambda}$ in $\Delta E = h\nu = \frac{hc}{\lambda} = hc\frac{1}{\lambda}$ yields

$$\Delta E = \left(6.63 \text{ x } 10^{-34} \text{ J}\cdot\text{s}\right)\left(3.00 \text{ x } 10^8 \text{ m}\cdot\text{s}^{-1}\right)\left(1.097 \text{ x } 10^7 \text{ m}^{-1}\right) = \mathbf{2.18 \text{ x } 10^{-18} \text{ J}}$$

(c) There is a limit to the energy difference, since the $n = \infty$ level corresponds to the electron being infinitely far from the nucleus.

(d) The energy associated with removing an electron from an atom in the gas phase is known as the ionization energy of the atom; see Section 7.3. The energy that was calculated in part c correspondes to removing the electron from one atom of H. The value that is commonly tabulated for the ionization energy of H corresponds to the energy that is needed to remove the electron from one mole of H atoms, namely

$$\left(2.18 \text{ x } 10^{-18} \frac{\text{J}}{\text{atom}}\right)\left(6.022 \text{ x } 10^{23} \frac{\text{atoms}}{\text{mol}}\right) = 1.314 \text{ x } 10^{6} \frac{\text{J}}{\text{mol}} \text{ or } 1314 \frac{\text{kJ}}{\text{mol}}$$

6.64 ***See Sections 2.2, 5.7 and 6.3 and Examples 5.10 and 6.5.***

(a) Solving $p = m\nu = \frac{h}{\lambda}$ for ν gives $\nu = \frac{h}{m\lambda}$

$$\nu = \frac{6.63 \text{ x } 10^{-34} \text{ kg}\cdot\text{m}^2\cdot\text{s}^{-1}}{\left(9.11 \text{ x } 10^{-31} \text{ kg}\right)\left(100 \text{ pm} \times \frac{10^{-12} \text{ m}}{1 \text{ pm}}\right)} = \mathbf{7.28 \text{ x } 10^{6} \text{ m}\cdot\text{s}^{-1}}$$

(b) Solving $p = m\nu = \frac{h}{\lambda}$ for ν gives $\nu = \frac{h}{m\lambda}$

$$\nu = \frac{6.63 \text{ x } 10^{-34} \text{ kg}\cdot\text{m}^2\cdot\text{s}^{-1}}{\left(1.67 \text{ x } 10^{-27} \text{ kg}\right)\left(100 \text{ pm} \times \frac{10^{-12} \text{ m}}{1 \text{ pm}}\right)} = \mathbf{3.97 \text{ x } 10^{3} \text{ m}\cdot\text{s}^{-1}}$$

For neutrons at 300 K,

$$\bar{\mu}_{rms} = \sqrt{\frac{3RT}{M}} = \sqrt{\frac{(3)\left(8.314 \text{ kg}\cdot\text{m}^2\cdot\text{s}^{-2}\cdot\text{mol}^{-1}\cdot\text{K}^{-1}\right)(300 \text{ K})}{\left[\left(1.675 \text{ x } 10^{-27} \frac{\text{kg}}{\text{neutron}}\right)\left(6.022 \text{ x } 10^{23} \frac{\text{neutrons}}{\text{mol}}\right)\right]}} = \mathbf{2.73 \text{ x } 10^{3} \text{ m}\cdot\text{s}^{-1}}$$

A velocity higher than the rms speed at 300 K is needed to obtain a wavelength of 100 pm.

Chapter 7: Periodic Trends

7.4 ***See Section 7.1, Figure 7.1, Table 7.1 and Example 7.1.***

(a) Ge $[Ar]4s^23d^{10}4p^2$ [Ar] ↑↓ (4s) ↑↓ ↑↓ ↑↓ ↑↓ ↑↓ (3d) ↑ ↑ __ (4p)

(b) S $[Ne]3s^23p^4$ [Ne] ↑↓ (3s) ↑↓ ↑ ↑ (3p)

(c) Rb $[Kr]5s^1$ [Kr] ↑ (5s)

7.6 ***See Section 7.1 and Figure 7.1.***

(a) The element which has 13 total electrons and is the first element in the p block in the third period ($3p^1$) is Al.

(b) The element which has 11 total electrons and is the first element in the s block in the third period ($3s^1$) is Na.

(c) The element which has 25 total electrons and is the fifth element in the 3d block ($3d^5$) is Mn.

7.8 ***See Section 7.1 and Figure 7.1.***

The element which is a member of the halogen group (VIIA, 17) and a member of the fourth period is Br.

7.10 ***See Section 7.1, Table 7.1 and Figure 7.1.***

(a) Ca $[Ar]4s^2$ [Ar] ↑↓ (4s)

(b) Tc $[Kr]5s^24d^5$ [Kr] ↑↓ (5s) ↑ ↑ ↑ ↑ ↑ (4d)

(c) In $[Kr]5s^24d^{10}5p^1$ [Kr] ↑↓ (5s) ↑↓ ↑↓ ↑↓ ↑↓ ↑↓ (4d) ↑ __ __ (5p)

7.12 ***See Section 7.1, Table 7.1 and Figure 7.1.***

(a) Y $[Kr]5s^24d^1$ [Kr] ↑↓ (5s) ↑ __ __ __ __ (4d) 1 unpaired

(b) Se $[Ar]5s^24d^{10}5p^4$ [Ar] ↑↓ (5s) ↑↓ ↑↓ ↑↓ ↑↓ ↑↓ (4d) ↑↓ ↑ ↑ (5p) 2 unpaired

(c) Cd $[Kr]5s^24d^{10}$ [Kr] ↑↓ (5s) ↑↓ ↑↓ ↑↓ ↑↓ ↑↓ (4d) 0 unpaired

7.14 *See Section 7.1, Figure 7.1, Table 7.1 and Example 7.1.*

(a) Al $[Ne]3s^2 3p^1$ Valence electrons $3s^2 3p^1$

(b) Cs $[Xe]6s^1$ Valence electron $6s^1$

(c) As $[Ar]4s^2 3d^{10} 4p^3$ Valence electrons $4s^2 4p^3$

7.16 *See Section 7.1 and Figure 7.1.*

A valence shell configuration of ns^2 corresponds to Group IIA (2).

7.18 *See Section 7.1 and Figure 7.1.*

(a) Group IA, ns^1 (b) Group IVA, $ns^2 np^2$ (c) Group VIIA, $ns^2 np^5$

7.22 *See Section 7.1, Figure 7.1 and Example 7.2.*

(a) S^{2-} has 18 electrons, two more than S: $[Ne]3s^2 3p^6$ or [Ar].

(b) Mn^{2+} has 23 electrons, two fewer than Mn: $[Ar]3d^5$.

Note: The electrons of highest *n* value, the $4s^2$ electrons, are removed first in forming Mn^{2+} from Mn.

(c) Ge^{2+} has 30 electrons, two fewer than Ge: $[Ar]4s^2 3d^{10}$.

Note: The electrons of highest ℓ value, the $4p^2$ electrons, are removed in forming Ge^{2+} from Ge, not the $4s^2$ electrons.

7.24 *See Section 7.1, Figure 7.1 and Table 7.1.*

(a) The 1+ cation would be formed from an element having a total of 14 electrons and a $3s^2 3p^2$ valence configuration. The element is Si.

(b) The 1+ cation would be formed from an element having a total of 12 electrons and a $3s^2$ valence configuration. The element is Mg.

(c) The 1+ cation would be formed from an element having a total of 26 electrons and a $4s^2 3d^6$ valence configuration. The element is Fe.

7.26 *See Section 7.1, Figure 7.1, Table 7.1 and Example 7.2.*

(a) Y^{3+} [Kr] 0 unpaired electrons

(b) Ni^{2+} $[Ar]3d^8$ [Ar] $\underline{\uparrow\downarrow}\ \underline{\uparrow\downarrow}\ \underline{\uparrow\downarrow}\ \underline{\uparrow}\ \underline{\uparrow}$ (3d) 2 unpaired electrons

(c) Cl^- [Ar] 0 unpaired electrons

7.28 *See Section 7.1, Figure 7.1, Table 7.1 and Example 7.2.*

(a) Fe^{3+} $[Ar]3d^5$ (b) Cr^{3+} $[Ar]3d^3$

Note: The electrons of highest *n* value, the 4s electrons, are removed first in forming these cations from their respective neutral atoms.

7.30 *See Section 7.1 and Example 7.3.*

Se has 34 electrons and the electron configuration $[Ar]4s^2 3d^{10} 4p^4$. As^- and Br^+ would have the same number of electrons as Se and also have identical electron configurations.

7.32 *See Section 7.1, Table 7.1 and Example 7.2.*

Li^{3+}, B^{3+}, and N^{3+} would have no unpaired electrons. Li^{3+} would have zero electrons, B^{3+} would have $1s^2$ and N^{3+} would have $1s^2 2s^2$.

7.34 *See Section 7.1, Table 7.1 and Example 7.2.*

The maximum number of unpaired electrons that is possible for the d orbitals is 5. The transition metal in the fourth period which forms a 2+ ion having 5 unpaired electrons is Mn.

Mn^{2+} $[Ar]3d^5$ $\underline{\uparrow}\ \underline{\uparrow}\ \underline{\uparrow}\ \underline{\uparrow}\ \underline{\uparrow}$

3d

7.36 *See Section 7.2 and Example 7.4.*

(a) Na is larger than Na^+. Cations are normally smaller than their corresponding neutral atoms, because the same numbers of protons are exerting an attractive force on fewer electrons and there are less electron-electron repulsions in the cations. Hence, Na^+ has a higher effective nuclear charge than Na and is smaller than Na.

(b) O^{2-} is larger than F^-. O^{2-} and F^- both have 10 electrons and are therefore isoelectronic. O^{2-} has fewer protons attracting the same number of electrons as F^- and therefore has a lower effective nuclear charge.

(c) Ni^{2+} is larger than Ni^{3+}. Ni^{3+} has the same number of protons exerting an attractive force on fewer electrons and also has less electron-electron repulsions. Hence, it has a higher effective nuclear charge than Ni^{2+} and is smaller than Ni^{2+}.

7.38 *See Section 7.2 and Figures 7.8, 9 and 10.*

(a) O < B < Li Size generally decreases as effective nuclear charge increases from left to right in a given period.

(b) N < C < Si Size generally decreases as effective nuclear charge increases from left to right in a given period and generally increases as more subshells are added going down a given group.

(c) S < As < Sn Size generally decreases as effective nuclear charge increases from left to right in a given period and generally increases as more subshells are added going down a given group. The increase in size associated with going down groups usually dominates the decrease in size associated with going across periods.

7.40 *See Section 7.1 and Figures 7.8, 9 and 10.*

(a) $Be^{2+} < Be < Li$ Cations are normally smaller than their corresponding neutral atoms, and size generally decreases as effective nuclear charge increases from left to right in a given period. More extensive discussion of these trends is given in the solutions for problems 7.36a and 7.38a.

(b) $Cl < S < S^{2-}$ Size generally decreases as effective charge increases from left to right in a given period, and anions are normally larger than their corresponding neutral atoms.

(c) $N < C < Si$ See Solution for 7.38a.

7.46 *See Section 7.3 and Figures 7.12 and 7.13.*

(a) Cl has a higher ionization energy than Si. Cl has a higher effective nuclear charge than Si and a smaller size than Si. Hence, Cl exerts a greater attractive force on its outer electraons than Si.

(b) Na has a higher ionization energy than Rb. The electron that is removed from Rb is in an orbital that is larger in size and therefore further from the nucleus. This increase in size dominates the increase in effective nuclear charge that occurs from Na to Rb.

(c) F^- would have a higher ionization energy than O^{2-}. F^- and O^{2-} are isoelectronic. However, F^- has more protons attracting the same number of electrons and therefore has a higher effective nuclear charge and smaller size.

7.48 *See Section 7.3 and Figures 7.12 and 13.*

(a) Cl has a higher ionization energy than Ge. Ionization energies generally increase with increasing effective nuclear charge and decreasing size going across a given period. Ionization energies generally decrease with increasing size going down a given group. Hence, Cl has a higher ionization energy than Ge for several reasons.

(b) F has a higher ionization energy than B. Ionization energies generally increase with increasing effective nuclear charge and decreasing size going across a given period.

(c) Al^{3+} has a higher ionization energy than Na^+. Al^{3+} and Na^+ are isoelectronic. However, Al^{3+} has a higher effective nuclear charge and a smaller size causing it to have a higher ionization energy than Na^+.

7.50 *See Section 7.3 and Figures 7.12 and 13.*

(a) $O^{2-} < O < F$ This order is one of increasing effective nuclear charge and decreasing size.

(b) $Si < C < N$ This order is one of increasing effective nuclear charge and decreasing size.

Note: See solution for 7.38a.

(c) $Sr < Ru < Te$ This order is one of increasing effective nuclear charge and decreasing size.

7.56 *See Section 7.3.*

Be^+ has $[He]2s^1$ as its electron configuration, and B^+ has $[He]2s^2$ as its electron configuration. B^+ has the higher effective nuclear charge and smaller size and therefore has the higher ionization energy.

Note: The situation is analogous to discussion of the first ionization energies of Li and Be.

7.72 *See Section 7.1 and Example 7.2.*

Ca^{2+} has 18 electrons, two fewer than Ca: [Ar].

Hence, the energy level diagram for Ca^{2+} is:

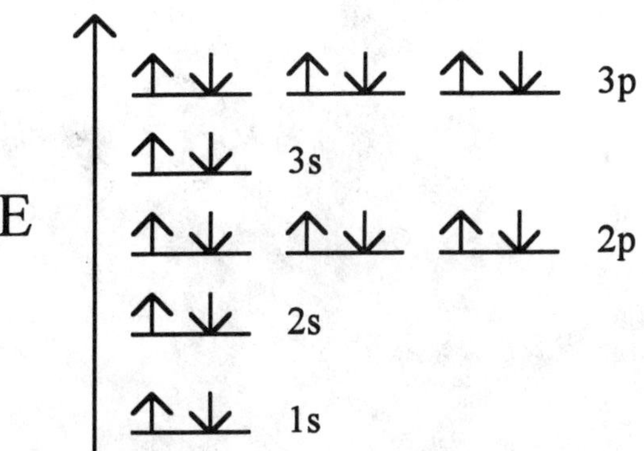

7.74 *See Section 7.1 and Example 7.2.*

Pd^{2+} has 44 electrons, two fewer than Pd: $[Kr]4d^8$. If Pd had $[Kr]5s^24d^8$ as its ground state electron configuration, Pd^{2+} would still have $[Kr]4d^8$ as its electron configuration, since electrons would be removed from the subshell having the highest *n* value. Hence, the fact that the palladium atom is an exception does not influence the electron configuration of Pd^{2+}.

7.76 *See Section 7.1 and Example 7.2.*

The formula for chromium(IV) oxide is CrO_2.
A Cr^{4+} cation would have an $[Ar]3d^2$ electon configuration with two unpaired electrons.

7.78 *See Section 7.5.*

Unbalanced:	NaCl	$\xrightarrow{\text{elect}}$	Na + Cl_2	Start with Cl_2.
Step 1:	$\underline{2}$NaCl	$\xrightarrow{\text{elect}}$	$\underline{2}$Na + Cl_2.	Balances Na, Cl.

Strategy: *g NaCl → mol NaCl → mol Cl_2 → L Cl_2 at STP*

$$? \text{ L } Cl_2 \text{ at STP} = 2.44 \text{ g NaCl} \times \frac{1 \text{ mol NaCl}}{58.45 \text{ g NaCl}} \times \frac{1 \text{ mol } Cl_2}{2 \text{ mol NaCl}} \times \frac{22.4 \text{ L } Cl_2}{1 \text{ mol } Cl_2} = \mathbf{0.468 \text{ L } Cl_2}$$

7.80 *See Section 7.1, Table 7.1 and Example 7.2.*

Fe^{2+} has 24 electrons, two fewer than Fe: $[Ar]3d^6$

Fe^{3+} has 23 electrons, three fewer than Fe: $[Ar]3d^5$

7.82 *See Section 7.2 and 3.*

The effective nuclear charge increases with the increasing number of protons in the isoelectronic series S^{2-}, K^+ and Ca^{2+} causing the following trends:

Size: $Ca^{2+} < K^+ < S^{2-}$ IE: $S^{2-} < K^+ < Ca^{2+}$

Chapter 8: Chemical Bonds

8.2 ***See Section 8.1, Figure 8.1 and Example 8.1.***

(a) $Na\cdot$ (b) $:\ddot{\underset{\cdot}{F}}:$ (c) $:\ddot{\underset{\cdot\cdot}{O}}:^{2-}$ (d) Mg^{2+}

8.4 ***See Section 8.1 and Figure 8.1.***

(a) $\cdot A \cdot$ corresponds to Group IIA (2): Be.

(b) $\cdot\dot{\underset{\cdot}{A}}\cdot$ corresponds to Group IVA (14): C.

(c) $\cdot\dot{\underset{\cdot}{A}}\cdot^{-}$ corresponds to Group IIIA (13): B^-.

8.6 ***See Section 8.2 and Example 8.2.***

(a) $\cdot Ba \cdot + 2\ :\ddot{\underset{\cdot}{Br}}: \rightarrow Ba^{2+} + 2\ :\ddot{\underset{\cdot\cdot}{Br}}:^{-}$

(b) $2\ K\cdot + :\dot{\underset{\cdot}{S}}: \rightarrow 2\ K^{+} + :\ddot{\underset{\cdot\cdot}{S}}:^{2-}$

8.10 ***See Section 8.2 and Example 8.3.***

(a) The lattice energy of LiCl is greater than that of LiBr because the smaller anion size of Cl^- causes the coulombic attractions to be stronger in LiCl than LiBr.

(b) The lattice energy of Na_2O is greater than that of NaF because Na_2O has 1+ and 2- coulombic attractions compared to 1+ and 1- coulombic attractions for NaF.

8.12 ***See Section 8.1 and Examples 8.2 and 3.***

(a) Lithium having one valence electron forms Li^+ and oxygen having six valence electrons forms O^{2-} as they react to form Li_2O. Similarly, sodium having one valence electron forms Na^+ and sulfur having six valence electrons forms S^{2-} as they react to form Na_2S. Since the smaller sizes of Li^+ and O^{2-} compared to Na^+ and S^{2-} lead to stronger coulombic attractions, Li_2O has the higher lattice energy.

(b) Potassium having one valence electron forms K^+ and chlorine having seven valence electrons forms Cl^- as they react to form KCl. Similarly, magnesium having two valence electrons forms Mg^{2+} and fluorine having seven valence electrons forms F^- as they react to form MgF_2. Since Mg^{2+} has a higher charge than K^+ and is smaller than K^+ and since F^- is smaller than Cl^-, MgF_2 has stronger coulombic attractions than KCl and the higher lattice energy.

8.14 *See Section 7.2.*

(a) $Ca^{2+} < K^{+} < S^{2-}$ (b) $Na^{+} < F^{-} < O^{2-}$

Note: The effective nuclear charge decreases with the decreasing number of protons in these isoelectronic series of ions causing the size to increase in the orders given.

8.18 *See Section 8.3 and Figure 8.1.*

(a) CF_4 Carbon has four valence electrons and each fluorine has seven valence electrons. Carbon obtains an octet by forming four covalent bonds whereas each fluorine obtains an octet by forming one covalent bond.

(b) NI_3 Nitrogen has five valence electrons and each iodine has seven valence electrons. Nitrogen obtains an octet by forming three covalent bonds whereas each iodine obtains an octet by forming one covalent bond.

(c) Cl_2O Oxygen has six valence electrons and each chlorine has seven valence electrons. Oxygen obtains an octet by forming two covalent bonds whereas each chlorine obtains an octet by forming one covalent bond.

8.20 *See Section 8.3 and Examples 8.4, 5 and 6.*

(a) H_2S

H—S—H

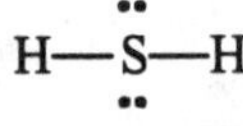

Total valence electrons $= [2 \times 1(\mathrm{H}) + 1 \times 6(\mathrm{S})] = 8$.
Four electrons remain after assigning two single bonds, and four unshared electrons are needed to give each atom a noble gas configuration (4 for S). Eight electrons are used in writing the Lewis structure.

(b) H_2CO

H—C—O (with H bonded to C)

H—C=O (with H bonded to C; two lone pairs on O)

Total valence electrons $= [2 \times 1(\mathrm{H}) + 1 \times 4(\mathrm{C}) + 1 \times 6(\mathrm{O})] = 12$.
Six electrons remain after assigning three single bonds, and eight unshared electrons are needed to give each atom a noble gas configuration (2 for C and 6 for O). Hence, two electrons (8-6) must be used to form an additional bond. The additional bond must be between C and O, since H obtains a noble gas configuration by forming just one bond. Twelve electrons are used in writing the Lewis structure.

(c) PF_3

F—P—F (with F bonded to P)

:F—P—F: (with :F: bonded to P; lone pairs on each F and one on P)

Total valence electrons $= [1 \times 5(\mathrm{P}) + 3 \times 7(\mathrm{F})] = 26$.
Twenty electrons remain after assigning three single bonds, and twenty unshared electrons are needed to give each atom a noble gas configuration (2 for P and 6 for each F). Twenty six electrons are used in writing the Lewis structure.

8.22 *See Section 8.3.*

(a) The H-S bonds are single bonds having a bond order of 1.

(b) The H-C bonds are single bonds having a bond order of 1. The C=O double bond has a bond order of 2.

(c) The F-P bonds are single bonds having a bond order of 1.

8.24 *See Section 8.3 and Examples 8.4, 5 and 6.*

(a) AsH_3

Total valence electrons $= [1 \times 5(As) + 3 \times 1(H)] = 8$.

Two electrons remain after assigning three single bonds, and two unshared electrons are needed to give each atom a noble gas configuration (2 for As). Eight electrons are used in writing the Lewis structure.

```
H—As—H
   |
   H

   ••
H—As—H
   |
   H
```

(b) NO^+

Total valence electrons $= [1 \times 5(N) + 1 \times 6(O) - 1(\text{charge})] = 10$.

Eight electrons remain after assigning a single bond, and twelve unshared electrons are needed to give each atom a noble gas configuration (6 for N and 6 for O). Hence, four electrons (12-8) must be used to form two additional bonds between N and O making the bond a triple bond. Ten electrons are used in writing the Lewis structure.

```
N—O +

           +
[ :N≡O: ]
```

(c) CF_3OH

Total valence electrons $= [1 \times 4(C) + 3 \times 7(F) + 1 \times 6(O) + 1 \times 1(H)] = 32$.

Twenty two electrons remain after assigning five single bonds, and twenty two electrons are needed to give each atom a noble gas configuration (6 for each F and 4 for O). Thirty two electrons are used in writing the Lewis structure.

```
    F
    |
F—C—O—H
    |
    F

    ••
   :F:
••  |    ••
:F—C—O—H
••  |    ••
   :F:
    ••
```

8.26 *See Section 8.3 and Examples 8.4, 5 and 6.*

(a) CH_3CHO

Total valence electrons $= [2 \times 4(C) + 4 \times 1(H) + 1 \times 6(O)] = 18$.

Six electrons remain after assigning six single bonds, and eight unshared electrons are needed to give each atom a noble gas configuration (6 for O and 2 for second C). Hence, two electrons (8-6) must be used to form an additional bond. The additional bond must be between C and O, since H attains a noble gas configuration by forming just one bond. Eighteen electrons are used in writing the Lewis structure.

```
  H   O
  |   |
H—C—C—H
  |
  H

  H  :O:
  |   ||
H—C—C—H
  |
  H
```

(b) NH_2OH

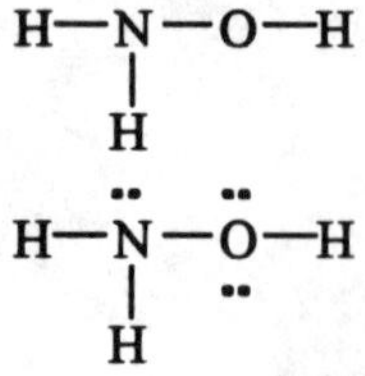

H—N—O—H
|
H

H—N̈—Ö—H (with lone pairs)
|
H

Total valence electrons $= [1 \times 5(N) + 3 \times 1(H) + 1 \times 6(O) = 14]$.

Six electrons remain after assigning four single bonds, and six unshared electrons are needed to give each atom a noble gas configuration (2 for N and 4 for O). Fourteen electrons are used in writing the Lewis structure.

(c) CH_3CHCH_2

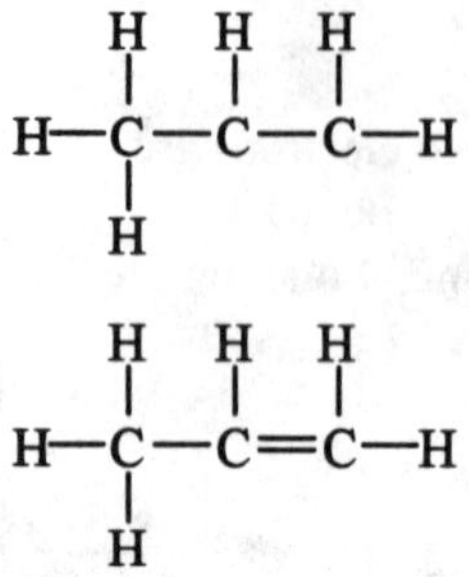

H—C(H)(H)—C(H)—C(H)—H

H—C(H)(H)—C(H)=C(H)—H

Total valence electrons $= [3 \times 4(C) + 6 \times 1(H)] = 18$.

Two electrons remain after assigning eight single bonds, and four unshared electrons are needed to give each atom a noble gas configuration (2 each for second and third C atoms). Hence, two electrons (4-2) must be used to form an additional bond. The additional bond must be between the second and third C atoms, since the first C atom has an octet and H attains a noble gas configuration by forming just one bond. Eighteen electrons are used in writing the Lewis structure.

(d) N_2H_2

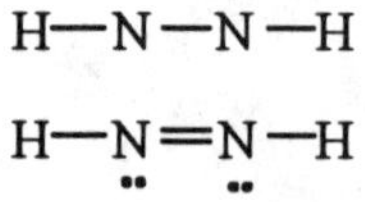

H—N—N—H

H—N̤=N̤—H

Total valence electrons $= [2 \times 5(N) + 2 \times 1(H)] = 12$.

Six electrons remain after assigning three single bonds, and eight unshared electrons are needed to give each atom a noble gas configuration (4 for each N). Hence, two electrons (8-6) must be used to form an additional bond. The additional bond must be formed between the nitrogen atoms, since hydrogen attains a noble gas configuration by forming just one bond. Fourteen electrons are used in writing the Lewis structure.

8.28 ***See Section 8.3 and Examples 8.4, 5 and 6.***

(a) C_2H_2

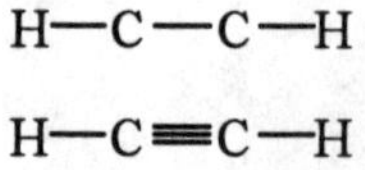

H—C—C—H

H—C≡C—H

Total valence electrons $= [2 \times 4(C) + 2 \times 1(H)] = 10$.

Four electrons remain after assigning three single bonds, and eight unshared electrons are needed to give each atom a noble gas configuration (4 for each C). Hence, four electrons (8-4) must be used to form two additional bonds. The additional bonds must be formed between the carbon atoms, since hydrogen attains a noble gas configuration by forming just one bond. Ten electrons are used in writing the Lewis structure.

(b) HOCl

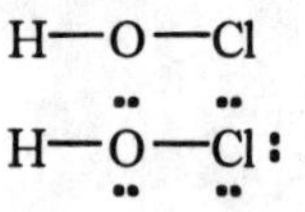

H—O—Cl

H—Ö—C̈l: (with lone pairs)

Total valence electrons $= [1 \times 1(H) + 6 \times 1(O) + 7 \times 1(Cl)] = 14$.

Ten electrons remain after assigning two single bonds, and ten unshared electrons are needed to give each atom a noble gas configuration (4 for O and 6 for Cl). Ten electrons are used in writing the Lewis structure.

(c) CH_2CHCN

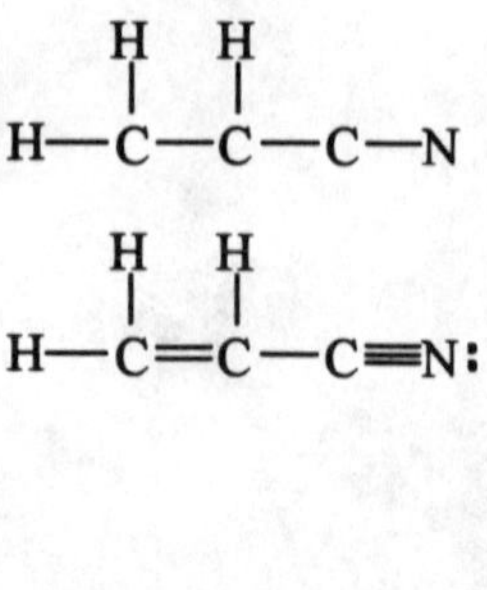

H—C(H)—C(H)—C—N

H—C(H)=C(H)—C≡N:

Total valence electrons $= [3 \times 4(C) + 3 \times 1(H) + 1 \times 5(N)] = 20$.

Eight electrons remain after assigning six single bonds, and fourteen unshared electrons are needed to give each atom a noble gas configuration (2 each for first and second C, 4 for third C and 6 for N). Hence, six electrons (14-8) must be used to form three additional bonds. One additional bond is placed between the first and second carbon atoms to give these atoms a noble gas configuration, and two additional bonds are placed between the third carbon atom and the nitrogen atom. Twenty electrons are used in writing the Lewis structure.

(d) H_2O_2

H—O—O—H

H—Ö—Ö—H (with two unshared pairs on each O)

Total valence electrons $= [2 \times 1(H) + 2 \times 6(O)] = 14$.
Eight electrons remain after assigning three single bonds, and eight unshared electrons are needed to give each atom a noble gas configuration (4 for each O). Fourteen electrons are used in writing the Lewis structure.

8.32	*See Section 8.1 and Figure 8.8.*

(a) Br (b) Cl (c) N (d) O (e) Ge

Electronegativity values decrease down the periodic table and also to the left across the table.

8.34	*See Section 8.4, Figure 8.8 and Example 8.7.*

(a) The order of increasing electronegativity is C < N < O (2.5 < 3.0 < 3.5). The C-O bond is more polar than the N-O bond due to the greater difference in electronegativity. The polarity is $\overset{\mapsto}{C-O}$.

(b) The order of increasing electronegativity is Ge = Si < C (1.9 = 1.9 < 2.5). The Ge-C bond is more polar than the Si-Ge bond due to the greater difference in electronegativity. The polarity is $\overset{\mapsto}{Ge-C}$.

(c) The order of increasing electronegativity is H < S < O (2.1 < 2.5 < 3.5). The H-O bond is more polar than the H-S bond due to the greater difference in electronegativity. The polarity is $\overset{\mapsto}{H-O}$.

(d) The order of increasing electronegativity is Si < B < C (1.8 < 2.0 < 2.5). The B-C bond is more polar than the B-Si bond due to the greater difference in electronegativity. The polarity is $\overset{\mapsto}{B-C}$.

8.36	*See Section 8.4 and Examples 8.7.*

N_2 has identical nuclei bonded together and is therefore nonpolar. BrF is more polar than ClF due to the greater difference in electronegativity (1.2 vs. 1.0). BrF also has a greater distance between the two electrical charges of +q and -q than ClF. Hence, BrF has the largest dipole moment.

8.40	*See Section 8.5 and Examples 8.8, 9 and 10.*

Total valence electrons $= [3 \times 1(H) + 1 \times 4(C) + 1 \times 5(N)] = 12$. Four electrons remain after assigning four single bonds in either connectivity. Six unshared electrons are needed to give each atom a noble gas configuration. Hence, two electrons (6-4) must be used to form one additional bond between carbon and nitrogen. The appropriate Lewis structures associated with the given connectivities are:

A: H—C=N̈—H (H on C)

B: H—C̈=N—H (H on N)

Formal Charges:	A	B
H	$1-0-\left(\frac{1}{2}\right)(2)=0$	$1-0-\left(\frac{1}{2}\right)(2)=0$
C	$4-0-\left(\frac{1}{2}\right)(8)=0$	$4-2-\left(\frac{1}{2}\right)(6)=-1$
N	$5-2-\left(\frac{1}{2}\right)(6)=0$	$5-0-\left(\frac{1}{2}\right)(8)=+1$

When these formal charges are included in the Lewis structures, we obtain:

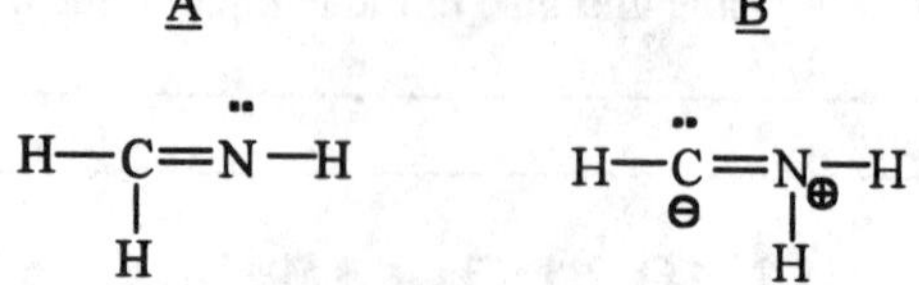

Structure A has the smaller formal charges and is therefore the more favorable arrangement.

8.42 ***See Section 8.6 and Example 8.12.***

(a) NO_2^-, O-N-O

Total valence electrons $= \left[1\times 5(\mathrm{N})+2\times 6(\mathrm{O})+1(\text{charge})\right]=18$. Fourteen electrons remain after assigning two single bonds, and sixteen unshared electrons are needed to give each atom a noble gas configuration (6 for each O and 4 for N). Hence, two electrons (16-14) must be used to form one additional bond. This gives the following as possible resonance forms:

A: $[:\ddot{\underset{..}{O}}—\ddot{N}=\ddot{\underset{..}{O}}]^-$ B: $[\ddot{\underset{..}{O}}=\ddot{N}—\ddot{\underset{..}{O}}:]^-$

Formal Charges:	A	B
O left	$6-6-\left(\frac{1}{2}\right)(2)=-1$	$6-4-\left(\frac{1}{2}\right)(4)=0$
N	$5-2-\left(\frac{1}{2}\right)(6)=0$	$5-2-\left(\frac{1}{2}\right)(6)=0$
O right	$6-4-\left(\frac{1}{2}\right)(4)=0$	$6-6-\left(\frac{1}{2}\right)(2)=-1$

When these formal charges are included in the Lewis structures, we obtain

A: $\ominus:\ddot{\underset{..}{O}}—\ddot{N}=\ddot{\underset{..}{O}}$ B: $\ddot{\underset{..}{O}}=\ddot{N}—\ddot{\underset{..}{O}}:\ominus$

Structures A and B have equal formal charges and are therefore equally important.

(b) ClCN, Cl-C-N

Total valence electrons $= \left[1\times 7(\mathrm{Cl})+1\times 4(\mathrm{C})+1\times 5(\mathrm{N})\right]=16$. Twelve electons remain after assigning two single bonds, and sixteen electrons are needed to give each atom a noble gas configuration (6 for Cl, 4 for C and 6 for N).

Hence, four electrons (16-12) must be used to form two additional bonds. This gives the following as possible resonance forms:

A	B	C
$\ddot{\underset{..}{Cl}}=C=\ddot{\underset{..}{N}}$	$:\ddot{\underset{..}{Cl}}-C\equiv N:$	$\ddot{Cl}\equiv C-\ddot{\underset{..}{N}}:$

Formal Charges:

	A	B	C
Cl	$7-4-\left(\frac{1}{2}\right)(4)=+1$	$7-6-\left(\frac{1}{2}\right)(2)=0$	$7-2-\left(\frac{1}{2}\right)(6)=+2$
C	$4-0-\left(\frac{1}{2}\right)(8)=0$	$4-\left(\frac{1}{2}\right)(8)=0$	$4-\left(\frac{1}{2}\right)(8)=0$
N	$5-4-\left(\frac{1}{2}\right)(4)=-1$	$5-2-\left(\frac{1}{2}\right)(6)=0$	$5-6-\left(\frac{1}{2}\right)(2)=-2$

When these formal charges are included in the Lewis structures, we obtain

A	B	C
$\oplus\ \ddot{\underset{..}{Cl}}=C=\ddot{\underset{..}{N}}\ \ominus$	$:\ddot{\underset{..}{Cl}}-C\equiv N:$	$\oplus\oplus\ddot{Cl}\equiv C-\ddot{\underset{..}{N}}:\ominus\ominus$

Structure B has the lowest formal charges and is therefore most important. Structure C has high formal charges and isn't very important.

8.44 ***See Section 8.6 and Example 8.12.***

(a) CH_2N_2

H—C—N—N (with H below C)

Total valence electrons $=\left[1\times 4(\mathrm{C})+2\times 1(\mathrm{H})+2\times 5(\mathrm{N})\right]=16$.
Eight electrons remain after assigning four single bonds, and twelve unshared electrons are needed to give each atom a noble gas configuration (2 for C, 4 for first N and 6 for second N). Hence, four electrons (12-8) must be used to form two additional bonds. This gives the following possible resonance forms:

A	B
$H-\underset{\underset{H}{\vert}}{C}=N=\ddot{N}:$	$H-\underset{\underset{H}{\vert}}{\ddot{C}}-N\equiv N:$

Formal Charges:

	A	B
H	$1-0-\left(\frac{1}{2}\right)(2)=0$	$1-0-\left(\frac{1}{2}\right)(2)=0$
C	$4-0-\left(\frac{1}{2}\right)(8)=0$	$4-2-\left(\frac{1}{2}\right)(6)=-1$
N left (first)	$5-0-\left(\frac{1}{2}\right)(8)=+1$	$5-0-\left(\frac{1}{2}\right)(8)=+1$
N right (second)	$5-4-\left(\frac{1}{2}\right)(4)=-1$	$5-2-\left(\frac{1}{2}\right)(6)=0$

When these formal charges are included in the Lewis structures, we obtain

A: H—C(—H)=N(⊕)=N̈(⊖):

B: H—C̈(⊖)(—H)—N(⊕)≡N:

Structure A is favored, since it places the -1 formal charge on the more electronegative N atom and not the less electronegative C atom.

(b) OCN^-

[O—C—N]$^-$

Total valence electrons $= \left[1\times 6(\mathrm{O}) + 1\times 4(\mathrm{C}) + 1\times 5(\mathrm{N}) + 1(\mathrm{charge})\right] = 16.$

Twelve electrons remain after assigning two single bonds, and sixteen unshared electrons are needed to give each atom a noble gas configuration (6 for O, 4 for C and 6 for N). Hence four electrons (16-12) must be used to form two additional bonds. This gives the following possible resonance forms:

A: [Ö=C=N̈:]$^-$

B: [:O≡C—N̈:]$^-$

C: [:Ö—C≡N:]$^-$

Formal Charges:

	A	B	C
O	$6-4-\left(\frac{1}{2}\right)(4)=0$	$6-2-\left(\frac{1}{2}\right)(6)=+1$	$6-6-\left(\frac{1}{2}\right)(2)=-1$
C	$4-0-\left(\frac{1}{2}\right)(8)=0$	$4-0-\left(\frac{1}{2}\right)(8)=0$	$4-0-\left(\frac{1}{2}\right)(8)=0$
N	$5-4-\left(\frac{1}{2}\right)(4)=-1$	$5-6-\left(\frac{1}{2}\right)(2)=-2$	$5-2-\left(\frac{1}{2}\right)(6)=0$

When these formal charges are included in the Lewis structures, we obtain

A: Ö=C=N̈(⊖):

B: :O(⊕)≡C—N̈(⊖⊖):

C: :Ö(⊖)—C≡N:

Structure C is favored, since it places the -1 formal charge on the more electronegative O atom and not the less electronegative N atom.

8.46 ***See Section 8.6 and Example 8.12.***

(a) N_3^-

[N—N—N]$^-$

Total valence electrons $= \left[3\times 5(\mathrm{N}) + 1(\mathrm{charge})\right] = 16.$

Twelve electrons remain after assigning two single bonds and sixteen unshared electrons are needed to give each atom a noble gas configuration (6 for left and right N and 4 for middle N). Hence, four electrons (16-12) must be used to form two additional bonds. This gives the following possible resonance forms:

A	B	C
$[:\ddot{N}=N=\ddot{N}:]^-$	$[:N\equiv N-\ddot{N}:]^-$	$[:\ddot{N}-N\equiv N:]^-$

Formal Charges:

	A	B	C
N left	$5-4-\left(\frac{1}{2}\right)(4)=-1$	$5-2-\left(\frac{1}{2}\right)(6)=0$	$5-6-\left(\frac{1}{2}\right)(2)=-2$
N middle	$5-0-\left(\frac{1}{2}\right)(8)=+1$	$5-0-\left(\frac{1}{2}\right)(8)=+1$	$5-0-\left(\frac{1}{2}\right)(8)=+1$
N right	$5-4-\left(\frac{1}{2}\right)(4)=-1$	$5-6-\left(\frac{1}{2}\right)(2)=-2$	$5-2-\left(\frac{1}{2}\right)(6)=0$

When these formal charges are included in the Lewis structures, we obtain

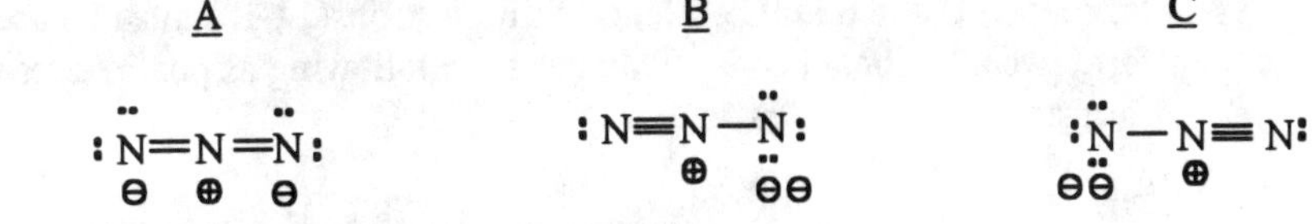

Structure A is favored, since it avoids having a -2 formal charge on one atom.

(b) CO_3^{2-}

Total valence electrons $=[1\times 4(\text{C})+3\times 6(\text{O})+2(\text{charge})]=24$. Eighteen electrons remain after assigning three single bonds, and twenty unshared electrons are needed to give each atom a noble gas configuration (6 for each O and 2 for C). Hence, two electrons (20-18) must be used to form one additional bond. This gives the following possible resonance forms:

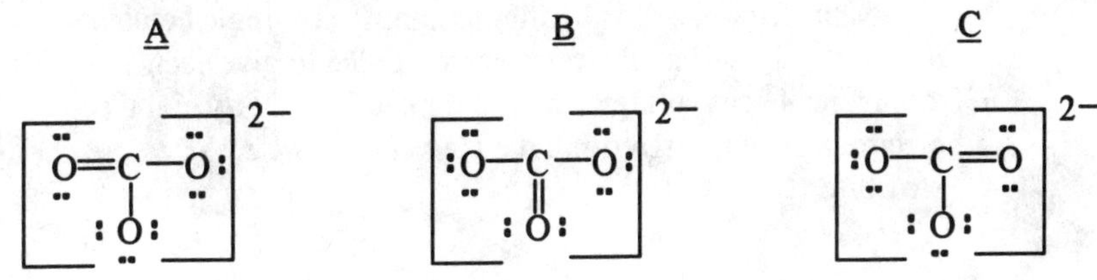

Formal Charges:

	A	B	C
O left	$6-4-\left(\frac{1}{2}\right)(4)=0$	$6-6-\left(\frac{1}{2}\right)(2)=-1$	$6-6-\left(\frac{1}{2}\right)(2)=-1$
C	$4-0-\left(\frac{1}{2}\right)(8)=0$	$4-0-\left(\frac{1}{2}\right)(8)=0$	$4-0-\left(\frac{1}{2}\right)(8)=0$
O middle	$6-6-\left(\frac{1}{2}\right)(2)=-1$	$6-4-\left(\frac{1}{2}\right)(4)=0$	$6-6-\left(\frac{1}{2}\right)(2)=-1$
O right	$6-6-\left(\frac{1}{2}\right)(2)=-1$	$6-6-\left(\frac{1}{2}\right)(2)=-1$	$6-4-\left(\frac{1}{2}\right)(4)=0$

When these formal charges are included in the Lewis structures, we obtain

A B C

Structures A, B, and C have equal formal charges and are therefore equally important.

8.48 ***See Section 8.6 and Example 8.12.***

CH_3NCO

Total valence electrons $= [2 \times 4(\mathrm{C}) + 3 \times 1(\mathrm{H}) + 1 \times 5(\mathrm{N}) + 1 \times 6(\mathrm{O})] = 22$.

Ten electrons remain after assigning six single bonds, and fourteen unshared electrons are needed to give each atom a noble gas configuration (4 for N, 4 for second C and 6 for O). Hence, four electrons(14-10) must be used to form two additional bonds. This gives the following as possible resonance forms:

The first structure is likely to be the most important and the last structure is likely to be the least important, since it places a positive formal charge on the most electronegative element in the molecule.

8.50 ***See Section 8.7 and Examples 8.13 and 14.***

(a) SeF_6

Total valence electrons $= [1 \times 6(\mathrm{Se}) + 6 \times 7(\mathrm{F})] = 48$.

Thirty six electrons remain after assigning six single bonds, and thirty six unshared electrons are needed to give each atom a noble gas configuration (6 for each F). Forty eight electrons are used in writing the **electron rich** Lewis structure.

(b) BBr_3

Total valence electrons $= [1 \times 3(\mathrm{B}) + 3 \times 7(\mathrm{Br})] = 24$.

Eighteen electrons remain after assigning three single bonds, and eighteen unshared electrons are needed to give each atom a noble gas configuration (6 for each Br). Eighteen electrons are shown in the **electron deficient** Lewis structure. However, some contribution from double bonded resonance forms is likely.

(c) NO_2

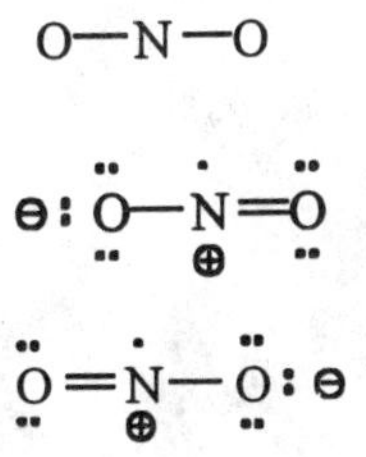

O—N—O

Total valence electrons $= [1 \times 5(N) + 2 \times 6(O)] = 17$.

Thirteen electrons remain after assigning two single bonds, and sixteen unshared electrons are needed to give each atom a noble gas configuration (6 for each O and 4 for N). Hence, three electrons (16-13) are available to form additional bonds. Two of these are used to form one additional bond, and the odd electron is assigned to N. Seventeen electrons are used in writing the **odd electron** Lewis structure.

8.54 ***See Section 8.7 and Example 8.15.***

(a) SeO_2

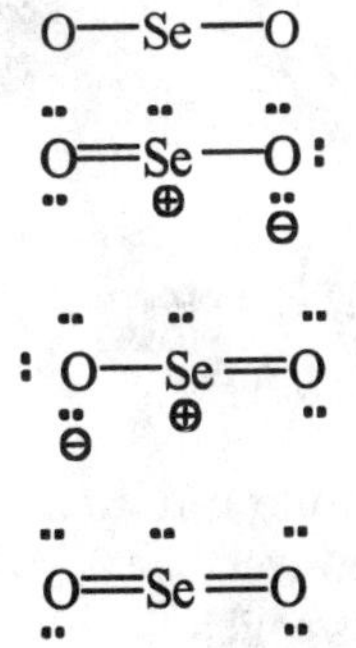

O—Se—O

Total valence electrons $= [1 \times 6(Se) + 2 \times 6(O)] = 18$.

Fourteen electrons remain after assigning two single bonds, and sixteen unshared electrons are needed to give each atom a noble gas configuration (6 for each O and 4 for Se). Hence, two electrons (16-14) must be used to form one additional bond. This gives two equally important resonance forms. However, eighteen electrons can also be used to draw a Lewis structure which shows an expanded valence shell for Se and has formal charges of zero for all atoms. This is likely to be the most important Lewis structure.

O=Se=O

(b) SO_3

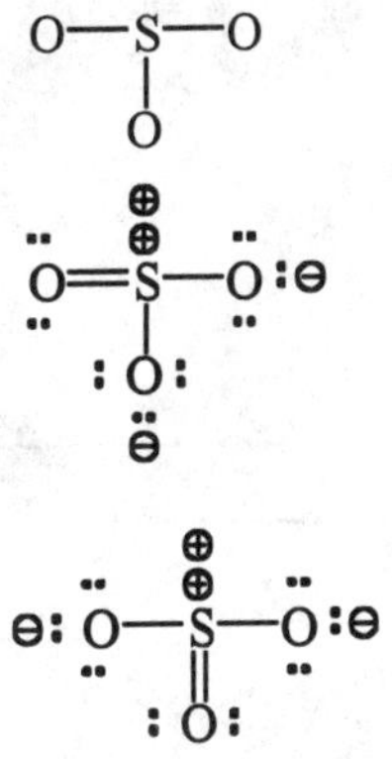

Total valence electrons $= [1 \times 6(S) + 3 \times 6(O)] = 24$.

Eighteen electrons remain after assigning three single bonds, and twenty unshared electrons are needed to give each atom a noble gas configuration (6 for each O and 2 for S). Hence, two electrons must be used to form an additional bond. This gives three equally important resonance forms. However, twenty four electrons can also be used to draw three resonance forms having two double bonds and one form having three double bonds. All of these involve an expanded valence shell for S and reduced formal charges. However, only the last structure has formal charges of zero for all atoms and is therefore likely to be the most important Lewis structure.

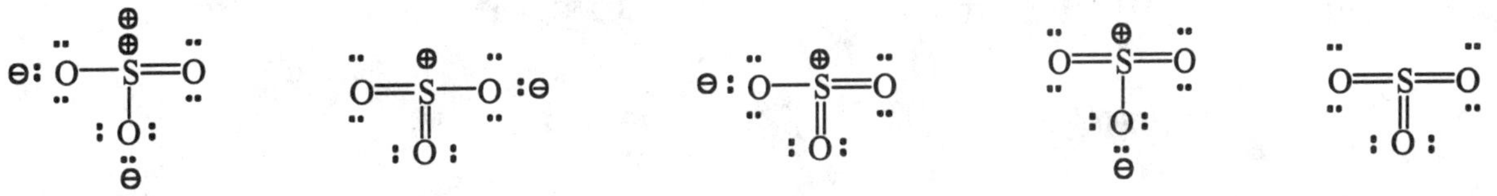

8.56 ***See Section 8.7 and Example 8.15.***

(a) ClO_2^-, $[O-Cl-O]^-$

Total valence electrons $= [1\times7(Cl)+2\times6(O)+1(charge)] = 20$.

Sixteen electrons remain after assigning two single bonds, and sixteen unshared electrons are needed to give each atom a noble gas configuration (6 for each O and 4 for C). Sixteen electrons can be used to draw a Lewis structure which obeys the octet rule. However, sixteen electrons can also be used to draw three resonance forms which have an expanded valence shell for Cl and reduced formal charges. However, the last one shown places the -1 formal charge on the less electronegative atom and isn't likely to be important. The forms with just one double bond are therefore likely to be the most important resonance forms.

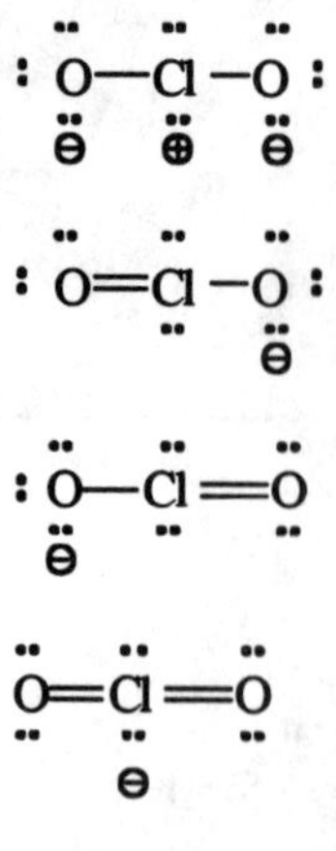

(b) HSO_3^-

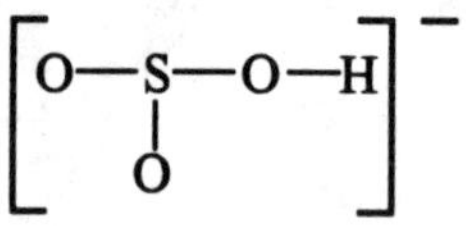

Total valence electrons $= [1\times1(H)+1\times6(S)+3\times6(O)+1(charge)] = 26$.

Eighteen electrons remain after assigning four single bonds, and eighteen unshared electrons are needed to give each atom a noble gas configuration (6 for each terminal O, 2 for S and 4 for other O). Eighteen electrons can be used to draw a Lewis structure which obeys the octet rule. However, eighteen electrons can also be used to draw three resonance forms which have an expanded valence shell for S. Two of these have reduced formal charges and are likely to be the most important resonance forms.

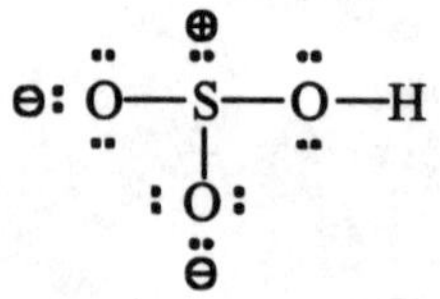

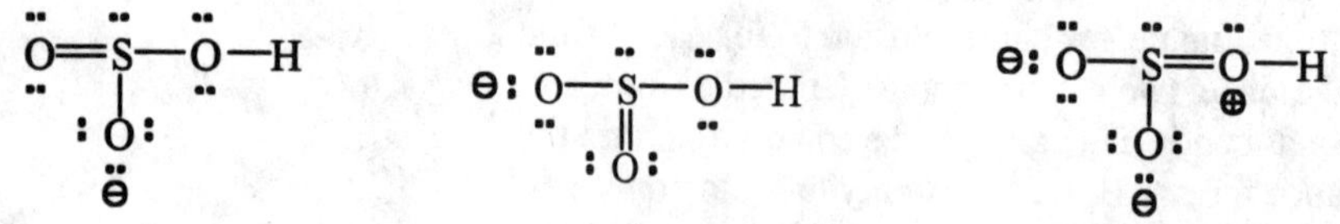

8.58 ***See Section 8.8 and Table 8.3.***

(a) $NH_3(g) \rightarrow N(g) + 3H(g)$ $\Delta H = 3D_{N-H}$ $\Delta H = (3\text{ mol})\left(389\frac{kJ}{mol}\right) = \mathbf{1167\ kJ}$

(b) $CH_3OH(g) \rightarrow C(g) + 4H(g) + O(g)$ $\Delta H = 3D_{C-H} + D_{C-O} + D_{O-H}$

$$\Delta H = (3\text{ mol})\left(414\frac{kJ}{mol}\right) + (1\text{ mol})\left(351\frac{kJ}{mol}\right) + (1\text{ mol})\left(463\frac{kJ}{mol}\right) = \mathbf{2056\ kJ}$$

Note: Lewis strutctures can be used to determine what bonds are broken.

8.60 *See Section 8.8, Table 8.3 and Example 8.16.*

(a) $2H_2(g) + O_2(g) \rightarrow 2H_2O(g)$

$$\Delta H_{rxn} = [2D_{H-H} + D_{O=O}] - [4D_{O-H}]$$

$$\Delta H_{rxn} = \left[(2\ mol)\left(436\frac{kJ}{mol}\right) + (1\ mol)\left(498\frac{kJ}{mol}\right)\right] - \left[(4\ mol)\left(463\frac{kJ}{mol}\right)\right] = \mathbf{-482\ kJ}$$

(b) $2CO(g) + O_2(g) \rightarrow 2CO_2(g)$

$$\Delta H_{rxn} = [2D_{C\equiv O} + D_{O=O}] - [4D_{C=O}]$$

$$\Delta H_{rxn} = \left[(2\ mol)\left(1,072\frac{kJ}{mol}\right) + (1\ mol)\left(498\frac{kJ}{mol}\right)\right] - \left[(4\ mol)\left(799\frac{kJ}{mol}\right)\right] = \mathbf{-554\ kJ}$$

Note: Lewis strutctures can be used to determine what bonds are broken and what bonds are formed.

8.62 *See Section 8.3 and Examples 8.4, 5 and 6.*

(a) $C_2H_4(g) + 3O_2(g) \rightarrow 2CO_2(g) + 2H_2O(g)$

$$\Delta H_{rxn} = [4D_{C-H} + D_{C=C} + 3D_{O=O}] - [4D_{C=O} + 4D_{O-H}]$$

$$\Delta H_{rxn} = \left[(4\ mol)\left(414\frac{kJ}{mol}\right) + (1\ mol)\left(611\frac{kJ}{mol}\right) + (3\ mol)\left(498\frac{kJ}{mol}\right)\right]$$

$$-\left[(4\ mol)\left(799\frac{kJ}{mol}\right) + (4\ mol)\left(463\frac{kJ}{mol}\right)\right] = \mathbf{-1287\ kJ}$$

(b) $H_2CO(g) + H_2(g) \rightarrow H_3COH\ (CH_3OH)$

$$\Delta H_{rxn} = [2D_{C-H} + D_{C=O} + D_{H-H}] - [3D_{C-H} + D_{C-O} + D_{O-H}]$$

$$\Delta H_{rxn} = \left[(2\ mol)\left(414\frac{kJ}{mol}\right) + (1\ mol)\left(799\frac{kJ}{mol}\right) + (1\ mol)\left(436\frac{kJ}{mol}\right)\right]$$

$$-\left[(3\ mol)\left(414\frac{kJ}{mol}\right) + (1\ mol)\left(351\frac{kJ}{mol}\right) + (1 mol)\left(463\frac{kJ}{mol}\right)\right] = \mathbf{7\ kJ}$$

Note: Lewis strutctures can be used to determine what bonds are broken and what bonds are formed.

8.66 ***See Section 8.3 and Examples 8.4, 5 and 6.***

S_2N_2

Total valence electrons $= [2 \times 6(S) + 2 \times 5(N)] = 22$.

Fourteen electrons remain after assigning four single bonds, and sixteen unshared electrons are needed to give each atom a noble gas configuration. Hence, two electrons (16-14) must be used to form one additional bond. This bond can exist in any one of the four bonding positions. Twenty two electrons are used drawing the Lewis structure which obeys the octet rule. However, twenty two electrons can also be used to draw a Lewis strucutre which shows an expanded valence shell for sulfur and has formal charges of zero for all atoms.

8.68 ***See Section 8.7 and Example 8.15.***

SF_4CH_2

Total valence electrons $= [1 \times 6(S) + 4 \times 7(F) + 1 \times 4(C) + 2 \times 1(H)] = 40$.

Twenty six electrons remain after assigning seven single bonds, and twenty six unshared electrons are needed to give each attached atom a noble gas configuration (6 for each F and 2 for C). Twenty six electrons can be used to draw a single bonded structure. However, twenty six electrons can also be used to draw a strucutre with one double bond and reduced formal charges. This structure is also compatible with the unusually short C-S bond distance of 155 pm.

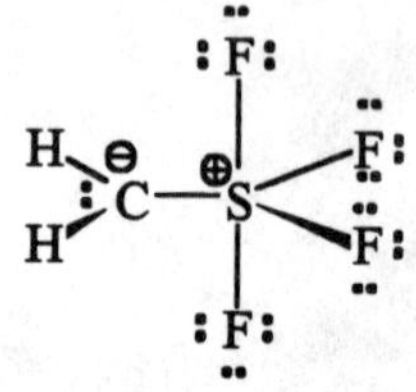

Note: The S atom already has an expanded valence shell with the single bonded structure. However, electron withdrawal from S by the highly electronegative fluorine atoms probably contributes to forming a C-S double bond.

8.70 ***See Section 8.6 and Example 8.12.***

N_2O Total valence electrons $= [2 \times 5(N) + 1 \times 6(O)] = 16$. Twelve electrons remain after assigning two single bonds in either connectivity, and sixteen electrons are needed to give each atom a noble gas configuration (6 for each terminal atom and 4 for the central atom). Hence, four electrons (16-12) must be used to form two additional bonds. This gives the following as possible resonance forms:

A: $:N{\equiv}N{-}\ddot{\underset{..}{O}}:$ B: $:\ddot{N}{=}N{=}\ddot{O}:$

Formal Charges:

	A	B
N left	$5-2-\left(\frac{1}{2}\right)(6)=0$	$5-4-\left(\frac{1}{2}\right)(4)=-1$
N right	$5-0-\left(\frac{1}{2}\right)(8)=+1$	$5-0-\left(\frac{1}{2}\right)(8)=+1$
O	$6-6-\left(\frac{1}{2}\right)(2)=-1$	$6-4-\left(\frac{1}{2}\right)(4)=0$

C: $:\ddot{N}{=}O{=}\ddot{N}:$ D: $:N{\equiv}O{-}\ddot{\underset{..}{N}}:$ E: $:\ddot{\underset{..}{N}}{-}O{\equiv}N:$

	C	D	E
N left	$5-4-\left(\frac{1}{2}\right)(4)=-1$	$5-2-\left(\frac{1}{2}\right)(6)=0$	$5-6-\left(\frac{1}{2}\right)(2)=-2$
N right	$5-4-\left(\frac{1}{2}\right)(4)=-1$	$6-0-\left(\frac{1}{2}\right)(8)=+2$	$6-0-\left(\frac{1}{2}\right)(8)=+2$
O	$6-0-\left(\frac{1}{2}\right)(8)=+2$	$5-6-\left(\frac{1}{2}\right)(2)=-2$	$5-2-\left(\frac{1}{2}\right)(6)=0$

When these formal charges are included in the Lewis structures, we obtain

A	B	C	D	E
$:N{\equiv}\overset{\oplus}{N}{-}\ddot{O}:^{\ominus}$	$:\ddot{N}^{\ominus}{=}N^{\oplus}{=}\ddot{O}:$	$:\ddot{N}^{\ominus}{=}O^{\oplus\oplus}{=}\ddot{N}:^{\ominus}$	$:N{\equiv}O^{\oplus\oplus}{-}\ddot{N}:^{\ominus\ominus}$	$:\ddot{N}^{\ominus\ominus}{-}O^{\oplus\oplus}{\equiv}N:$

Structures A and B have lower formal charges and also avoid having a positive formal charge on the more electronegative O atom, as occurs in C, D and E. Structure A has its negative formal charge on the most electronegative element in the molecule and is therefore likely to be the most important resonance form. These considerations indicate a N-N-O connectivity is more likely than a N-O-N connectivity.

8.72 ***See Section 8.3 and 4, Figure 8.8 and Examples 8.4, 5, 6 and 7.***

BrNO

Total valence electrons $= [1\times 7(\mathrm{Br})+1\times 5(\mathrm{N})+1\times 6(\mathrm{O})]=18$.

Br—N—O

Fourteen electrons remain after assigning two single bonds, and sixteen electrons are needed to give each atom a noble gas configuration (6 for Br, 4 for N and 6 for O). Hence, two electrons (16-14) must be used to form an additional bond between N and O. Eighteen electrons are used in writing the Lewis structure.

$:\ddot{\underset{..}{Br}}{-}\ddot{N}{=}\ddot{\underset{..}{O}}$

According to Figure 8.8, the electronegativity difference between N and O is greater than that between N and Br. Hence, the N-O bond should be the more polar bond. However, caution should be used in assuming the values given in Figure 8.8 are applicable to multiply bonded atoms.

8.74 ***See Section 8.8, Table 8.3 and Example 8.6.***

$HCN(g) + 2H_2(g) \rightarrow H_3CNH_2(g)$

$$\Delta H_{rxn} = \left[D_{C-H} + D_{C\equiv N} + 2D_{H-H}\right] - \left[3D_{C-H} + D_{C-N} + 2D_{N-H}\right]$$

$$\Delta H_{rxn} = \left[(1\text{ mol})\left(414\frac{kJ}{mol}\right) + (1\text{ mol})\left(891\frac{kJ}{mol}\right) + (2\text{ mol})\left(436\frac{kJ}{mol}\right)\right]$$

$$-\left[(3\text{ mol})\left(414\frac{kJ}{mol}\right) + (1\text{ mol})\left(293\frac{kJ}{mol}\right) + (2\text{ mol})\left(389\frac{kJ}{mol}\right)\right] = \mathbf{-136\ kJ}$$

Note: Lewis strutctures can be used to determine what bonds are broken and what bonds are formed.

Chapter 9: Molecular Structure

9.2 ***See Section 9.1 and Figure 9.1.***

(a) trigonal planar (b) tetrahedral (c) tetrahedral (d) trigonal bipyramidal

9.4 ***See Section 9.1 and Figure 9.6.***

(a) CF_4 (b) CS_2 (c) AsF_5 (d) CF_2O (e) NH_4^+

(a) tetrahedral

(b) linear

(c) trigonal bipyramidal

(d) trigonal planar

(e) tetrahedral

9.6 ***See Section 9.1 and Figures 9.1 and 6.***

(a) SeO_2 (b) N_2O (c) H_3O^+ (d) IF_5 (e) SCl_4

(a) electron rich

(a)	(b)	(c)	(d)	(e)
trigonal planar	linear	tetrahedral	octahedral	trigonal bipyramidal
bent	linear	trigonal pyramidal	square pyramidal	see-saw

9.8 ***See Section 9.1 and Figures 9.1 and 6.***

(a) BCl_3 NCl_3 (b) OF_2 SF_6

BCl_3	NCl_3	OF_2	SF_6
trigonal planar electron pair geometry	tetrahedral electron pair geometry	tetrahedral electron pair geometry	octahedral electron pair geometry
120° bond angles	109° bond angles	109° bond angles	90°&180° bond angles

NCl_3 has smaller bond angles than BCl_3.

SF_6 has smaller bond angles than OF_2.

9.10 ***See Section 9.1, Figures 9.1 and 6 and Examples 9.2 and 3.***

(a) H_3CCCH

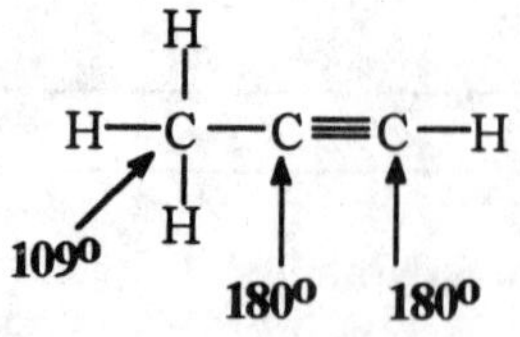

H—C—C≡C—H (C bonded to H above and below)

109° at first C; 180° at second C; 180° at third C

(b) Br_2CCH_2

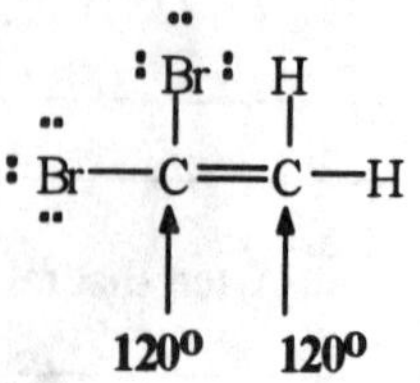

:Br—C=C—H with :Br: and H above

120° ; 120°

(c) H_3CNH_2

H—C—N—H with H's

109° ; 109°

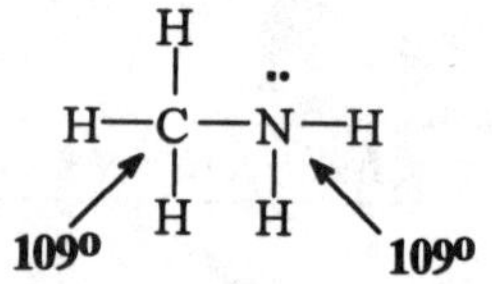

9.12 ***See Section 9.1, Figures 9.1 and Examples 9.2 and 3.***

(a) C_2H_4O

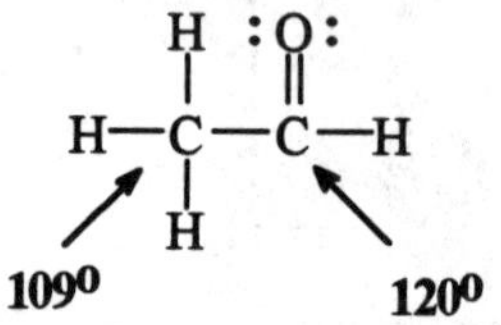

H—C—C—H with H, :O: above, H below

109° ; 120°

(b) C_3H_6O

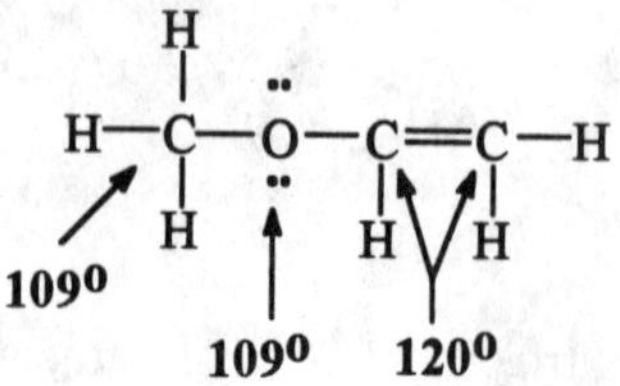

H—C—O—C=C—H with H's

109° ; 109° ; 120°

(c) XeF_2

:F—Xe—F:

180°

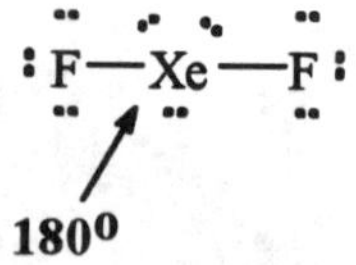

9.16 ***See Section 9.2 and Example 9.4.***

(a) BeF_2

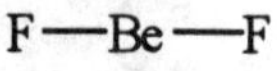

F—Be—F

symmetrical

nonpolar

(b) SF_6

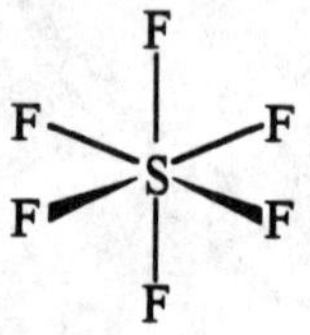

symmetrical

nonpolar

(c) SiH_4

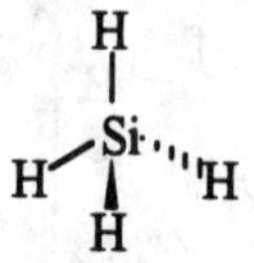

symmetrical

nonpolar

(d) ClCN

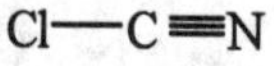

Cl—C≡N

unsymmetrical

polar

9.18 ***See Section 9.2 and Example 9.4.***

(a) XeO_2

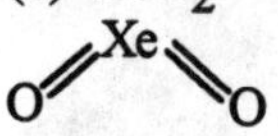

O=Xe=O

unsymmetrical

polar

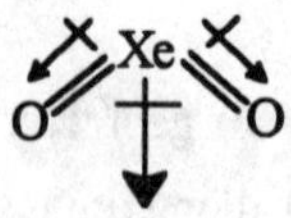

(b) I_2

I—I

symmetrical

nonpolar

(c) NO

N=O

unsymmetrical

polar

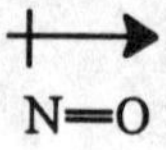

N=O

(d) PCl_5

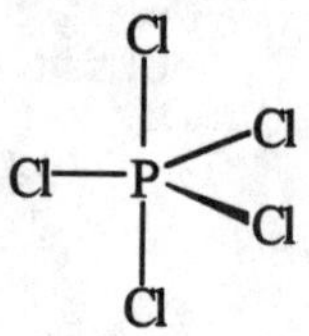

symmetrical

nonpolar

9.20	*See Section 9.2 and Example 9.4.*

(a) NCl_3

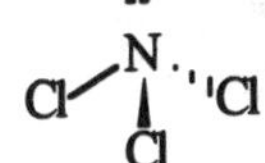

unsymmetrical

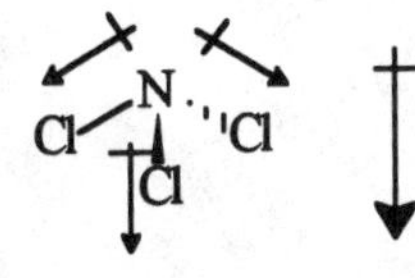

polar

(b) CBr_4

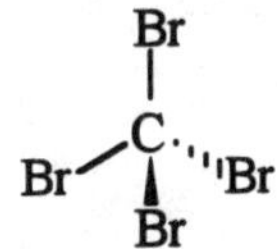

symmetrical

nonpolar

(c) BeI_2

I—Be—I

symmetrical

nonpolar

9.22	*See Section 9.3 and Example 9.6.*

F [He] ↑↓ (2s) ↑↓ ↑↓ ↑ (2p)

Cl [Ne] ↑↓ (3s) ↑↓ ↑↓ ↑ (3p)

The partially filled 3p orbital of Cl overlaps with the partially filled 2p orbital of F to form the bond in ClF.

9.24	*See Section 9.3.*

(a) 120°, sp^2 hybrids (b) 90°, sp^3d^2 (c) 180°, sp hybrids

9.26	*See Section 9.3 and Examples 9.7 and 8.*

(a) CF_4

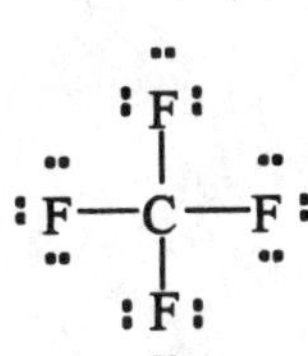

tetrahedral

sp^3 for C

(b) $SbCl_6^-$

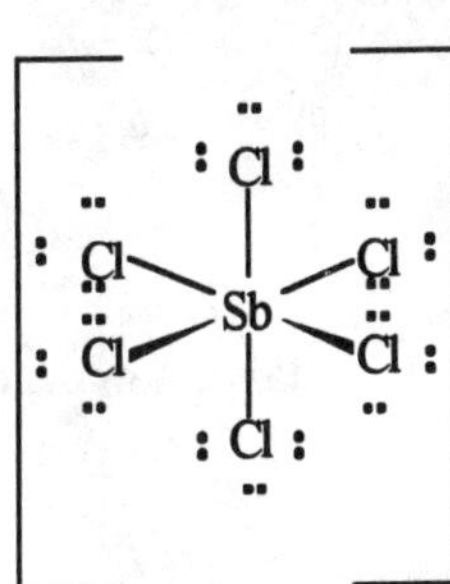

octahedral

sp^3d^2 for Sb

(c) AsF_5

trigonal bipyramidal

sp^3d for As

(d) SiH_4

tetrahedral

sp^3 for Si

(e) NH_4^+

tetrahedral

sp^3 for N

9.28	*See Section 9.3 and Examples 9.7 and 8.*

(a) N_2O

:N≡N—O:

linear

sp for central N

(b) $SnCl_2$

:Cl—Sn—Cl:

trigonal planar

sp^2 for Sn

(c) I_3^-

[:I—I—I:]⁻

trigonal bipyramidal

sp^3d for central I

(d) SeO_2

O=Se=O

trigonal planar

sp^2 for Se

9.30 ***See Section 9.3 and Examples 9.7 and 8.***

(a) CO_3^{2-} | (b) CH_2F_2 | (c) H_2CO

trigonal planar | tetrahedral | trigonal planar

sp^2 for C | sp^3 for C | sp^2 for C

9.32 ***See Section 9.3 and Examples 9.7 and 8.***

SeF_4

trigonal bipyramidal

sp^3d for Se

F [He] ↑↓ (2s) ↑↓ ↑↓ ↑ (2p)

A sp^3d orbital from Se containing one electron overlaps with a 2p orbital of F containing one electron to form a Se-F bond in SeF_4. The lone pair of electrons is in a sp^3d orbital of Se.

9.34 ***See Section 9.3 and Examples 9.7 and 8.***

(a) OF_2

tetrahedral

sp^3 for O

F [He] ↑↓ (2s) ↑↓ ↑↓ ↑ (2p)

A sp^3 orbital from O containing one electron overlaps with a 2p orbital from F containing one electron to form a O-F bond in OF_2. The lone pairs of electrons are in sp^3 orbitals of O.

(b) NH_3

tetrahedral

sp^3 for N

H ↑ (1s)

A sp^3 orbital from N containing one electron overlaps with a 1s orbital of H containing one electron to form a N-H bond in NH_3. The lone pair of electrons is in a sp^3 orbital of N.

(c) BCl_3

trigonal planar

sp^2 for B

Cl [Ne] ↑↓ (3s) ↑↓ ↑↓ ↑ (3p)

A sp^2 orbital from B containing one electron overlaps with a 3p orbital of Cl containing one electron to form a B-Cl bond in BCl_3.

9.38	*See Section 9.3 and Example 9.8.*

$SbCl_5$ trigonal bipyramid

sp^3d for Sb

Cl [Ne] ↑↓ (3s) ↑↓ ↑↓ ↑ (3p)

A sp^3d orbital of Sb containing one electron overlaps with a 3p orbital of Cl containing one electron to form a Sb-Cl bond in $SbCl_5$.

$SbCl_6^-$ octahedral

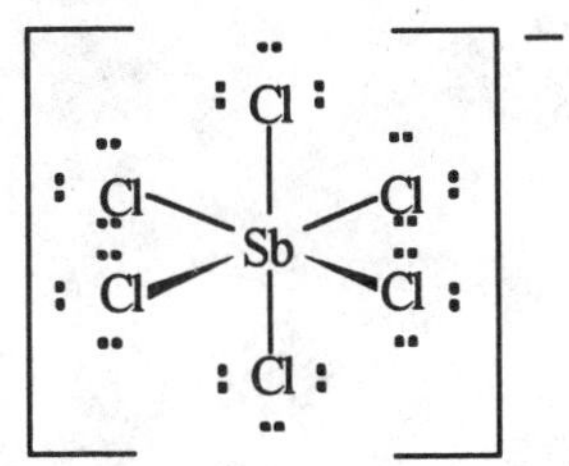

sp^3d^2 for Sb

Cl^- [Ne] ↑↓ (3s) ↑↓ ↑↓ ↑↓ (3p) or [Ar]

Sb ([Kr] $5s^24d^65p^3$) has five valence electrons, and there are six sp^3d^2 hybrid orbitals. This leaves an empty sp^3d^2 orbital of Sb to overlap with a 3p orbital of Cl^- containing two electrons to form a coordinate covalent Sb-Cl bond in forming $SbCl_6^-$.

9.42	*See Section 9.4 and Figures 9.24 through 31.*

(a) 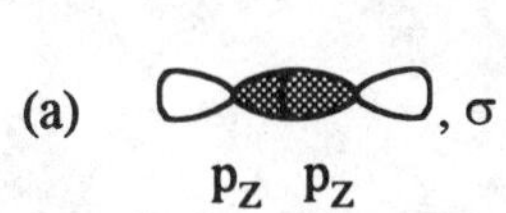 , σ

p_z p_z

(b) 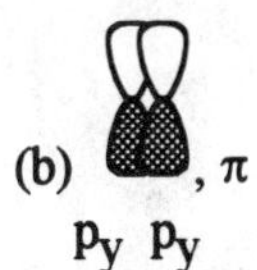 , π

p_y p_y

(c) 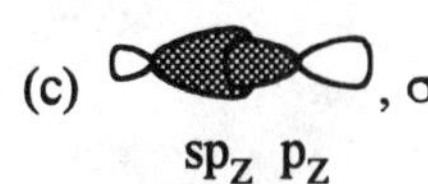 , σ

sp_z p_z

9.44	*See Sections 9.3 and 4 and Example 9.9.*

H_3CCN

tetrahedral about C_1 and therefore sp^3 for C_1

linear about C_2 and therefore sp for C_2

Bond	Orbital Overlaps	Bond Type
C-H	sp^3-1s	σ
C_1-C_2	sp^3-sp	σ
C_2≡N	sp-p_z	σ
	p_y-p_y	π
	p_x-p_x	π

There are a total of five σ bonds and two π bonds in H_3CCN.

9.48	*See Sections 9.3 and 4.*

(a)

The single bonded carbon atoms are tetrahedral and sp^3.

The double bonded carbon atoms are trigonal planar and sp^2.

(b)

The C is trigonal planar and sp^2.

(c)

The N is tetrahedral and sp^3.

The C_1 is tetrahedral and sp^3.

The C_2 is trigonal planar and sp^2.

The O of C_2-O-H is tetrahedral and sp^3.

9.50	*See Section 9.3 and 4.*

sp^2 for N

sp^2 for N

Note: A top view of a trigonal planar arrangement is shown here for convenience. A side view is often shown like that shown in 9.48b above.

9.52	*See Sections 9.3 and 4.*

(a)

The C is tetrahedral and sp^3.

The N is tetrahedral and sp^3.

(b)

The C_1 is tetrahedral and sp^3.

The C_2 and C_3 are linear and sp.

9.54	***See Sections 9.3 and 4.***

H_2CCHCN

H—C_1=C_2−C_3≡N:
 | |
 H H

The C_1 is trigonal planar and sp^2.

The C_2 is trigonal planar and sp^2.

The C_3 is linear and sp.

9.56	***See Section 9.5, Figure 9.37 and Example 9.11.***

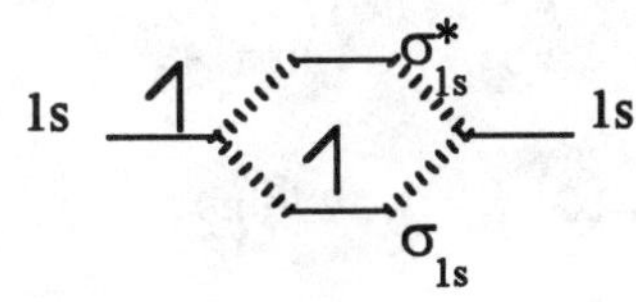

H H_2^+ H^+

For H_2^+, the electron configuration is $(\sigma_{1s})^1$, there is one unpaired electron, the bond order is $\frac{1}{2}[1-0]=0.5$, and it is predicted to be stable.

9.58	***See Section 9.5, Figure 9.37 and Example 9.11.***

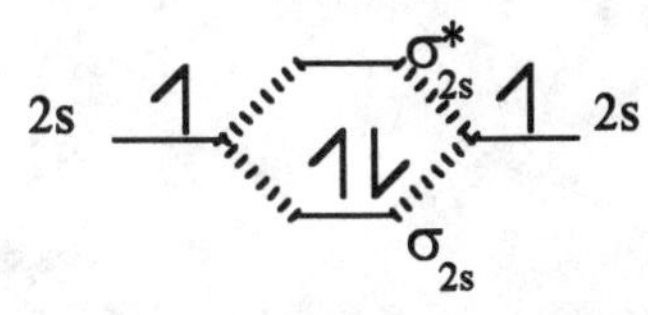

Li Li_2 Li

For Li_2, the electron configuration is $(\sigma_{2s})^2$, there are no unpaired electrons, the bond order is $\frac{1}{2}[2-0]=1$, and it is predicted to be stable.

9.60	***See Section 9.5 and Example 9.12.***

(a) C_2^+ has (4 + 4 - 1) = 7 valence electrons : $(\sigma_{2s})^2(\sigma^*{}_{2s})^2(\pi_{2p})^3$.

The bond order is $\frac{1}{2}[5-2]=1.5$, and there is one unpaired electron in a π_{2p} orbital.

(b) N_2^- has (5 + 5 + 1) = 11 valence electrons: $(\sigma_{2s})^2(\sigma^*{}_{2s})^2(\pi_{2p})^4(\sigma_{2p})^2(\pi^*{}_{2p})^1$.

The bond order is $\frac{1}{2}[8-3]=2.5$, and there is one unpaired electron in a $\pi^*{}_{2p}$ orbital.

(c) Be_2^- has (2 + 2 + 1) = 5 valence electrons: $(\sigma_{2s})^2(\sigma^*{}_{2p})^2(\pi_{2p})^1$.

The bond order is $\frac{1}{2}[3-2]=0.5$, and there is one unpaired electron in a π_{2p} orbital.

9.62	***See Section 9.5.***

(a) O_2^- and F_2^+ have 13 valence electrons. A bond order of $\frac{1}{2}[8-5]=1.5$ indicates these species should be stable.

(b) C_2^{2-}, N_2, and O_2^{2+} have 10 valence electrons. A bond order of $\frac{1}{2}[8-2]=3.0$ indicates these species should be stable.

(c) Li_2^{2-}, Be_2, and B_2^{2+} have 4 valence electrons. A bond order of $\frac{1}{2}[2-2]=0$ indicates these species would not be stable.

9.64 ***See Section 9.5.***

Specie	Number of valence electrons	Electron Configuration	Bond Order	Number of unpaired electrons
(a) CN	9	$(\sigma_{2s})^2(\sigma^*_{2s})^2(\pi_{2p})^4(\sigma_{2p})^1$	$\frac{1}{2}[7-2]=2.5$	1 in σ_{2p}
(b) CO^-	11	$(\sigma_{2s})^2(\sigma^*_{2s})^2(\pi_{2p})^4(\sigma_{2p})^2(\pi^*_{2p})^1$	$\frac{1}{2}[8-3]=2.5$	1 in π^*_{2p}
(c) BeB^-	6	$(\sigma_{2s})^2(\sigma^*_{2s})^2(\pi_{2p})^2$	$\frac{1}{2}[4-2]=1.0$	2 in π_{2p}
(d) BC^+	6	$(\sigma_{2s})^2(\sigma^*_{2s})^2(\pi_{2p})^2$	$\frac{1}{2}[4-2]=1.0$	2 in π_{2p}

9.66 ***See Secion 9.5 and Figure 9.48.***

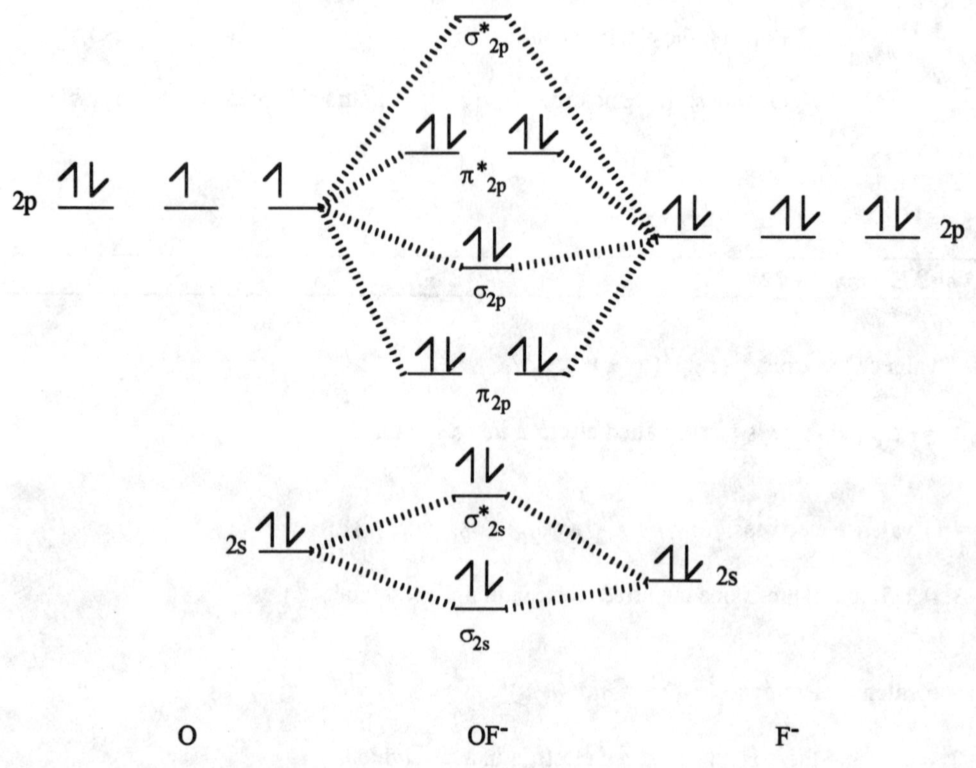

The bond order for OF^- is $\frac{1}{2}[8-6]=1$.

9.70 ***See Sections 9.1, 3 and 4 and Example 9.7.***

N_2O_5

Each N is trigonal planar with 120° bond angles and sp^2.

The central O is tetrahedral with 109° bond angles and sp^3.

9.72	*See Section 8.1.*

ethylene oxide, C_2H_4O

```
     ..
     O
    / \
H—C—C—H
   |   |
   H   H
```

ethylene glycol, $HOCH_2CH_2OH$

```
        H   H
  ..    |   |   ..
H—O—C—C—O—H
  ..    |   |   ..
        H   H
```

acrylonitrile, CH_2CHCN

```
H—C=C—C≡N:
   |   |
   H   H
```

9.74	*See Sections 9.1 and 9.3.*

A bond angle of 109° is associated with a tetrahedral electron pair geometry and sp^3 hybrids. A bond angle of 120° is associated with a trigonal planar geometry and sp^2 hybrids. The lone pair of electrons with the sp^2 hybrids would be located in the unhybridized atomic p orbital.

9.76	*See Sections 9.1-4 and Example 9.10.*

$COCl_2$

```
 ..
:Cl\      ..
 ..  C=O
 ..  /    ..
:Cl
 ..
```

The C has a trigonal planar electron pair geometry and sp^2 hybrids. The molecule is unsymmetrical and therefore polar.

9.78	*See Section 9.5.*

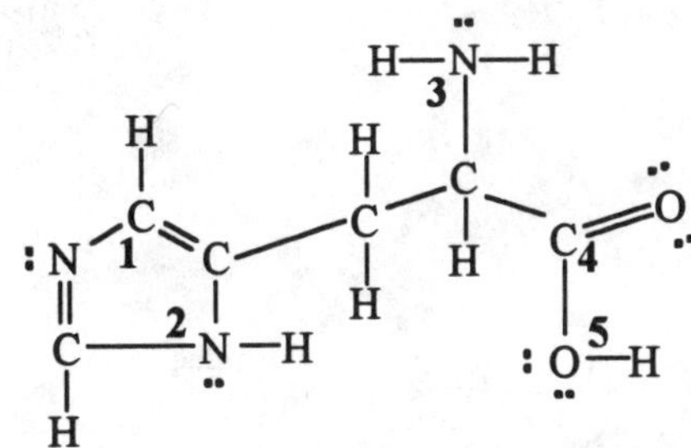

Bond ∠ 1: trigonal planar and therefore 120°.
Bond ∠ 2: tetrahedral and therefore 109°.
Bond ∠ 3: tetrahedral and therefore 109°.
Bond ∠ 4: trigonal planar and therefore 120°.
Bond ∠ 5: tetrahedral and therefore 109°.

9.80 ***See Section 9.5 and Figure 9.48.***

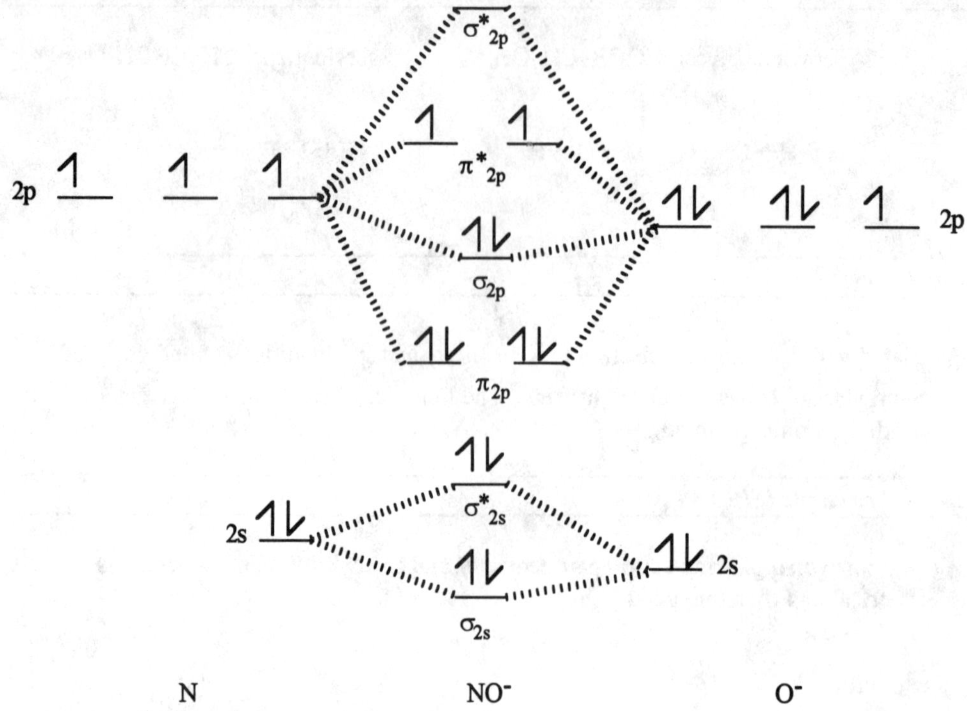

The bond order for $NO^- = \frac{1}{2}[8-4] = 2$.

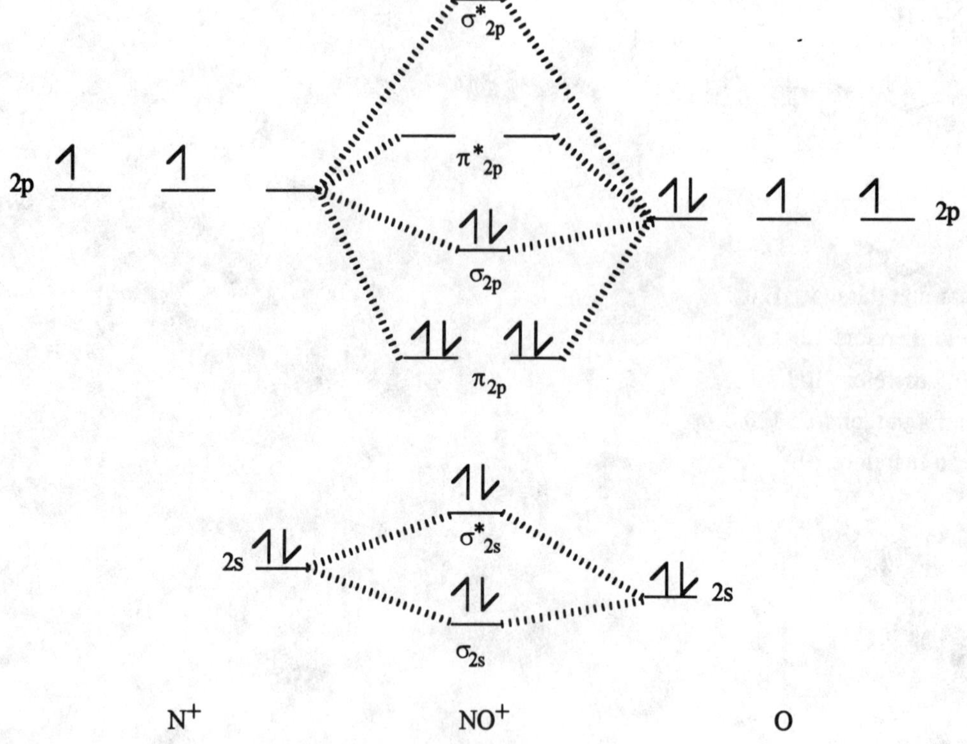

The bond order for $NO^+ = \frac{1}{2}[8-2] = 3$.

Chapter 10: Liquids and Solids

10.2 ***See Section 10.1.***

	Molecule	Comment	Most Important Intermolecular Forces
(a)	C_3H_8	virtually nonpolar	London dispersion forces
(b)	$HO(CH_2)_2OH$	polar with O-H bonds	hydrogen bonding
(c)	C_6H_{12}	virtually nonpolar	London dispersion forces
(d)	PH_3O	polar	dipole-dipole and London dispersion forces
(e)	NO	slightly polar	London dispersion forces
(f)	NH_2OH	polar with N-H and O-H bonds	hydrogen bonding

10.4 ***See Section 10.1.***

(a) C_2H_4 and CH_4 are both symmetrical nonpolar molecules. C_2H_4 has more electrons and a larger size. C_2H_4 is therefore more polarizable and has the higher boiling point (-103.7°C vs. -161°C).

(b) Cl_2 is symmetrical and nonpolar whereas ClF is unsymmetrical and polar. Cl_2 has more electrons and a larger size. Cl_2 is therefore more polarizable and has the higher boiling point (-35°C vs. -100.8°C).

(c) S_2F_2 and S_2Cl_2 are both unsymmetrical trigonal pyramidal molecules having one sulfur atom as the central atom. S_2F_2 would be expected to be more polar than S_2Cl_2 based on the higher electronegativity of F. However, S_2Cl_2 has more electrons and a larger size. S_2Cl_2 is therefore more polarizable and has the higher boiling point (135.6°C vs. -38.4°C).

(d) NH_3 is a polar molecule which can hydrogen bond to other NH_3 molecules whereas PH_3 cannot participate in hydrogen bonding. NH_3 has the higher boiling point (-33°C vs. -88°C).

(e) CH_3I and CHI_3 are both unsymmetrical polar molecules. CHI_3 has more electrons and a larger size. CHI_3 is therefore more polarizable and has the higher boiling point (218°C vs. 42.4°C).

(f) BBr_2I and $BBrI_2$ are both unsymmetrical polar molecules. $BBrI_2$ has more electrons and a larger size. $BBrI_2$ is therefore more polarizable and has the higher boiling point (180°C vs. 125°C).

10.8 ***See Section 10.1.***

	Molecule	Molecular Shape	Comment	Type(s) of Forces
(a)	N_2O	linear (NNO)	unsymmetrical, polar	dipole-dipole and London dispersion
(b)	CH_4	tetrahedral	symmetrical, nonpolar	London dispersion
(c)	NH_3	trigonal pyramidal	unsymmetrical, polar with N-H bonds	hydrogen bonds and London dispersion
(d)	SO_2	v-shaped	unsymmetrical, polar	dipole-dipole and London dispersion

10.14 *See Section 10.2 and 5.*

	Property	Change with Increasing Strength of Intermolecular Forces
(a)	enthalpy of fusion	increase
(b)	melting point	increase
(c)	surface tension	increase
(d)	viscosity	increase
(e)	enthalpy of vaporization	increase
(f)	boiling point	increase

10.16 *See Section 10.2.*

The adhesive forces between the polar sites on glass (largely SiO_2) and the dipoles of the water molecules are sufficiently strong to overcome the cohesive forces between the water molecules and the forces of gravity and thus suspend liquid on the walls of the buret.

10.18 *See Sections 10.1 and 2.*

In each case, the substance having the stronger intermolecular forces of attraction will have the larger enthalpy of vaporization.

	Molecules	Comment	Substance with Larger ΔH_{vap}
(a)	C_3H_8 and CH_4	C_3H_8 is larger and more polarizable.	C_3H_8
(b)	I_2 and ICl	I_2 is larger and more polarizable.	I_2
(c)	S_2Cl_2 and S_2F_2	S_2Cl_2 is larger and more polarizable.	S_2Cl_2
(d)	H_2Se and H_2O	H_2O has hydrogen bonding.	H_2O
(e)	CH_2Cl_2 and CH_3Cl	CH_2Cl_2 is larger and more polarizable.	CH_2Cl_2
(f)	NOF and NOCl	NOCl is larger and more polarizable.	NOCl

10.20 *See Section2 10.1 and 2.*

C_2H_5OH has polar O-H bonds and thus experiences hydrogen bonding whereas CH_3-O-CH_3 does not. Hence, C_2H_5OH is expected to have the higher surface tension.

10.24 *See Sections 10.1 and 2.*

	Substance	Type of Solid	Most Important Forces
(a)	H_2O	molecular, polar	hydrogen bonds
(b)	C_6H_6	molecular, nonpolar	London dispersion
(c)	$CaCl_2$	ionic	cation-anion attractions
(d)	SiO_2	covalent network	Si-O covalent bonds
(e)	Fe	metallic	metal cations-sea of electrons, metallic bonds
(f)	SiC	covalent network	Si-C covalent bonds

10.26 *See Sections 10.1, 3 and 5.*

	Substance	Comment
increasing boiling point ↓	N_2	nonpolar with least number of electrons and smallest size
	CO_2	nonpolar having more electrons and larger size
	H_2O	polar with hydrogen bonds
	KCl	ionic with ions of smaller charges
	CaO	ionic with ions of higher charges

10.28 *See Section 10.3.*

$$1\ NH_4^+ \text{ ion at center of unit cell} \quad = \mathbf{1\ NH_4^+\ ion/unit\ cell}$$

$$\left(8\ Cl^- \text{ ions at corners}\right) \times \frac{1}{8} \quad = \mathbf{1\ Cl^-\ ion/unit\ cell}$$

10.30 *See Section 10.3.*

$$\left[1\ \text{W at center} + \left(8\ \text{W at corners}\right) \times \frac{1}{8}\right] = \mathbf{2\ W\ atoms\ /\ body\text{-}centered\ unit\ cell}$$

10.32 *See Section 10.3 and Example 10.3.*

Density is mass per unit volume. Mass per unit cell depends on the number of atoms/unit cell.

$$\left[\left(8\ \text{Ni atoms at corners}\right) \times \frac{1}{8} + \left(6\ \text{Ni atoms in faces}\right) \times \frac{1}{2}\right] = 4\ \text{Ni atoms/ unit cell}$$

$$?\ \text{mass per unit cell} = \frac{4\ \text{Ni atoms}}{1\ \text{unit cell}} \times \frac{1\ \text{mol Ni}}{6.022\ \text{x}\ 10^{23}\ \text{Ni atoms}} \times \frac{58.69\ \text{g Ni}}{1\ \text{mol Ni}} = 3.898\ \text{x}\ 10^{-22}\ \text{g/ unit cell}$$

$$?\ \text{volume of unit cell} = \left(351\ \text{pm} \times \frac{1\ \text{m}}{10^{12}\ \text{pm}} \times \frac{10^2\ \text{cm}}{1\ \text{m}}\right)^3 = 4.32\ \text{x}\ 10^{-23}\ \text{cm}^3\ /\ \text{unit cell}$$

$$\text{Density of Ni} = \frac{3.898\ \text{x}\ 10^{-22}\ \text{g/ unit cell}}{4.32\ \text{x}\ 10^{-23}\ \text{cm}^3\ /\ \text{unit cell}} = \mathbf{9.02\ g/cm^3}$$

10.34 *See Section 10.3 and Examples 10.3 and 4.*

(a) Volume = (length of edge)^3. Hence, the given data must be used to calculate the volume of the unit cell. Since density = mass/ volume and the density is given, the face-centered cubic array data must be used to calculate the number of atoms per unit cell and then the mass per unit cell to determine the volume of the unit cell.

$$\left[\left(8\ \text{Ar atoms at corners}\right) \times \frac{1}{8} + \left(6\ \text{Ar atoms in faces}\right) \times \frac{1}{2}\right] = 4\ \text{Ar atoms/ unit cell}$$

$$?\ \text{mass per unit cell} = \frac{4\ \text{Ar atoms}}{1\ \text{unit cell}} \times \frac{1\ \text{mol Ar}}{6.022\ \text{x}\ 10^{23}\ \text{Ar atoms}} \times \frac{39.95\ \text{g Ar}}{1\ \text{mol Ar}} = 2.654\ \text{x}\ 10^{-22}\ \text{g}$$

$$? \text{ volume of unit cell} = 2.654 \times 10^{-22} \text{ g} \times \frac{1 \text{ cm}^3}{1.65 \text{ g}} = 1.61 \times 10^{-22} \text{ cm}^3$$

$$\text{length of edge} = \sqrt[3]{\text{volume}} = \sqrt[3]{1.61 \times 10^{-22} \text{ cm}^3} = \mathbf{5.44 \times 10^{-8} \text{ cm}}$$

(b) The length of a face diagonal of a cube is equal to $\sqrt{2}a$, where a is the length of the edge, and is also equal to

4r, where r is the radius of an atom. Hence, $r_{Ar} = \frac{\sqrt{2}a}{4} = \frac{\sqrt{2}(5.44 \times 10^{-8} \text{ cm})}{4} = \mathbf{1.92 \times 10^{-8} \text{ cm}}$

10.36 See Section 10.3.

$$\left[1 \text{ Ti}^{4+} \text{ at center} + \left(8 \text{ Ti}^{4+} \text{ at corners}\right) \times \frac{1}{8}\right] = 2 \text{ Ti}^{4+} / \text{unit cell}$$

Since the formula is TiO_2, there must be **4 O^{2-}/unit cell.**

10.38 *See Section 10.3 and Example 10.3.*

Density is mass per unit volume. Mass per unit cell depends on the number of ions of each type per unit cell.

1 Cs^+ at center of unit cell = 1 Cs^+ ion/unit cell and $\left(8 \text{ I}^- \text{ at corners of unit cell}\right) \times \frac{1}{8} = 1 \text{ I}^-$ ion/unit cell

Assuming the mass of an electron is small compared to mass of an atom,

$$? \text{ g per unit cell} = \left[\left(\frac{1 \text{ Cs}^+ \text{ ion}}{\text{unit cell}} \times \frac{1 \text{ mol Cs}^+}{6.022 \times 10^{23} \text{ Cs}^+ \text{ ions}} \times \frac{132.9 \text{ g Cs}^+}{1 \text{ mol Cs}^+}\right)\right.$$

$$\left. + \left(\frac{1 \text{ I}^- \text{ ion}}{\text{unit cell}} \times \frac{1 \text{ mol I}^-}{6.022 \times 10^{23} \text{ I}^- \text{ ions}} \times \frac{126.9 \text{ g I}^-}{1 \text{ mol I}^-}\right)\right] = 4.314 \times 10^{-22} \text{ g/unit cell}$$

$$? \text{ volume of unit cell} = \left(445 \text{ pm} \times \frac{1 \text{ m}}{10^{12} \text{ pm}} \times \frac{10^2 \text{ cm}}{1 \text{m}}\right)^3 = 8.81 \times 10^{-23} \text{ cm}^3 / \text{unit cell}$$

$$\text{Density of CsI} = \frac{4.314 \times 10^{-22} \text{ g/unit cell}}{8.81 \times 10^{-23} \text{ cm}^3 / \text{unit cell}} = \mathbf{4.90 \text{ g/cm}^3}$$

10.40 ***See Section 10.3 and Example 10.3.***

The number of atoms per unit cell can be determined from the mass per unit cell. The mass per unit cell can be determined from the volume of the unit cell and the density.

$$? \text{ volume of unit cell} = \left(389 \text{ pm} \times \frac{1 \text{ m}}{10^{12} \text{ pm}} \times \frac{10^{2} \text{ cm}}{1 \text{ m}}\right)^{3} = 5.89 \times 10^{-23} \text{ cm}^3/\text{unit cell}$$

$$? \text{ g per unit cell} = 5.89 \times 10^{-23} \text{ cm}^3 \times 12.02 \text{ g Pd/cm}^3 = 7.08 \times 10^{-22} \text{ g Pd/unit cell}$$

$$? \text{ Pd atoms per unit cell} = \frac{7.08 \times 10^{-22} \text{ g Pd}}{\text{unit cell}} \times \frac{1 \text{ mol Pd}}{106.4 \text{ g Pd}} \times \frac{6.022 \times 10^{23} \text{ Pd atoms}}{1 \text{ mol Pd}} = 4.00 \text{ Pd atoms/unit cell}$$

The cubic unit cell which has four atoms per unit cell is the face-centered cubic unit cell

$$\left(8 \text{ corners} \times \frac{1}{8}\right) + \left(6 \text{ faces} \times \frac{1}{2}\right) = 4.$$

Hence, Pd must exist in the face-centered cubic unit cell.

10.42 ***See Section 10.4.***

Only (a) KBr and (d) sugar have the highly ordered structures needed to produce well-defined x-ray diffraction patterns.

10.44 ***See Section 10.4 and Example 10.6.***

Solving $n\lambda = 2d\sin\theta$ for λ gives $\lambda = \dfrac{2d\sin\theta}{n}$

$$\lambda = \frac{2(232 \text{ pm})(\sin 13.4°)}{1} = \mathbf{108 \text{ pm}}$$

10.46 ***See Section 10.5.***

(a) The vapor pressure does not change because the fraction of molecules with enough kinetic energy to escape from the liquid does not change.

(b) The vapor pressure increases because the fraction of molecules with enough kinetic energy to escape from the liquid increases.

(c) The vapor pressure does not change because the fraction of molecules with enough kinetic energy to escape from the liquid does not change.
Note: The initial increase in vapor pressure vanishes as the system returns to equilibrium.

10.48 ***See Section 10.5.***

(a) Assuming all 1.50 g of water are vaporized,

Known Quantities: $n = 1.50\ g\ H_2O \times \dfrac{1\ mol\ H_2O}{18.0\ g\ H_2O} = 0.0833\ mol$ $V = 1.00\ L$ $T = 30 + 273 = 303\ K$

Solving PV = nRT for P gives $P = \dfrac{nRT}{V}$ $P = \dfrac{(0.0833\ mol)\left(0.0821\dfrac{L \cdot atm}{mol \cdot K}\right)(303\ K)}{1.00\ L} = 2.07\ atm$

or $2.07\ atm \times \dfrac{760\ torr}{atm} = \mathbf{1.57 \times 10^3\ torr}$.

(b) At 30°C, the vapor pressure of water is **31.8 torr**.

(c) Since P is directly proportional to n and n is directly proportional to mass,

$? \ g\ H_2O$ actually evaporated at $30^\circ C = 31.8\ torr \times \dfrac{1.50\ g\ H_2O}{1.57 \times 10^3\ torr} = \mathbf{0.0304\ g\ H_2O}$.

10.54 ***See Section 10.5 and Example 10.8.***

q for raising temp. of 10.0 g of ice from -10°C to 0°C:

$$q_1 = nC\Delta t = 10.0\ g \times \frac{1\ mol}{18.0\ g} \times \frac{37.3\ J}{mol\ ^\circ C} \times \left[0^\circ C - \left(-10^\circ C\right)\right] = 207\ J$$

q for melting 10.0 g of ice: $q_2 = n\Delta H = 10.0\ g \times \dfrac{1\ mol}{18.0\ g} \times 6.01 \times 10^3\ \dfrac{J}{mol} = 3.34 \times 10^3\ J$

q for raising temp. of 10.0 g of liquid from 0°C to 95°C:

$$q_3 = nC\Delta t = 10.0\ g \times \frac{1\ mol}{18.0\ g} \times \frac{75.4\ J}{mol\ ^\circ C} \times \left[95^\circ C - 0^\circ C\right] = 3.98 \times 10^3\ J$$

$$q_{total} = q_1 + q_2 + q_3 = 207\ J + 3.34 \times 10^3\ J + 3.98 \times 10^3\ J = \mathbf{7.53 \times 10^3\ J}$$

10.56 ***See Section 10.5 and Table 10.6.***

$\Delta H_{fus} < \Delta H_{vap} < \Delta H_{sub}$

10.58 ***See Section 10.5.***

Assuming all three changes occur at the same temperature, $\Delta H_{sub} = \Delta H_{fus} + \Delta H_{vap}$.
Solving for ΔH_{fus} gives $\Delta H_{fus} = H_{sub} - \Delta H_{vap}$. Hence, ΔH_{fus} = 60.2 kJ/mol - 45.5 kJ/mol = **14.7 kJ/mol.**

10.60 *See Sections 10.5 and 6 and Exercise 10.59.*

(a) The heating curve expected when heat is added to the sample at constant pressure starting with solid at point B of the phase diagram shown in Exercise 10.59 is:

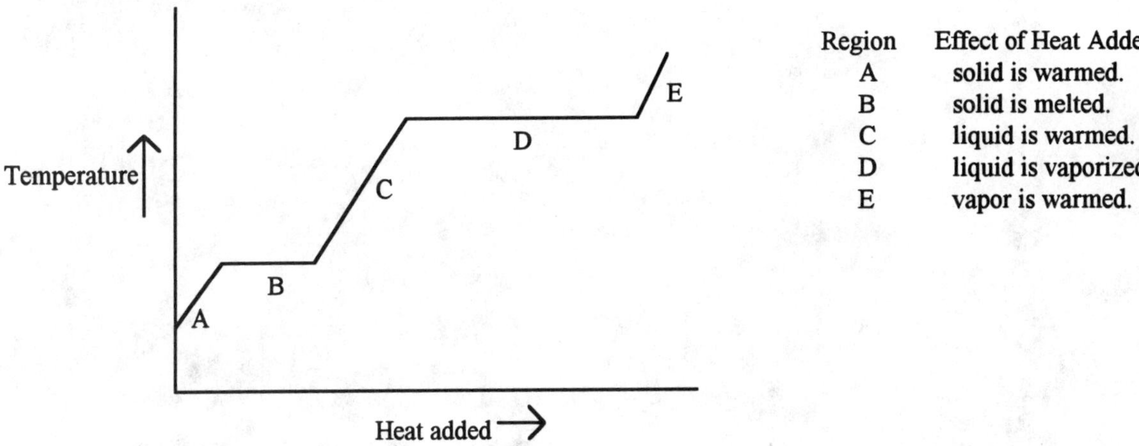

Region	Effect of Heat Added
A	solid is warmed.
B	solid is melted.
C	liquid is warmed.
D	liquid is vaporized.
E	vapor is warmed.

(b) As the pressure is lowered at constant temperature, starting at point A, solid will be converted to liquid and eventually liquid will be converted to vapor.

(c) The positive slope of the solid-liquid equilibrium line indicates an increase in pressure at a constant temperature will cause liquid to be converted to solid. This means the solid phase is more dense than the liquid phase of this substance.
(An increase in pressure at constant temperature always favors a decrease in volume, as was noted in Chapter 5.)

10.62 *See Section 10.6.*

(a) Addition of heat at constant pressure will convert solid to gas.

(b) A sudden increase in pressure will cause some ice to melt, since the density of the solid is less than that of the liquid. This will cause the temperature to decrease until equilibrium is re-established..

(c) An increase in pressure at constant temperature will convert some of the vapor to the liquid.

10.66 *See Section 10.5 and Example 10.8.*

(a) q for lowering temp. of 10.0 g Freon-11 from 23.8°C to 0°C:

$q = nC\Delta t \qquad q = 10.0\ \text{g} \times \frac{1\ \text{mol}}{18.0\ \text{g}} \times \frac{0.870\ \text{J}}{\text{mol}\ ^\circ\text{C}} \times \left[0^\circ\ \text{C} - 23.8^\circ\ \text{C}\right] = \text{-207 J}$

Hence, **207 J** must be removed.

(b) Solving $q = m\Delta H$ for m gives $m = \frac{q}{\Delta H}$ $\qquad m = \frac{207\ \text{J}}{180\ \text{J}\cdot\text{g}^{-1}} = \mathbf{1.15\ g}$

Chapter 11: Solutions

11.2 ***See Section 11.1 and Examples 11.1 and 2.***

(a) $\text{mass percent } C_6H_5CO_2H = \dfrac{1.20 \text{ g } C_6H_5CO_2H}{751.2 \text{ g soln}} \times 100\% = \mathbf{0.160\%\ C_6H_5CO_2H}$

(b) $? \text{ mol } C_6H_5CO_2H = 1.20 \text{ g } C_6H_5CO_2H \times \dfrac{1 \text{ mol } C_6H_5CO_2H}{122.0 \text{ g } C_6H_5CO_2H} = 0.00984 \text{ mol } C_6H_5CO_2H$

$? \text{ mol } H_2O = 750.0 \text{ g } H_2O \times \dfrac{1 \text{ mol } H_2O}{18.0 \text{ g } H_2O} = 41.70 \text{ mol } H_2O$

$? \text{ mol total of soln} = 0.00984 \text{ mol} + 41.70 \text{ mol} = 41.71 \text{ mol total of soln}$

$\chi_{C_6H_5CO_2H} = \dfrac{\text{mol } C_6H_5CO_2H}{\text{mol total of soln}}$ $\chi_{C_6H_5CO_2H} = \dfrac{0.00984 \text{ mol } C_6H_5CO_2H}{41.71 \text{ mol total of soln}} = \mathbf{2.36 \times 10^{-4}}$

(c) $\text{molality} = \dfrac{\text{mol } C_6H_5CO_2H}{\text{kg } H_2O}$ $\text{molality} = \dfrac{0.00984 \text{ mol } C_6H_5CO_2H}{0.750 \text{ kg } H_2O} = \mathbf{0.0131\ m\ C_6H_5CO_2H}$

11.4 ***See Section 11.1 and Example 11.1.***

$? \text{ mol } H_2O_2 = 250. \text{ g soln} \times \dfrac{3.0 \text{ g } H_2O_2}{100 \text{ g soln}} \times \dfrac{1 \text{ mol } H_2O_2}{34.0 \text{ g } H_2O_2} = \mathbf{0.22\ mol\ H_2O_2}$

11.6 ***See Section 11.1 and Examples 11.1 and 2.***

(a) $? \text{ g } H_3BO_3 \text{ in } 155 \text{ g soln} = 155 \text{ g soln} \times \dfrac{1.00 \text{ g } H_3BO_3}{100.00 \text{ g soln}} = 1.55 \text{ g } H_3BO_3$

Dissolve 1.55 g H_3BO_3 in sufficient water to give 155 g soln, 153.45 g H_2O.

(b) $? \text{ mol } H_3BO_3 \text{ in } 155 \text{ g soln} = 1.55 \text{ g } H_3BO_3 \times \dfrac{1 \text{ mol } H_3BO_3}{61.8 \text{ g } H_3BO_3} = 0.0251 \text{ mol } H_3BO_3$

$\text{molality} = \dfrac{\text{mol } H_3BO_3}{\text{kg } H_2O}$ $\text{molality} = \dfrac{0.0251 \text{ mol } H_3BO_3}{0.15345 \text{ kg } H_2O} = \mathbf{0.163\ m\ H_3BO_3}$

11.8 ***See Section 11.1 and Example 11.4.***

$? \text{ g } AgNO_3 \text{ in } 100.00 \text{ g soln} = 100.00 \text{ g soln} \times \dfrac{0.10 \text{ g } AgNO_3}{100.00 \text{ g soln}} = 0.10 \text{ g } AgNO_3$

$$? \text{ mol AgNO}_3 \text{ in } 100.00 \text{ g soln} = 0.10 \text{ g AgNO}_3 \times \frac{1 \text{ mol AgNO}_3}{169.9 \text{ g AgNO}_3} = 5.9 \times 10^{-4} \text{ mol AgNO}_3$$

$$? \text{ g H}_2\text{O in } 100.00 \text{ g soln} = 100.00 \text{ g soln} - 0.10 \text{ g AgNO}_3 = 99.90 \text{ g H}_2\text{O}$$

$$\text{molality} = \frac{\text{mol AgNO}_3}{\text{kg H}_2\text{O}} \qquad \text{molality} = \frac{5.9 \times 10^{-4} \text{ mol AgNO}_3}{0.09990 \text{ kg H}_2\text{O}} = \mathbf{5.9 \times 10^{-3} \text{ m AgNO}_3}$$

11.10 *See Section 11.1 and Example 11.4.*

A 0.75 m NaOCl solution contains 0.75 mol NaOCl per kg of water.

$$? \text{ mol H}_2\text{O in one kg H}_2\text{O} = 1{,}000 \text{ g H}_2\text{O} \times \frac{1 \text{ mol H}_2\text{O}}{18.02 \text{ g H}_2\text{O}} = 55.49 \text{ mol H}_2\text{O}$$

$$? \text{ mol total of solution} = 0.75 \text{ mol NaOCl} + 55.49 \text{ mol H}_2\text{O} = 56.24 \text{ mol total of soln}$$

$$\chi_{\text{NaOCl}} = \frac{0.75 \text{ mol NaOCl}}{56.24 \text{ mol total of soln}} = \mathbf{0.013}$$

11.12 *See Section 11.1 and Examples 11.4 and 5.*

(a) $$? \text{ g C}_{12}\text{H}_{22}\text{O}_{11} \text{ in } 100.0 \text{ g soln} = 100.0 \text{ g soln} \times \frac{10.0 \text{ g C}_{12}\text{H}_{22}\text{O}_{11}}{100.0 \text{ g soln}} = 10.0 \text{ g C}_{12}\text{H}_{22}\text{O}_{11}$$

$$? \text{ mol C}_{12}\text{H}_{22}\text{O}_{11} \text{ in } 100.0 \text{ g soln} = 10.0 \text{ g C}_{12}\text{H}_{22}\text{O}_{11} \times \frac{1 \text{ mol C}_{12}\text{H}_{22}\text{O}_{11}}{342.0 \text{ g C}_{12}\text{H}_{22}\text{O}_{11}} = 0.0292 \text{ mol C}_{12}\text{H}_{22}\text{O}_{11}$$

$$? \text{ g H}_2\text{O in } 100.0 \text{ g soln} = 100.0 \text{ g soln} - 10.0 \text{ g C}_{12}\text{H}_{22}\text{O}_{11} = 90.0 \text{ H}_2\text{O}$$

$$\text{molality} = \frac{\text{mol C}_{12}\text{H}_{22}\text{O}_{11}}{\text{kg H}_2\text{O}} \qquad \text{molality} = \frac{0.0292 \text{ mol C}_{12}\text{H}_{22}\text{O}_{11}}{0.090 \text{ kg H}_2\text{O}} = \mathbf{0.324 \text{ m C}_{12}\text{H}_{22}\text{O}_{11}}$$

(b) $$? \text{ L soln having mass of } 100.0 \text{ g} = 100.0 \text{ g soln} \times \frac{1 \text{ mL}}{1.038 \text{ g soln}} \times \frac{1 \text{ L}}{10^3 \text{ mL}} = 9.634 \times 10^{-2} \text{ L}$$

$$\text{molarity} = \frac{\text{mol C}_{12}\text{H}_{22}\text{O}_{11}}{\text{L soln}} \qquad \text{molarity} = \frac{\dfrac{0.0292 \text{ mol C}_{12}\text{H}_{22}\text{O}_{11}}{100.0 \text{ g soln}}}{\dfrac{9.634 \times 10^{-2} \text{ L}}{100.0 \text{ g soln}}} = \mathbf{0.303 \; \mathit{M} \; C_{12}H_{22}O_{11}}$$

(c) $$? \text{ mol H}_2\text{O in } 100.0 \text{ g soln} = 90.0 \text{ g H}_2\text{O} \times \frac{1 \text{ mol H}_2\text{O}}{18.0 \text{ g H}_2\text{O}} = 5.00 \text{ mol H}_2\text{O}$$

$? \text{ mol total of soln for } 100.0 \text{ g soln} = 0.0292 \text{ mol } C_{12}H_{22}O_{11} + 5.00 \text{ mol } H_2O = 5.03 \text{ mol total of soln}$

$$\chi_{C_{12}H_{22}O_{11}} = \frac{\text{mol } C_{12}H_{22}O_{11}}{\text{mol total of soln}} \qquad \chi_{C_{12}H_{22}O_{11}} = \frac{0.0292 \text{ mol } C_{12}H_{22}O_{11}}{5.03 \text{ mol total of soln}} = \mathbf{5.81 \times 10^{-3}}$$

11.14 ***See Section 11.1 and Examples 11.4 and 5.***

(a) $? \text{ g for exactly 1 L soln} = 1 \text{ L soln} \times \frac{10^3 \text{ mL soln}}{1 \text{ L soln}} \times \frac{1.031 \text{ g soln}}{1 \text{ mL soln}} = 1{,}031 \text{ g soln}$

$$? \text{ g } H_3PO_4 \text{ in 1 L soln} = 1 \text{ L soln} \times \frac{0.631 \text{ mol } H_3PO_4}{1 \text{ L soln}} \times \frac{98.0 \text{ g } H_3PO_4}{1 \text{ mol } H_3PO_4} = 61.8 \text{ g } H_3PO_4$$

$$\text{mass percent } H_3PO_4 = \frac{61.8 \text{ g } H_3PO_4}{1{,}031 \text{ g soln}} \times 100\% = \mathbf{5.99\%\ H_3PO_4}$$

(b) $? \text{ g } H_2O \text{ in 1 L soln} = 1{,}031 \text{ g soln} - 61.8 \text{ g } H_3PO_4 = 969 \text{ g } H_2O$

$$? \text{ mol } H_2O \text{ in 1 L soln} = 969 \text{ g } H_2O \times \frac{1 \text{ mol } H_2O}{18.0 \text{ g } H_2O} = 53.8 \text{ mol } H_2O$$

$? \text{ mol total of soln in 1 L soln} = 0.631 \text{ mol } H_3PO_4 + 53.8 \text{ mol } H_2O = 54.4 \text{ mol total of soln}$

$$\chi_{H_3PO_4} = \frac{\text{mol } H_3PO_4}{\text{mol total of soln}} \qquad \chi_{H_3PO_4} = \frac{0.631 \text{ mol } H_3PO_4}{54.4 \text{ mol total of soln}} = \mathbf{0.0116}$$

(c) $\text{molality} = \frac{\text{mol } H_3PO_4}{\text{kg } H_2O} \qquad \text{molality} = \frac{0.631 \text{ mol } H_3PO_4}{0.969 \text{ kg } H_2O} = \mathbf{0.651\ m\ H_3PO_4}$

11.16 ***See Section 11.1 and Examples 11.4 and 5.***

(a) $? \text{ g of exactly 1 L soln} = 1 \text{ L soln} \times \frac{10^3 \text{ mL soln}}{1 \text{ L soln}} \times \frac{0.973 \text{ g soln}}{1 \text{ mL soln}} = 973 \text{ g soln}$

$$? \text{ g } NH_3 \text{ in 1 L soln} = 973 \text{ g soln} \times \frac{6.00 \text{ g } NH_3}{100.00 \text{ g soln}} = 58.4 \text{ g } NH_3$$

$$? \text{ mol } NH_3 \text{ in 1 L soln} = M_{NH_3} = \frac{58.4 \text{ g } NH_3}{1 \text{ L soln}} \times \frac{1 \text{ mol } NH_3}{17.0 \text{ g } NH_3} = \mathbf{3.44\ \mathit{M}\ NH_3}$$

$? \text{ g } H_2O \text{ in 1 L soln} = 973 \text{ g soln} - 58.4 \text{ g } NH_3 = 915 \text{ g } H_2O$

$$? \text{ mol } H_2O \text{ in 1 L soln} = 915 \text{ g } H_2O \times \frac{1 \text{ mol } H_2O}{18.0 \text{ g } H_2O} = 50.8 \text{ mol } H_2O$$

$? \text{ mol total of soln} = 3.44 \text{ mol } NH_3 + 50.8 \text{ mol } H_2O = 54.2 \text{ mol total of soln}$

$$\text{molality} = \frac{\text{mol NH}_3}{\text{kg H}_2\text{O}} \qquad \text{molality} = \frac{3.44 \text{ mol NH}_3}{0.915 \text{ kg H}_2\text{O}} = \mathbf{3.76\ m\ NH_3}$$

$$\chi_{\text{NH}_3} = \frac{\text{mol NH}_3}{\text{mol total of soln}} \qquad \chi_{\text{NH}_3} = \frac{3.44 \text{ mol NH}_3}{54.2 \text{ mol total of soln}} = \mathbf{0.0635}$$

(b) $$? \text{ g of exactly 1 L soln} = 1 \text{ L soln} \times \frac{10^3 \text{ mL soln}}{1 \text{ L soln}} \times \frac{0.936 \text{ g soln}}{1 \text{ mL soln}} = 936 \text{ g soln}$$

$$? \text{ g NH}_3 \text{ in 1 L soln} = 1 \text{ L soln} \times \frac{8.80 \text{ mol}}{1 \text{ L soln}} \times \frac{17.0 \text{ g NH}_3}{1 \text{ mol NH}_3} = 150 \text{ g NH}_3$$

$$? \text{ g H}_2\text{O in 1 L soln} = 936 \text{ g soln} - 150 \text{ g NH}_3 = 786 \text{ g H}_2\text{O}$$

$$? \text{ mol H}_2\text{O in 1 L soln} = 786 \text{ g H}_2\text{O} \times \frac{1 \text{ mol H}_2\text{O}}{18.0 \text{ g H}_2\text{O}} = 43.7 \text{ mol H}_2\text{O}$$

$$? \text{ mol total of soln} = 8.80 \text{ mol NH}_3 + 43.7 \text{ mol H}_2\text{O} = 52.5 \text{ mol total of soln}$$

$$\text{molality} = \frac{\text{mol NH}_3}{\text{kg H}_2\text{O}} \qquad \text{molality} = \frac{8.80 \text{ mol NH}_3}{0.786 \text{ kg H}_2\text{O}} = \mathbf{11.2\ m\ NH_3}$$

$$\text{mass percent NH}_3 = \frac{\text{g NH}_3}{\text{g soln}} \times 100\% \qquad \text{mass percent} = \frac{150 \text{ g NH}_3}{936 \text{ g soln}} 100\% = \mathbf{16.0\%\ NH_3}$$

$$\chi_{\text{NH}_3} = \frac{\text{mol NH}_3}{\text{mol total of soln}} \qquad \chi_{\text{NH}_3} = \frac{8.80 \text{ mol NH}_3}{52.52 \text{ mol total of soln}} = \mathbf{0.168}$$

(c) $$? \text{ g soln containing exactly 1 kg H}_2\text{O} = 1{,}000 \text{ g H}_2\text{O} + (8.02 \text{ mol NH}_3)\left(\frac{17.0 \text{ g NH}_3}{1 \text{ mol NH}_3}\right) = 1{,}136.3 \text{ g soln}$$

$$? \text{ L soln containing exactly 1 kg H}_2\text{O} = 1{,}136.3 \text{ g soln} \times \frac{1 \text{ mL soln}}{0.950 \text{ g soln}} \times \frac{1 \text{ L soln}}{10^3 \text{ mL soln}} = 1.20 \text{ L soln}$$

$$? \text{ mol H}_2\text{O in exactly 1 kg H}_2\text{O} = 1 \text{ kg H}_2\text{O} \times \frac{10^3 \text{ g H}_2\text{O}}{1 \text{ kg H}_2\text{O}} \times \frac{1 \text{ mol H}_2\text{O}}{18.0 \text{ g H}_2\text{O}} = 55.6 \text{ mol H}_2\text{O}$$

$$? \text{ mol total of soln when exactly 1 kg H}_2\text{O} = 8.02 \text{ mol NH}_3 + 55.6 \text{ mol H}_2\text{O} = 63.6 \text{ mol total of soln}$$

$$\text{molarity} = \frac{\text{mol NH}_3}{\text{L soln}} \qquad \text{molarity} = \frac{8.02 \text{ mol NH}_3}{1.20 \text{ L soln}} = \mathbf{6.68\ \mathit{M}\ NH_3}$$

$$\text{mass percent NH}_3 = \frac{\text{g NH}_3}{\text{g soln}} \times 100\% \qquad \text{mass percent NH}_3 = \frac{136.3 \text{ g NH}_3}{1{,}136.3 \text{ g soln}} \times 100\% = \mathbf{12.0\%\ NH_3}$$

$$\chi_{NH_3} = \frac{\text{mol } NH_3}{\text{mol total of soln}} \qquad \chi_{NH_3} = \frac{8.02 \text{ mol } NH_3}{63.6 \text{ mol total of soln}} = \mathbf{0.126}$$

(d) The quantity chosen for total mol of soln is arbitrary; 10.00 mol was chosen since this quantity is easy to work with.

$$? \text{ mol } NH_3 \text{ in } 10.00 \text{ total mol of soln} = 10.00 \text{ total mol of soln} \times \frac{0.0738 \text{ mol } NH_3}{1 \text{ mol total of soln}} = 0.738 \text{ mol } NH_3$$

$$? \text{ mol } H_2O \text{ in } 10.00 \text{ total mol of soln} = 10.00 \text{ mol total of soln} - 0.738 \text{ mol } NH_3 = 9.26 \text{ mol } H_2O$$

$$? \text{ g of } 10.00 \text{ total mol of soln} = (0.738 \text{ mol } NH_3)\left(\frac{17.0 \text{ g } NH_3}{1 \text{ mol } NH_3}\right) + (9.26 \text{ mol } H_2O)\left(\frac{18.0 \text{ g } H_2O}{1 \text{ mol } H_2O}\right)$$

$$= 12.6 \text{ g } NH_3 + 167 \text{ g } H_2O = 180 \text{ g soln}$$

$$? \text{ L of } 10.00 \text{ total mol of soln} = 180 \text{ g soln} \times \frac{1 \text{ mL soln}}{0.969 \text{ g soln}} \times \frac{1 \text{ L soln}}{10^3 \text{ mL soln}} = 0.186 \text{ L soln}$$

$$\text{molality} = \frac{\text{mol } NH_3}{\text{kg } H_2O} \qquad \text{molality} = \frac{0.738 \text{ mol } NH_3}{0.167 \text{ kg } H_2O} = \mathbf{4.42 \text{ m } NH_3}$$

$$\text{molarity} = \frac{\text{mol } NH_3}{\text{L soln}} \qquad \text{molarity} = \frac{0.738 \text{ mol } NH_3}{0.186 \text{ L soln}} = \mathbf{3.97}\ \boldsymbol{M}\ \mathbf{NH_3}$$

$$\text{mass percent} = \frac{\text{g } NH_3}{\text{g soln}} \times 100\% \qquad \text{mass percent} = \frac{12.6 \text{ g } NH_3}{180 \text{ g soln}} \times 100\% = \mathbf{7.00\%\ NH_3}$$

Summary Table:

	Density(g/cm^3)	Molality	Molarity	Mass % NH_3	Mole Fraction
(a)	0.973	**3.76**	**3.44**	6.00	**0.0635**
(b)	0.936	**11.2**	8.80	**16.0**	**0.168**
(c)	0.950	8.02	**6.68**	**12.0**	**0.126**
(d)	0.969	**4.42**	**3.97**	**7.00**	0.0738

11.18 ***See Section 11.1.***

$$? \text{ mol } C_6H_6 = 98.0 \text{ g } C_6H_6 \times \frac{1 \text{ mol } C_6H_6}{78.1 \text{ g } C_6H_6} = 1.25 \text{ mol } C_6H_6$$

$$? \text{ mol } C_6H_{14} = 12.0 \text{ g } C_6H_{14} \times \frac{1 \text{ mol } C_6H_{14}}{86.2 \text{ g } C_6H_{14}} = 0.139 \text{ mol } C_6H_{14}$$

$$? \text{ mol } C_8H_{18} = 20.0 \text{ g } C_8H_{18} \times \frac{1 \text{ mol } C_8H_{18}}{114.2 \text{ g } C_8H_{18}} = 0.175 \text{ mol } C_8H_{18}$$

$$? \text{ mol total of soln} = 1.25 \text{ mol } C_6H_6 + 0.139 \text{ mol } C_6H_{14} + 0.175 \text{ mol } C_8H_{18} = 1.56 \text{ mol total of soln}$$

$$\chi_{C_6H_6} = \frac{\text{mol } C_6H_6}{\text{mol total of soln}} \qquad \chi_{C_6H_6} = \frac{1.25 \text{ mol } C_6H_6}{1.56 \text{ mol total of soln}} = \mathbf{0.801}$$

11.20 ***See Section 11.2.***

As n increases, the overall polarity of the straight chain alcohols ($CH_3(CH_2)_nOH$) decreases. Hence, the solubility in water decreases as n increases.

11.22 ***See Section 11.2.***

Substances that have similar intermolecular forces have strong solute-solvent interactions and therefore tend to form solutions. Hence, we should obtain following order of increasing solubility in benzene (C_6H_6):
$H_2O < C_2H_5OH < C_6H_{14}$.

11.24 ***See Section 11.2 and Example 11.6.***

(a) Br_2 is nonpolar and therefore more soluble in nonpolar carbon tetrachloride than polar water.

(b) $CaCl_2$ is ionic and therefore more soluble in polar water than nonpolar benzene.

(c) $CHCl_3$ is slightly polar and therefore more soluble in the less polar diethyl ether than the hydrogen-bonded water.

(d) $C_2H_4(OH)_2$ has O-H bonds and is therefore more soluble in water than nonpolar benzene.

11.26 ***See Section 11.2.***

(a) $(CH_3)_2CO$ in water, hydrogen bonding

(b) IBr in $CHCl_3$, dipole-dipole and London dispersion

(c) $CaCl_2$ in water, ion-dipole

(d) Kr in CH_3OH, London dispersion

11.28 ***See Section 11.2.***

(a) $CH_3(CH_2)_{10}OH$ is less polar than $CH_3(CH_2)_2OH$ due to its longer CH_2 chain and is therefore more soluble than $CH_3(CH_2)_2OH$ in C_6H_{14}.

(b) CCl_4 is nonpolar and therefore more soluble than ionic $BaCl_2$ in C_6H_{14}.

(c) $Fe(C_5H_5)_2$ is molecular and nonpolar and therefore more soluble than ionic $FeCl_2$ in C_6H_{14}.

11.30 ***See Section 11.3.***

The solubility of the solute (KCl) increases with increasing temperature. The enthalpy of solution is positive, and the process is endothermic.

11.32 ***See Section 11.3.***

The solubility of nitrogen should decrease with increasing temperature, since the dissolving process is exothermic. The solubility of nitrogen should, however, increase with increasing pressure, as does the solubility of other gases. Hence,
(a) lower temperature and higher pressure yields higher solubility.
(b) same temperature and lower pressure yields lower solubility.

(c) lower temperature and same pressure yields higher solubility.
(d) higher temperature and same pressure yields lower solubility.
(e) higher temperature and lower pressure yields lower solubility.

11.34 ***See Section 11.3 and Example 11.7.***

(a) Solving C = kP for k gives $k = \frac{C}{P}$ $\qquad k = \frac{9.38 \text{ x } 10^{-3} \text{ molal}}{0.200 \text{ atm} \times 760 \text{ torr/atm}} = \mathbf{6.17 \text{ x } 10^{-5} \text{ molal/torr}}$

(b) Substituting in C = kP gives $C = (6.17 \text{ x } 10^{-5} \text{ molal/torr})(300 \text{ torr}) = 1.85 \text{ x } 10^{-2} \text{ molal}$

Since 1.85 x 10^{-2} molal corresponds to 1.85 x 10^{-2} mol C_2H_2/kg H_2O,

$$? \text{ g } C_2H_2 = 1.85 \text{ x } 10^{-2} \text{ mol } C_2H_2 \times \frac{26.0 \text{ g } C_2H_2}{1 \text{ mol } C_2H_2} = \mathbf{0.481 \text{ g } C_2H_2}$$

11.38 ***See Section 11.3.***

The solubility of nitrous oxide should decrease with increasing temperature, since the dissolving process is exothermic. The solubility of nitrous oxide should, however, increase with increasing pressure, as does the solubility of other gases. Hence,
(a) lower temperature and same pressure yields higher solubility; i.e. larger than 0.055 molal.
(b) higher temperature and same pressure yields lower solubility; i.e. smaller than 0.055 molal.
(c) same temperature and lower pressure yields lower solubility; i.e. smaller than 0.055 molal.
(d) higher temperature and lower pressure yields lower solubility; i.e. smaller than 0.055 molal.

11.40 ***See Section 11.3.***

The solubility of the solute ($Ca(OH)_2$) decreases with increasing temperature. The enthalpy of solution is negative, and the process is exothermic.

11.42 ***See Section 11.3 and Figure 11.10.***

The solubility of NH_4Cl increases more rapidly with increasing temperature indicating it has a more positive and thus more endothermic enthalpy of solution.

11.44 ***See Section 11.4 and Example 11.8.***

$\Delta P = \chi_{solute} P^{\circ}_{solv}$ and $P_{solv} = \chi_{solv} P^{\circ}_{solv}$

$$? \text{ mol } C_6H_5OH = 10.0 \text{ g } C_6H_5OH \times \frac{1 \text{ mol } C_6H_5OH}{94.1 \text{ g } C_6H_5OH} = 0.106 \text{ mol } C_6H_5OH$$

$$? \text{ mol } CHCl_3 = 95.0 \text{ g } CHCl_3 \times \frac{1 \text{ mol } CHCl_3}{119.4 \text{ g } CHCl_3} = 0.796 \text{ mol } CHCl_3$$

$$? \text{ mol total of soln} = 0.106 \text{ mol } C_6H_5OH + 0.796 \text{ mol } CHCl_3 = 0.902 \text{ mol total of soln}$$

$$\chi_{solute} = \chi_{C_6H_5OH} = \frac{0.106 \text{ mol } C_6H_5OH}{0.902 \text{ mol total of soln}} = 0.118 \qquad \chi_{solv} = 1 - \chi_{solute} \qquad \chi_{solv} = 1 - 0.118 = 0.882$$

$\Delta P = \chi_{C_6H_5OH} P^{\circ}_{CHCl_3}$ $\Delta P = (0.118)(360 \text{ torr}) = \mathbf{42.5\ torr}$

$P_{solv} = \chi_{CHCl_3} P^{\circ}_{CHCl_3}$ $P_{solv} = (0.882)(360 \text{ torr}) = \mathbf{318\ torr}$ or $P_{solv} = P^{\circ}_{CHCl_3} - \Delta P = \mathbf{318\,torr}$

11.46 ***See Section 11.4, Table 11.3 and Examples 11.10 and 12.***

$\Delta T_f = mk_f$ and $\Delta T_b = mk_b$ $? \text{ mol urea} = 2.00 \text{ g urea} \times \frac{1 \text{ mol urea}}{60.1 \text{ g urea}} = 0.0333 \text{ mol urea}$

$\text{molality} = \frac{\text{mol urea}}{\text{kg water}}$ $\text{molality} = \frac{0.0333 \text{ mol urea}}{0.0250 \text{ kg water}} = 1.33 \text{ m urea}$

$\Delta T_f = mk_f$ $\Delta T_f = (1.33 \text{ m})(1.86^{\circ}\text{C/m}) = 2.47^{\circ}\text{C}$, $T_f = 0.00^{\circ}\text{C} - 2.47^{\circ}\text{C} = \mathbf{-2.47^{\circ}C}$

$\Delta T_b = mk_b$ $\Delta T_b = (1.33 \text{ m})(0.512^{\circ}\text{C/m}) = 0.681^{\circ}\text{C}$, $T_f = 100.00^{\circ}\text{C} + 0.681^{\circ}\text{C} = \mathbf{100.68^{\circ}C}$

11.48 ***See Section 11.4, Table 11.3 and Examples 11.10 and 12.***

Solving $\Delta T_b = mk_b$ for k_b gives $k_b = \frac{\Delta T_b}{m}$ Using the data of 11.47 gives

$$? \text{ mol } C_6H_5OH = 0.500 \text{ g } C_6H_5OH \times \frac{1 \text{ mol } C_6H_5OH}{94.1 \text{ g } C_6H_5OH} = 0.00531 \text{ mol } C_6H_5OH$$

$\text{molality} = \frac{\text{mol } C_6H_5OH}{\text{kg } C_6H_{12}}$ $\text{molality} = \frac{0.00531 \text{ mol } C_6H_5OH}{0.0120 \text{ kg } C_6H_{12}} = 0.442 \text{ m } C_6H_5OH$

$k_b = \frac{\Delta T_b}{m}$ $k_b = \frac{81.94^{\circ}\text{C} - 80.72^{\circ}\text{C}}{0.442 \text{ m}} = \mathbf{2.76^{\circ}\ C/m}$

11.50 ***See Section 11.4 and Example 11.13.***

Solving $\Delta T_f = mk_f$ for m gives $m = \frac{\Delta T_f}{k_f}$ $m = \frac{6.50^{\circ}\text{C} - 0.83^{\circ}\text{C}}{20.2^{\circ}\text{C/molal}} = 0.281 \text{ m}$

$$? \text{ mol solute} = 12.0 \text{ g solvent} \times \frac{1 \text{ kg solvent}}{10^3 \text{ g solvent}}\ \frac{0.281 \text{ mol solute}}{1 \text{ kg solvent}} = 3.37 \times 10^{-3} \text{ mol solute}$$

$\text{molar mass} = \frac{\text{g solute}}{\text{mol solute}}$ $\text{molar mass} = \frac{0.350 \text{ g}}{3.37 \times 10^{-3} \text{ mol}} = \mathbf{104\ g/mol}$

11.52 ***See Section 11.5 and Example 11.15.***

The boiling point of a solution depends on the total molal concentration of solute particles. This concentration is given by the product of the van't Hoff factor times the molal concentration of the solute.

Compound	Present As	*i*	m	$i \times m$
LiBr	Li^+ & Br^-	2	0.02	0.04
sucrose	molecules	1	0.03	0.03

Compound	Present As	i	m	$i \times$ m
$MgSO_4$	Mg^{2+} & SO_4^{2-}	2	0.03	0.06
$CaCl_2$	Ca^{2+} & $2Cl^-$	3	0.03	0.09

Hence, the boiling points of these solutions should increase in the order:

0.03 m sucrose < 0.02 m LiBr < 0.03 m $MgSO_4$ < 0.03 m $CaCl_2$

11.54 ***See Sections 11.4 and 5 and Example 11.5.***

Solving $\Delta T_f = imk_f$ for im, the total molality of solute particles, gives $im = \frac{\Delta T_f}{k_f}$ $\quad im = \frac{2.01^\circ \text{C}}{1.86^\circ \text{C/m}} = 1.08 \text{ m}$

Assuming an ideal i value of 2.0 for NaCl gives $\text{m} = \frac{1.08 \text{ m}}{2.0} = 0.54 \text{ m NaCl}$

This corresponds to 0.54 mol NaCl/kg H_2O and

$$? \text{ g NaCl per kg H}_2\text{O} = \frac{0.54 \text{ mol NaCl}}{\text{kg H}_2\text{O}} \times \frac{58.45 \text{ g NaCl}}{1 \text{ mol NaCl}} = \mathbf{31.6 \text{ g NaCl/ kg H}_2\text{O}}$$

11.56 ***See Sections 11.4 and 5.***

(a) Assuming no dissociation of K_2SO_4 occurs, the total expected molarity of K_2SO_4 solute particles is 0.029 M.

Substituting in $\Pi = M\text{RT}$ gives $\Pi = \left(0.029 \frac{\text{mol}}{\text{L}}\right)\left(0.0821 \frac{\text{L}\cdot\text{atm}}{\text{mol}\cdot\text{K}}\right)(298 \text{ K}) = 0.71 \text{ atm}$

$i = \frac{\text{measured value}}{\text{expected value}}$ $\qquad i = \frac{1.79 \text{ atm}}{0.71 \text{ atm}} = 2.5$

(b) At higher concentrations i will be smaller, since more extensive ion association will occur.

11.58 ***See Section 11.4.***

Known Quantities:

$$? \text{ g enzyme in exactly 1 L soln} = 1 \text{ L soln} \times \frac{10^3 \text{ mL soln}}{1 \text{ L soln}} \times \frac{1.00 \text{ g soln}}{1 \text{ mL soln}} \times \frac{10.0 \text{ g enzyme}}{100.00 \text{ g soln}} = 100. \text{ g enzyme}$$

$\Pi = 13.3 \text{ torr} \times \frac{1 \text{ atm}}{760 \text{ torr}} = 1.75 \times 10^{-2} \text{ atm}$ $\qquad T = 298 \text{ K}$

Solving $\Pi = \frac{\text{nRT}}{\text{V}}$ for n gives $\text{n} = \frac{\Pi \text{V}}{\text{RT}}$. Assuming a volume of exactly 1 L and substituting yields

$$\text{n} = \frac{(1.75 \times 10^{-2} \text{ atm})(1 \text{ L})}{\left(0.0821 \frac{\text{L}\cdot\text{atm}}{\text{mol}\cdot\text{K}}\right)(298 \text{ K})} = 7.15 \times 10^{-4} \text{ mol}$$

$$\text{molar mass} = \frac{\text{g solute}}{\text{mol solute}} \qquad \text{molar mass} = \frac{100.\ \text{g enzyme}}{7.15 \times 10^{-4}\ \text{mol enzyme}} = \mathbf{1.40 \times 10^{5}\ g/mol}$$

11.60 ***See Sections 11.4 and 5 and the Insight into Chemistry entitled Reverse Osmosis.***

The direction of net transport of water molecules depends on whether the pressure that is exerted on the membrane is greater on the pure water side or on the NaCl solution side. The pressure that is exerted on the pure water side will be equal to the pressure that is exerted on the NaCl side when there is no outside force acting on the system and the system is allowed to reach equilibrium. This pressure is given by Π. Hence, Π must be calculated and compared to the applied pressure to determine the direction of net transport of water molecules.

Known Quantities:
Assuming ideal behavior, total molarity of NaCl solute particles $= 2.0 \times 0.010\ M = 0.20\ M$ $\quad$ T = 298 K

Substituting in $\Pi = M\text{RT}$ gives $\Pi = \left(0.020\frac{\text{mol}}{\text{L}}\right)\left(0.0821\frac{\text{L}\cdot\text{atm}}{\text{mol}\cdot\text{K}}\right)(298\ \text{K}) = 0.489\ \text{atm or } 372\ \text{torr}$

Since Π is less than the 500 torr that is applied to the salt solution side, net transport of water molecules to the pure water side will occur. Reverse osmosis will occur.

11.62 ***See Section 11.6 and Example 11.6.***

(a) $? \text{ mol } C_6H_{12} = 25.0\ \text{g}\ C_6H_{12} \times \frac{1\ \text{mol}\ C_6H_{12}}{84.0\ \text{g}\ C_6H_{12}} = 0.298\ \text{mol}\ C_6H_{12}$

$? \text{ mol } C_6H_{14} = 44.0\ \text{g}\ C_6H_{14} \times \frac{1\ \text{mol}\ C_6H_{14}}{86.0\ \text{g}\ C_6H_{14}} = 0.512\ \text{mol}\ C_6H_{14}$

$? \text{ mol total of soln} = 0.298\ \text{mol}\ C_6H_{12} + 0.512\ \text{mol}\ C_6H_{14} = 0.810 \text{mol total of soln}$

$$\chi_{C_6H_{12}} = \frac{\text{mol}\ C_6H_{12}}{\text{mol total of soln}} \qquad \chi_{C_6H_{12}} = \frac{0.298\ \text{mol}\ C_6H_{12}}{0.810\ \text{mol total of soln}} = \mathbf{0.368}$$

$$\chi_{C_6H_{14}} = 1 - \chi_{C_6H_{12}} \qquad \chi_{C_6H_{14}} = 1 - 0.368 = \mathbf{0.632}$$

(b) $P_{C_6H_{12}} = \chi_{C_6H_{12}} P^{\circ}_{C_6H_{12}}$ $\qquad P_{C_6H_{12}} = (0.368)(150\ \text{torr}) = \mathbf{55.2\ torr}$

$P_{C_6H_{14}} = \chi_{C_6H_{14}} P^{\circ}_{C_6H_{14}}$ $\qquad P_{C_6H_{14}} = (0.632)(313\ \text{torr}) = \mathbf{198\ torr}$

(c) $\chi_{C_6H_{12}(g)} = \frac{55.2\ \text{torr}}{55.2\ \text{torr} + 198\ \text{torr}} = \mathbf{0.218}$ $\qquad \chi_{C_6H_{14}(g)} = 1 - \chi_{C_6H_{12}(g)}$ $\qquad \chi_{C_6H_{14}(g)} = 1 - 0.218 = \mathbf{0.782}$

11.64 *See Section 11.6 and Example 11.6.*

(a) $P_{HCO_2H} = \chi_{HCO_2H} P^\circ_{HCO_2H}$, $\quad P_{H_2O} = \chi_{H_2O} P^\circ_{H_2O}$, $\quad P_T = P_{HCO_2H} + P_{H_2O}$

Arbitrarily assuming a 100.0 g mixture of HCO_2H and H_2O,

$$? \text{ mol } HCO_2H = 100.0 \text{ g soln} \times \frac{77.5 \text{ g } HCO_2H}{100.0 \text{ g soln}} \times \frac{1 \text{ mol } HCO_2H}{46.0 \text{ g } HCO_2H} = 1.68 \text{ mol } HCO_2H$$

$$? \text{ mol } H_2O = 100.0 \text{ g soln} \times \frac{22.5 \text{ g } H_2O}{100.0 \text{ g soln}} \times \frac{1 \text{ mol } H_2O}{18.0 \text{ g } H_2O} = 1.25 \text{ mol } H_2O$$

? mol total of soln = 1.68 mol HCO_2H + 1.25 mol H_2O = 2.93 mol total of soln

$$\chi_{HCO_2H} = \frac{1.68 \text{ mol } HCO_2H}{2.93 \text{ mol total of soln}} = 0.573 \qquad \chi_{H_2O} = 1 - 0.573 = 0.427$$

$$P_{HCO_2H} = \chi_{HCO_2H} P^\circ_{HCO_2H} \qquad P_{HCO_2H} = (0.573)(917 \text{ torr}) = \mathbf{525 \text{ torr}}$$

$$P_{H_2O} = \chi_{H_2O} P^\circ_{H_2O} \qquad P_{H_2O} = (0.427)(974 \text{ torr}) = \mathbf{416 \text{ torr}}$$

$$P_T = P_{HCO_2H} + P_{H_2O} \qquad P_T = 525 \text{ torr} + 416 \text{ torr} = \mathbf{941 \text{ torr}}$$

(b) The actual vapor pressure of the azeotropic mixture at 107.1°C is 760 torr, the prevailing atmospheric pressure. Since the actual vapor pressure is less than that calculated (760 torr vs. 941 torr), the deviation from Raoult's Law is negative.

(c) Negative deviation implies the intermolecular forces between HCO_2H and H_2O are stronger than the average of those in the pure substances.

11.66 *See Section 11.3 and Example 11.7.*

(a) Solving $C = kP$ for k gives $k = \frac{C}{P}$. Substituting the given data yields

$$k = \frac{\frac{0.240 \text{ g } CO_2}{100 \text{ mL}}}{1.00 \text{ atm}} = 2.40 \times 10^{-3} \text{ g } CO_2 \cdot mL^{-1} \cdot atm^{-1}$$

Using this value of k to calculate C when P = 4.00 atm yields

$$C = (2.40 \times 10^{-3} \text{ g } CO_2 \cdot mL^{-1} \cdot atm^{-1})(4.00 \text{ atm}) = 9.60 \times 10^{-3} \text{ g } CO_2 \cdot mL^{-1}$$

$$\text{Hence, ? g } CO_2 \text{ in 12 oz can} = 12 \text{ oz} \times \frac{28.35 \text{ mL}}{1 \text{ oz}} \times 9.60 \times 10^{-3} \frac{\text{g } CO_2}{\text{mL}} = \mathbf{3.27 \text{ g } CO_2}$$

(b) Neglecting the equilibrium pressure of CO_2 in the atmosphere as the pressure of CO_2 remaining, we can assume that all of the CO_2 is expelled. This gives

$$?\text{L } CO_2 \text{ expelled when measured at STP} = 3.27 \text{ g } CO_2 \times \frac{1 \text{ mol } CO_2}{44.0 \text{ g } CO_2} \times \frac{22.4 \text{ L } CO_2}{1 \text{ mol } CO_2} = \mathbf{1.66 \text{ L } CO_2}$$

11.68 ***See Sections 5.4 and 11.4 and Examples 5.6 and 11.13.***

(a) Known Quantities: $m = 0.262 \text{ g}$ $\quad P = 745 \text{ torr} \times \frac{1 \text{ atm}}{760 \text{ torr}} = 0.980 \text{ atm}$

$$V = 51.0 \text{ mL} \times \frac{1 \text{ L}}{10^3 \text{ mL}} = 0.0510 \text{ L} \qquad T = 25 + 273 = 298 \text{ K}$$

Solving PV = nRT for n gives $n = \frac{PV}{RT}$ $\quad n = \frac{(0.980 \text{ atm})(0.0510 \text{ L})}{\left(0.0821 \frac{\text{L}\cdot\text{atm}}{\text{mol}\cdot\text{K}}\right)(298 \text{ K})} = 2.04 \times 10^{-3} \text{ mol}$

and using $\mathcal{M} = \frac{m}{n}$ yields $\quad \mathcal{M} = \frac{0.262 \text{ g}}{2.04 \times 10^{-3} \text{ mol}} = \mathbf{128 \text{ g/mol}}$

(b) Solving $\Delta T_f = mk_f$ for m gives $m = \frac{\Delta T_f}{k_f}$ $\quad m = \frac{0^\circ\text{C} - (-0.61^\circ\text{C})}{1.86^\circ \text{ C/molal}} = 0.33 \text{ m}$

$$? \text{ mol solute} = 12.0 \text{ g solvent} \times \frac{1 \text{ kg solvent}}{10^3 \text{ g solvent}} \times \frac{0.33 \text{ mol solute}}{1 \text{ kg solvent}} = 4.0 \times 10^{-3} \text{ mol solute}$$

$\text{molar mass} = \frac{\text{g solute}}{\text{mol solute}}$ $\quad \text{molar mass} = \frac{0.262 \text{ g}}{4.0 \times 10^{-3} \text{ mol}} = \mathbf{66 \text{ g/mol}}$

(c) The key to understanding this difference is to note that the number of moles of solute particles that are obtained in solution is almost twice the number of moles of gas. The gaseous substance must therefore act as an electrolyte in water and have an *i* value of approximately 2. This could occur if the gaseous compound acts as an acid in water: $HX(g) + H_2O \rightarrow H_3O^+ + X^-$. (See Section 3.1).

11.70 ***See Section 11.4 and 5 and Example 11.3.***

(a) Solving $\Delta T_f = mk_f$ for m gives $m = \frac{\Delta T_f}{k_f}$

For C_6H_6: $m = \frac{0.205^\circ \text{C}}{4.90^\circ \text{C/m}} = \mathbf{0.0418 \text{ m}}$ $\qquad$ For H_2O: $m = \frac{0.159^\circ \text{C}}{1.86^\circ \text{C/m}} = \mathbf{0.0855 \text{ m}}$

(b) Assuming 100.0 g soln for each solution and recognizing that 0.50 g is solute and 99.50 g is solvent gives

$$? \text{ mol } CH_3COOH = 0.50 \text{ g } CH_3COOH \times \frac{1 \text{ mol } CH_3COOH}{60.0 \text{ g } CH_3COOH} = 8.3 \times 10^{-3} \text{ mol } CH_3COOH$$

$$\text{molality} = \frac{\text{mol } CH_3COOH}{\text{kg solvent}} \qquad \text{molality} = \frac{8.3 \times 10^{-3} \text{ mol } CH_3COOH}{0.995 \text{ kg solvent}} = 8.3 \times 10^{-2} \text{ m}$$

The experimental molality that is obtained with benzene is approximately one-half that calculated and thus corresponds to an *i* value of one-half:

$$im = 0.0418 \text{ molal and m} = 8.3 \times 10^{-2} \text{ molal yields } \boldsymbol{i} \cong \tfrac{1}{2}$$

The experimental molality that is obtained with water is approximately equal to that calculated and thus corresponds to *an i* value of one:

$$im = 0.0855 \text{ molal and m} = 8.3 \times 10^{-2} \text{ molal yields } \boldsymbol{i} \cong 1$$

(c) CH_3COOH forms a hydrogen-bonded dimer in nonpolar benzene (C_6H_6) causing it to give one-half as many solute particles as might be expected.

Hence, the effective molality of CH_3COOH is approximately one-half that in water.

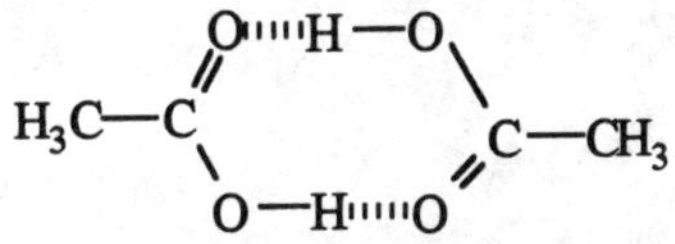

Note: The observation that the experimental molality for CH_3COOH in water is slightly higher than calculated can be attributed to the observation that CH_3COOH remains monomeric and acts as a weak acid in water.

Chapter 12: Chemical Equilibrium

12.6 ***See Section 12.1.***

(a) $PCl_5(g) \rightleftarrows PCl_3(g) + Cl_2(g)$

$$K_p = \frac{P_{PCl_3} P_{Cl_2}}{P_{PCl_5}}$$

(b) $2NO_2(g) \rightleftarrows 2NO(g) + O_2(g)$

$$K_p = \frac{(P_{NO})^2 P_{O_2}}{(P_{NO_2})^2}$$

(c) $2SO_3(g) \rightleftarrows 2SO_2(g) + O_2(g)$

$$K_p = \frac{(P_{SO_2})^2 P_{O_2}}{(P_{SO_3})^2}$$

(d) $H_2(g) + I_2(g) \rightleftarrows 2HI(g)$

$$K_p = \frac{(P_{HI})^2}{P_{H_2} P_{I_2}}$$

12.8 ***See Section 12.1 and Example 12.1.***

(a) $HCl(g) + \frac{1}{4}O_2(g) \rightleftarrows \frac{1}{2}Cl_2(g) + \frac{1}{2}H_2O(g)$

$$K_c = \frac{[Cl_2]^{\frac{1}{2}}[H_2O]^{\frac{1}{2}}}{[HCl][O_2]^{\frac{1}{4}}}$$

(b) $\frac{1}{2}N_2O_4(g) \rightleftarrows NO_2(g)$

$$K_c = \frac{[NO_2]}{[N_2O_4]^{\frac{1}{2}}}$$

(c) $N_2O_4(g) \rightleftarrows N_2(g) + 2O_2(g)$

$$K_c = \frac{[N_2][O_2]^2}{[N_2O_4]}$$

(d) $\frac{1}{2}O_2(g) + SO_2(g) \rightleftarrows SO_3(g)$

$$K_c = \frac{[SO_3]}{[O_2]^{\frac{1}{2}}[SO_2]}$$

12.10 ***See Section 12.1 and Example 12.2.***

$CO(g) + \frac{1}{2}O_2 \rightleftarrows CO_2(g)$

$$[CO] = \frac{0.050 \text{ mol}}{5.0 \text{ L}} = 0.010\ M$$

$$[O_2] = \frac{0.025 \text{ mol}}{5.0 \text{ L}} = 0.0050\ M$$

$$[CO_2] = \frac{4.95 \text{ mol}}{5.0 \text{ L}} = 0.99\ M$$

$$K_c = \frac{[CO_2]}{[CO][O_2]^{\frac{1}{2}}}$$

$$K_c = \frac{(0.99)}{(0.010)(0.0050)^{\frac{1}{2}}} = \mathbf{1.4 \times 10^3}$$

12.12 ***See Section 12.1 and Example 12.2.***

$PCl_5(g) \rightleftarrows PCl_3(g) + Cl_2(g)$

$$[PCl_5] = \frac{1.0 \text{ mol}}{20.0 \text{ L}} = 0.050\ M \qquad K_c = \frac{[PCl_3][Cl_2]}{[PCl_5]} \qquad K_c = \frac{(0.10)(0.10)}{(0.050)} = \mathbf{0.20}$$

$$[PCl_3] = \frac{2.0 \text{ mol}}{20.0 \text{ L}} = 0.10\ M$$

$$[Cl_2] = \frac{2.0 \text{ mol}}{20.0 \text{ L}} = 0.10\ M$$

12.14 ***See Section 12.2 and Example 12.4.***

$$2H_2(g) + O_2(g) \rightleftarrows 2H_2O(g) \qquad K_1 = \frac{[H_2O]^2}{[H_2]^2[O_2]} = 6.1 \text{ x } 10^3$$

$$H_2(g) + \frac{1}{2}O_2(g) \rightleftarrows H_2O(g) \qquad K_2 = \frac{[H_2O]}{[H_2][O_2]^{\frac{1}{2}}}$$

Examination shows that $K_1 = K_2^2 = 6.1 \text{ x } 10^3$. Hence, $K_2 = \sqrt{6.1 \text{ x } 10^3} = \mathbf{78}$

12.16 ***See Sections 12.1 and 4 and Example 12.2.***

$2SO_2(g) + O_2(g) \rightleftarrows 2SO_3(g)$

Conc, *M*	SO_2	O_2	SO_3
initial	0.0076	0.0036	0.0000
Change	−0.0044	−0.0022	+0.0044
equilibirum	0.0032	0.0014	0.0044

$$K_c = \frac{[SO_3]^2}{[SO_2]^2[O_2]} \qquad K_c = \frac{(0.0044)^2}{(0.0032)^2(0.0014)} = \mathbf{1.4 \text{ x } 10^3}$$

Note: The change in concentration of SO_2 can be determined from the initial and equilibrium concentrations given for SO_2. The change in the concentrations of O_2 and SO_3 can be determined from the change for SO_2 and the coefficients in the balanced equation.

12.18 See Section 12.1 and 4.

$CO(g) + Cl_2(g) \rightleftarrows COCl_2(g)$

P, atm	CO	Cl_2	$COCl_2$
initial	0.96	1.02	0.00
Change	-y	-y	+y
equilibrium	0.96-y	1.02-y	y

Since, $P_T = P_{CO} + P_{Cl_2} + P_{COCl_2}$ and $P_T = 1.22$ at equlibrium,

$(0.96\ \text{atm} - y) + (1.02\ \text{atm} - y) + y = 1.22\ \text{atm}$ 1.98 - y = 1.22 atm y = 0.76 atm. This gives

P, atm	CO	Cl_2	$COCl_2$
initial	0.96	1.02	0.00
Change	-0.76	-0.76	+0.76
equilibrium	0.20	0.26	0.76

$$K_p = \frac{P_{COCl_2}}{P_{CO}P_{Cl_2}} \qquad K_p \frac{(0.76)}{(0.20)(0.26)} = \mathbf{15}$$

12.22 See Section 12.2 and Example 12.5.

$2HF(g) \rightleftarrows H_2(g) + F(g) \qquad K_C = 0.010$

The form of the reaction quotient expression is the same as the equilibrium constant expression. Hence, we obtain

$$Q = \frac{[H_2][F_2]}{[HF]^2}$$

K_C is the value of Q that is obtained at equilibrium. If Q is greater than K_C, spontaneous reaction occurs to the left to reach equilibrium. If Q is less than K_C, spontaneous reaction occurs to the right to reach equilibrium. The system always reacts spontaneously to make Q = K_C.

Converting the initial amounts to initial molar concentrations and calculating Q gives:

Part	[HF]	$[H_2]$	$[F_2]$	Q	Q vs. K_c	Direction of Spontaneous Reaction
(a)	5 x 10^{-6}	2.0 x 10^{-5}	2.0 x 10^{-5}	16	$Q > K_c$	←
(b)	2.0 x 10^{-5}	2.0 x 10^{-4}	2.0 x 10^{-4}	1 x 10^2	$Q > K_c$	←
(c)	1.05 x 10^{-3}	1.0 x 10^{-4}	9.0 x 10^{-5}	8 x 10^{-3}	$Q < K_c$	→
(d)	2.00 x 10^{-3}	5.0 x 10^{-5}	8.0 x 10^{-5}	1 x 10^{-3}	$Q < K_c$	→

12.24 See Section 12.3 and Examples 12.6, 7 and 8.

(a) Adding SO_3, a reactant, causes Q < K and spontaneous reaction to the right.
(b) Removing O_2, a product, causes Q < K and spontaneous reaction to the right.

(c) Increasing the volume of the container favors a net increase in the number of mole of gas. In this case, this causes spontaneous reaction to the right (1.5 mol gas for products vs. 1 mol gas for reactants).
(d) An increase in temperature favors the endothermic direction. In this case, this causes spontaneous reaction to the right.
(e) The pressure of materials other than reactants or products has no effect on the equilibrium. Hence, adding argon has no effect on this equilibrium.

12.26 ***See Section 12.1 and Example 12.3.***

$CO_2(g) \rightleftarrows CO(g) + \frac{1}{2}O_2(g)$ $\qquad \Delta n_g = \left[\left(1+\frac{1}{2}\right)-1\right] = \frac{1}{2}$

Substituting in $K_c = K_p(RT)^{-\Delta n}$ gives $K_c = 2.48(0.0821 \times 3{,}000)^{-\frac{1}{2}} = \mathbf{0.158}$

12.28 ***See Sections 12.1 and 4 and Examples 12.2 and 3.***

$2SO_2(g) + O_2(g) \rightleftarrows 2SO_3(g)$ $\qquad \Delta n_g = [2-(2+1)] = -1$

Conc, *M*	SO_2	O_2	SO_3
initial	1.0×10^{-3}	1.0×10^{-3}	0
Change	-5.0×10^{-4}	-2.5×10^{-4}	$+5.0 \times 10^{-4}$
equilibrium	5.0×10^{-4}	7.5×10^{-4}	5.0×10^{-4}

$$K_c = \frac{[SO_3]^2}{[SO_2]^2[O_2]} \qquad K_c = \frac{(5.0 \times 10^{-4})^2}{(5.0 \times 10^{-4})^2(7.5 \times 10^{-4})} = \mathbf{1.3 \times 10^3}$$

$$K_c = K_p(RT)^{-\Delta n} \qquad K_p = 1.3 \times 10^3(0.0821 \times 922)^{-1} = \mathbf{17}$$

Note: the change in concentration of O_2 can be determined from the initial and equilibrium concentrations given for O_2. The change in the concentrations of SO_2 and SO_3 can be determined from the change for O_2 and the coefficients in the balanced equation.

12.30 ***See Sections 12.1 and 4 and Example 12.2.***

$2NO_2(g) \rightleftarrows N_2O_4(g)$

(a) $? \text{ mol } NO_2 = \frac{0.010 \text{ mol } NO_2}{1 \text{ L}} \times 20.0 \text{ L} = \mathbf{0.20 \text{ mol } NO_2}$

(b) $? \text{ mol } N_2O_4 \text{ formed} = \text{mol } NO_2 \text{ reacted} \times \frac{1 \text{ mol } N_2O_4}{2 \text{ mol } NO_2 \text{ reacted}}$

$? \text{ mol } N_2O_4 \text{ formed} = (2.0-0.20) \text{ mol } NO_2 \text{ reacted} \times \frac{1 \text{ mol } N_2O_4}{2 \text{ mol } NO_2 \text{ reacted}} = \mathbf{0.90 \text{ mol } N_2O_4}$

$$[N_2O_4] = \frac{0.90 \text{ mol } N_2O_4}{20.0 \text{ L}} = \mathbf{0.045\ \mathit{M}\ N_2O_4}$$

(c) $K_c = \frac{[N_2O_4]}{[NO_2]^2}$ $\quad K_c = \frac{(0.045)}{(0.010)^2} = \mathbf{4.5 \times 10^2}$

12.32 ***See Section 12.4 and Example 12.9.***

$H_2(g) + I_2(g) \rightleftarrows 2HI(g)$

Conc, M	H_2	I_2	HI
initial	0.50	0.50	0.
Change	-y	-y	+2y
equilibrium	0.50-y	0.50-y	2y

$K_c = \frac{[HI]^2}{[H_2][I_2]}$ $\quad K_c = \frac{(2y)^2}{(0.50-y)(0.50-y)} = 36$

Taking the square root of both sides yields $\frac{2y}{0.50-y} = 6.0$ and y = 0.38.

Hence, $[\mathbf{H_2}] = [\mathbf{I_2}] = 0.50 - 0.38 = \mathbf{0.12\ \mathit{M}}$ and $[\mathbf{HI}] = 2(0.38) = \mathbf{0.76\ \mathit{M}}$.

Inserting these equilibrium concentrations into the equilibrium expression gives

$K_c = \frac{[HI]^2}{[H_2][I_2]}$ $\quad K_c = \frac{(0.76)^2}{(0.12)(0.12)} = 40$ $\quad$ and this is in good agreement with the given value of 36 considering the limitations of significant figures.

12.34 ***See Section 12.4 and Example 12.9.***

$SO_3(g) + NO(g) \rightleftarrows SO_2(g) + NO_2(g)$

Conc, *M*	SO_3	NO	SO_2	NO_2
initial	0.025	0.025	0.00	0.00
Change	-y	-y	+y	+y
equilibrium	0.025-y	0.025-y	y	y

$K_c = \frac{[SO_2[NO_2]]}{[SO_3][NO]}$ $\quad K_c = \frac{(y)(y)}{(0.025-y)(0.025-y)} = 0.50$

Taking the square root of both sides yields $\frac{y}{0.025-y} = 0.71$ and y = 0.010.

Hence, $[\mathbf{SO_3}]=[\mathbf{NO}]=0.025-0.010=\mathbf{0.015}\ \boldsymbol{M}$ and $[\mathbf{SO_2}]=[\mathbf{NO_2}]=\mathbf{0.010}\ \boldsymbol{M}$.

Inserting these equilibrium concentrations into the equilibrium expression gives

$$K_c=\frac{[SO_2][NO_2]}{[SO_3][NO]} \qquad K_c=\frac{(0.10)(0.10)}{(0.15)(0.15)}=.044$$

and this is in agreement with the given value of 0.50 considering the limitations of significant figures.

12.36 ***See Section 12.4 and Example 12.10.***

$COCl_2 \rightleftarrows CO+Cl_2$

Conc, *M*	$COCl_2$	CO	Cl_2
initial	0.250	0.00	0.00
Change	-y	+y	+y
equilibrium	0.250-y	y	y

$$K_c=\frac{[CO][Cl_2]}{[COCl_2]} \qquad K_c=\frac{(y)(y)}{(0.250-y)}=4.93\ x\ 10^{-3}$$

Rearranging yields $y^2+4.93\ x\ 10^{-3}\,y-1.23\ x\ 10^{-3}=0$.

Substituting into the quadratic formula gives $y=\dfrac{-4.93\ x\ 10^{-3}\pm\sqrt{(4.93\ x\ 10^{-3})^2-4(1)(-1.23\ x\ 10^{-3})}}{2(1)}$

and $y=\dfrac{-4.93\ x\ 10^{-3}\pm 7.03\ x\ 10^{-2}}{2}=3.27\ x\ 10^{-2}$ or $-3.76\ x\ 10^{-2}$.

The only reasonable value for y is $3.27\ x\ 10^{-2}$, since $-3.76\ x\ 10^{-2}$ predicts negative concentrations for CO and Cl_2. Hence, $[\mathbf{COCl_2}]=0.250-0.0327=\mathbf{0.217}\ \boldsymbol{M}$ and $[\mathbf{CO}]=[\mathbf{Cl_2}]=\mathbf{0.0327}\ \boldsymbol{M}$.

Inserting these equilibrium concentrations into the equilibrium expression gives

$$K_c=\frac{[CO][Cl_2]}{[COCl_2]} \qquad K_c=\frac{(0.0327)(0.0327)}{(0.217)}=4.93\ x\ 10^{-3}$$

and this is in good agreement with the given value of $4.93\ x\ 10^{-3}$.

12.38 ***See Section 12.4 and Example 12.10.***

$CO(g)+Cl_2(g) \rightleftarrows COCl_2(g)$

Conc, *M*	CO	Cl_2	$COCl_2$
initial	0.200	0.200	0.0
Change	-y	-y	+y
equilibrium	0.200-y	0.200-y	y

$$K_c=\frac{[COCl_2]}{[CO][Cl_2]} \qquad K_c=\frac{y}{(0.200-y)(0.200-y)}=7.52$$

Rearranging and combining terms yields $7.52y^2 - 4.01y + 0.301 = 0$.

Substituting into the quadratic formula gives $y = \dfrac{4.01 \pm \sqrt{(-4.01)^2 - 4(7.52)(0.301)}}{2(7.52)}$

and $y = \dfrac{4.01 \pm 2.65}{15.04} = 0.443$ or 0.0904.

The only reasonable value for y is 0.0904 , since 0.443 predicts negative concentrations for CO and CL_2.

Hence, $[CO] = [Cl_2] = 0.200 - 0.0904 = \mathbf{0.110}$ ***M*** and $[COCl_2] = \mathbf{0.0904}$ ***M***.

Inserting these equilibrium concentrations inot the equilibrium expression gives

$K_c = \dfrac{[COCl_2]}{[CO][Cl_2]}$ $\quad K_c = \dfrac{0.0904}{(0.110)(0.110)} = 7.47$ and this is in good agreement with the given value of 7.52.

12.40 ***See Section 12.4 and Example 12.10.***

$2HI(g) \rightleftarrows H_2(g) + I_2(g)$

Conc, *M*	HI	H_2	I_2
initial	0.0020	0.	0.
Change	-2y	+y	+y
equilibrium	0.0020-2y	y	y

$K_c = K_p(RT)^{-\Delta n}$ $\quad K_c = K_p(RT)^{-0}$ $\quad K_c = K_p$ $\quad K_c = \dfrac{[H_2][I_2]}{[HI]^2}$ $\quad K_c = \dfrac{(y)(y)}{(0.0020 - 2y)^2} = 0.050$

Rearranging yields $0.80y^2 + 4.0 \times 10^{-4} y - 2.0 \times 10^{-7} = 0$.

Substituting into the quadratic formula gives $y = \dfrac{-4.0 \times 10^{-4} \pm \sqrt{(4.0 \times 10^{-4})^2 - 4(0.80)(-2.0 \times 10^{-7})}}{2(0.80)}$

and $y = \dfrac{-4.0 \times 10^{-4} \pm 8.9 \times 10^{-4}}{1.6} = 3.1 \times 10^{-4}$ or -8.1×10^{-4}.

The only reasonable value of y is 3.1×10^{-4}, since -8.1×10^{-4} predicts negative concentrations for H_2 and I_2.

Hence, $[HI] = 0.0020 - (2)(3.1 \times 10^{-4}) = \mathbf{1.4 \times 10^{-3}}$ ***M*** and $[H_2] = [I_2] = \mathbf{3.1 \times 10^{-4}}$ ***M***.

Inserting these equilibrium concentrations into the equilibrium expression gives

$K_c = \frac{[H_2][I_2]}{[HI]^2}$ $K_c = \frac{(3.1 \times 10^{-4})(3.1 \times 10^{-4})}{(1.4 \times 10^{-3})^2} 0.049$ and this is in good agreement with the given value of 0.050.

12.42 ***See Section 12.4 and Example 12.10.***

$SO_2(g) + Cl_2(g) \rightleftarrows SO_2Cl_2(g)$

Conc, *M*	SO_2	Cl_2	SO_2Cl_2
initial	0.0100	0.0200	0.
Change	-y	-y	+y
equilibrium	0.0100-y	0.0200-y	y

$K_c = \frac{[SO_2Cl_2]}{[SO_2][Cl_2]}$ $K_c = \frac{y}{(0.0100 - y)(0.0200 - y)} = 89.3$

Rearranging and combining terms yields $89.3y^2 - 3.68y + 0.0179 = 0$.

Substituting into the quadratic formula gives $y = \frac{3.68 \pm \sqrt{(-3.68)^2 - 4(89.3)(0.0179)}}{2(89.3)}$

and $y = \frac{3.68 \pm 2.67}{178.6} = 3.56 \times 10^{-2}$ or 5.66×10^{-3}.

The only reasonable answer is 5.66×10^{-3}, since 3.56×10^{-2} predicts negative concentrations for SO_2 and Cl_2. Hence, $[SO_2] = 0.0100 - 5.66 \times 10^{-3} = \mathbf{4.34 \times 10^{-3}}$ ***M***, $[Cl_2] = 0.0200 - 5.66 \times 10^{-3} = \mathbf{1.43 \times 10^{-2}}$ ***M*** and $[SO_2Cl_2] = \mathbf{5.66 \times 10^{-3}}$ ***M***.

Inserting these equilibrium expressions into the equilibrium expression gives

$K_c = \frac{[SO_2Cl_2]}{[SO_2][Cl_2]}$ $K_c = \frac{(5.66 \times 10^{-3})}{(4.34 \times 10^{-3})(1.43 \times 10^{-2})} = 91.1$ and this value is in good agreement with the given value of 89.3.

12.44 ***See Section 12.4 and Example 12.10.***

$CO(g) + H_2(g) \rightleftarrows CH_2O(g)$

P, atm	CO	H_2	CH_2O
initial	2.00	3.00	0.0
Change	-y	-y	+y
equilibrium	2.00-y	3.00-y	+y

$K_p = \frac{P_{CH_2O}}{P_{CO}P_{H_2}}$ $K_p = \frac{y}{(2.00 - y)(3.00 - y)} = 5.60$

Rearranging and combining terms yields $5.60y^2 - 29.0y + 33.6 = 0$.

Substituting into the quadratic formula gives $y = \dfrac{29.0 \pm \sqrt{(-29.0)^2 - 4(5.60)(33.6)}}{2(5.60)}$

and $y = \dfrac{29.0 \pm 9.4.0}{11.2} = 3.43$ or 1.75.

The only reasonable value for y is 1.75, since 3.43 predicts negative pressures for CO and H_2.

Hence, $\mathbf{P_{CO}} = 2.00 - 1.75 = \mathbf{0.25\ atm}$, $\mathbf{P_{H_2}} = 3.00 - 1.75 = \mathbf{1.25\ atm}$ and $\mathbf{P_{CH_2O}} = \mathbf{1.75\ atm}$.

Inserting these equilibrium pressures into the equilibrium expression gives

$K_p = \dfrac{P_{CH_2O}}{P_{CO}P_{H_2}}$ $\quad K_p = \dfrac{1.75}{(0.25)(1.25)} = 5.6$ and this is in agreement with the given value of 5.6.

12.46 ***See Section 12.5 and Example 12.12.***

For $CaCO_3(s) \rightleftarrows CaO(s) + CO_2(g)$, $K_p = \mathbf{P_{CO_2}} = \mathbf{0.12\ atm}$ at 1000K.

12.50 ***See Section 12.6 and Example 12.13.***

(a) $Mg_2F(s) \rightleftarrows Mg^{2+}(aq) + 2F^-(aq)$ $\quad K_{sp} = [Mg^{2+}][F^-]^2$

(b) $Ca_3(PO_4)_2(s) \rightleftarrows 3Ca^{2+}(aq) + 2PO_4^{3-}(aq)$ $\quad K_{sp} = [Ca^{2+}]^3[PO_4^{3-}]^2$

(c) $Al_2(CO_3)_3(s) \rightleftarrows 2Al^{3+}(aq) + 3CO_3^{2-}(aq)$ $\quad K_{sp} = [Al^{3+}]^2[CO_3^{2-}]^3$

(d) $LaF_3(s) \rightleftarrows La^{3+}(aq) + 3F^-(aq)$ $\quad K_{sp} = [La^{3+}][F^-]^3$

12.52 ***See Section 12.6 and Example 12.14.***

$AgI(s) \rightleftarrows Ag^+(aq) + I^-(aq)$

Conc, *M*	AgI	Ag^+	I^-
initial	excess	0.	0.
Change	-9 x 10^{-9}	+9 x 10^{-9}	+9 x 10^{-9}
equilibrium	excess	9 x 10^{-9}	9 x 10^{-9}

$K_{sp} = [Ag^+][I^-]$ $\quad K_{sp} = (9 \times 10^{-9})(9 \times 10^{-9}) = \mathbf{8 \times 10^{-17}}$

12.54 *See Section 12.6, Tables 12.4 and 6 and Example 12.16.*

$BaSO_4(s) \rightleftarrows Ba^{2+}(aq) + SO_4^{2-}(aq)$

Conc, *M*	$BaSO_4$	Ba^{2+}	SO_4^{2-}
initial	excess	0.	0.
Change	-s	+s	+s
equilibrium	excess	s	s

$K_{sp} = [Ba^+][SO_4^{2-}]$ $K_{sp} = (s)(s) = 8.7 \text{ x } 10^{-11}$

Hence, $s^2 = 8.7 \text{ x } 10^{-11}$ and $s = 9.3 \text{ x } 10^{-6}$ *M*.

$$?\frac{\text{g BaSO}_4}{\text{L}} = \frac{9.3 \text{ x } 10^{-6} \text{ mol BaSO}_4}{1 \text{ L soln}} \times \frac{233.4 \text{ g BaSO}_4}{1 \text{ mol BaSO}_4} = \mathbf{2.2 \text{ x } 10^{-3}} \ \frac{\mathbf{g\ BaSO_4}}{\mathbf{L}}$$

12.56 *See Section 12.6, Table 12.6 and Example 12.16.*

$Ag_2WO_4(s) \rightleftarrows 2Ag^+(aq) + WO_4^{2-}(aq)$

Conc, *M*	Ag_2WO_4	Ag^+	WO_4^{2-}
initial	excess	0.	0.
Change	-s	+2s	+s
equilibrium	excess	2s	s

$K_{sp} = [Ag^+]^2[WO_4^{2-}]$ $K_{sp} = (2s)^2(s) = 5.5 \text{ x } 10^{-12}$

Hence, $4s^3 = 5.5 \text{ x } 10^{-12}$ and $\mathbf{s = 1.1 \text{ x } 10^{-4}}$ ***M***.

12.58 *See Section 12.7 and Example 12.18.*

$BaSO_4(s) \rightleftarrows Ba^{2+}(aq) + SO_4^{2-}(aq)$

(a)

Conc, *M*	$BaSO_4$	Ba^{2+}	SO_4^{2-}
initial	excess	0.	0.
Change	-s	+s	+s
equilibrium	excess	s	s

$K_{sp} = [Ba^{2+}][SO_4^{2-}]$ $K_{sp} = (s)(s) = 8.7 \text{ x } 10^{-11}$

Hence, $s^2 = 8.7 \text{ x } 10^{-11}$ and $\mathbf{s = 9.3 \text{ x } 10^{-6}}$ ***M***.

(b) This problem is solved in the same manner as part (a), but the initial concentration of SO_4^{2-} is 0.10 *M* from 0.10 *M* Na_2SO_4,

Conc, *M*	$BaSO_4$	Ba^{2+}	SO_4^{2-}
initial	excess	0.	0.10
Change	-s	+s	+s
equilibrium	excess	s	0.10+s

$K_{sp} = [Ba^{2+}][SO_4^{2-}]$ $K_{sp} = (s)(0.10+s) = 8.7 \text{ x } 10^{-11}$

Assuming s is small compared to 0.10 leads to (s)(0.10) = 8.7×10^{-11} and **s = 8.7×10^{-10}**.

Since s is indeed small compared to 0.10, our assumption is valid.

Note: We had good reason to believe s would be small compared to 0.10. The value of s that was obtained in the absence of Na_2SO_4 in part (a) was 9.3×10^{-6} *M*, and application of the Principle of Le Chatelier leads us to predict it will be even smaller in the presence of 0.10 *M* Na_2SO_4.

12.60 ***See Section 12.7 and Example 12.18.***

$Cu(IO_3)_2\,(s) \rightleftarrows Cu^{2+}\,(aq) + 2IO_3^-\,(aq)$

Conc, *M*	$Cu(IO_3)_2$	Cu^{2+}	IO_3^-
initial	excess	0.	0.
Change	-s	+s	+2s
equilibrium	excess	s	2s

$K_{sp} = [Cu^{2+}][IO_3^-]^2$ $\quad K_{sp} = (s)(2s)^2 = 7.4 \times 10^{-8}$

Hence, $4s^3 = 7.4 \times 10^{-8}$ and **s = 2.6×10^{-3} *M*.**

(b) This problem is solved in the same manner as part (a), but the initial concentration of IO_3^- is 0.10 *M* from 0.10 *M* $NaIO_3$,

Conc, *M*	$Cu(IO_3)_2$	Cu^{2+}	IO_3^-
initial	excess	0.	0.10
Change	-s	+s	+2s
equilibrium	excess	s	0.10+2s

$K_{sp} = [Cu^{2+}][IO_3^-]^2$ $\quad K_{sp} = (s)(0.10+2s)^2 = 7.4 \times 10^{-8}$

Assuming s is small compared to 0.10 leads to $(s)(0.10)^2 = 7.4 \times 10^{-8}$ and **s = 7.4×10^{-6}**.

Since s is indeed small compared to 0.10, our assumption is valid.

Note: We had good reason to believe s would be small compared to 0.10. The value of s that was obtained in the absence of $NaIO_3$ in part (a) was 2.6×10^{-3} *M*, and application of the Principle of Le Chatelier leads us to predict it will be even smaller in the presence of 0.10 *M* $NaIO_3$.

12.62 ***See Section 12.6 and Example 12.17.***

(a) We need to calculate Q and compare it to K_{sp} for $Ag_2SO_4\,(s) \rightleftarrows 2Ag^+\,(aq) + SO_4^{2-}\,(aq)$.

Hence, we need to calculate the concentrations of Ag^+ and SO_4^{2-} in the mixed soultion.

Solving $M(\text{con}) \times V(\text{con}) = M(\text{dil}) \times V(\text{dil})$ for *M*(dil) gives $M(\text{dil}) = \dfrac{M(\text{con}) \times V(\text{con})}{V(\text{dil})}$.

Hence, $[Ag^+]$ in mixed soln $= \dfrac{0.0010\,M \times 10\text{ mL}}{20\text{ mL}} = 0.00050\,M$

and $\left[SO_4^{2-}\right]$ in mixed soln $= \dfrac{0.0010\ M \times 10\ \text{mL}}{20\ \text{mL}} = 0.00050\ M$

$Q = \left[Ag^+\right]^2\left[SO_4^{2-}\right] \qquad Q = \left(5.0 \text{ x } 10^{-4}\right)^2\left(5.0 \text{ x } 10^{-4}\right) = 1.2 \text{ x } 10^{-10}$

Since K_{sp} = 1.2 x 10^{-5}, Q is less than K_{sp} and **no precipitate will form.**

(b) We need to calculate Q and compare it to K_{sp} for $Fe(OH)_2\ (s) \rightleftarrows Fe^{2+}\ (aq) + 2OH^-\ (aq)$.
Hence, we need to calculate the concentrations of Fe^{2+} and OH^- in the mixed solution, using the same procedure that was used in working part (a).

$\left[Fe^{2+}\right] = \dfrac{1.0 \text{ x } 10^{-6}\ M \times 10\ \text{mL}}{30\ \text{mL}} = 3.3 \text{ x } 10^{-7}\ M \qquad \left[OH^-\right] = \dfrac{2\left(3.0 \text{ x } 10^{-4}\ M\right) \times 20\ \text{mL}}{30\ \text{mL}} = 4.0 \text{ x } 10^{-4}\ M$

$Q = \left[Fe^{2+}\right]\left[OH^-\right]^2 \qquad Q = \left(3.3 \text{ x } 10^{-7}\right)\left(4.0 \text{ x } 10^{-4}\right)^2 = 5.3 \text{ x } 10^{-14}$

Since K_{sp} = 1.6 x 10^{-14}, Q is greater than K_{sp} and a **precipitate will form.**

12.64 ***See Section 12.6, Example 12.17 and Insights into Chemistry: Selective Precipitation.***

(a) The halide requiring the lowest concentration of Ag^+ for precipitation will precipitate first. Hence, we need to calculate the concentrations of Br^- and I^- in the mixed solution and then the concentrations of Ag^+ required to precipitate AgBr and AgI in the mixed solution.

Solving $M(\text{con}) \times V(\text{con}) = M(\text{dil}) \times V(\text{dil})$ for M(dil) gives $M(\text{dil}) = \dfrac{M(\text{con}) \times V(\text{con})}{V(\text{dil})}$.

Hence, $\left[Br^-\right]$ in mixed solution $= \dfrac{0.010\ M \times 20\ \text{mL}}{50\ \text{mL}} = 0.0040\ M$

and $\left[I^-\right]$ in mixed solution $= \dfrac{0.010\ M \times 30\ \text{mL}}{50\ \text{mL}} = 0.0060\ M$

Since Q must equal K_{sp} for precipitation to occur and K_{sp} for AgBr is $K_{sp} = \left[Ag^+\right]\left[Br^-\right] = 7.7 \text{ x } 10^{-13}$

and K_{sp} for AgI is $K_{sp} = \left[Ag^+\right]\left[I^-\right] = 8.5 \text{ x } 10^{-17}$,

$\left[Ag^+\right]$ for AgBr to ppt $= \dfrac{K_{sp}\ \text{AgBr}}{\left[Br^-\right]} \qquad \left[Ag^+\right]$ for AgBr to ppt $= \dfrac{7.7 \text{ x } 10^{-13}}{0.0040} = 1.9 \text{ x } 10^{-10}\ M$

$\left[Ag^+\right]$ for AgI to ppt $= \dfrac{K_{sp}\ \text{AgI}}{\left[I^-\right]} \qquad \left[Ag^+\right]$ for AgI to ppt $= \dfrac{8.5 \text{ x } 10^{-17}}{0.0060} = 1.4 \text{ x } 10^{-14}\ M$

AgI requires the lower $\left[Ag^+\right]$ for precipitation and will therefore precipitate first.

The concentration of Ag^+ is 1.9×10^{-10} *M* when AgBr begins to precipitate. Hence,

$$[I^-] = \frac{K_{sp} AgI}{[Ag^+]} \qquad [I^-] = \frac{8.5 \times 10^{-17}}{1.9 \times 10^{-10}} = 4.5 \times 10^{-7}\ M$$

Note: The $[I^-]$ when AgI just begins to precipitate is 0.0060 *M*. However, AgI continues to precipitate as $AgNO_3$

is added. The $[I^-]$ is thus reduced to 4.5×10^{-7} as the $[Ag^+]$ increases from 1.4×10^{-14} *M* to 1.9×10^{-10} *M*, the $[Ag^+]$ at which the more soluble AgBr just beings to precipitate.

12.66 ***See Section 12.6 and Example 12.7.***

Since Q must equal K_{sp} for precipitation to occur, and K_{sp} for AgCl is $K_{sp} = [Ag^+][Cl^-] = 1.8 \times 10^{-10}$,

$$[Ag^+] \text{ for AgCl to ppt} = \frac{K_{sp} AgCl}{[Cl^-]} \qquad [Ag^+] \text{ for AgCl to ppt} = \frac{1.8 \times 10^{-10}}{7.4 \times 10^{-4}} = \mathbf{2.4 \times 10^{-7}\ M}$$

Chapter 13: Solutions of Acids and Bases

13.2 *See Section 13.1, Tables 13.1, 6 and 7 and Example 13.1.*

To obtain the formula of a conjugate acid, add H^+. This gives:

Part	Base	Base Fromula	Conj. Acid Formula	Conj. Acid Name
(a)	hydrogen sulfate	HSO_4^-	H_2SO_4	sulfuric acid
(b)	water	H_2O	H_3O^+	hydronium ion
(c)	ammonia	NH_3	NH_4^+	ammonium ion
(d)	pyridine	C_5H_5N	$C_5H_5NH^+$	pyridinium ion

13.4 *See Section 13.1, Tables 13.1, 6 and 7 and Example 13.1.*

To obtain the formula of a conjugate base, remove H^+. This gives:

Part	Acid Formula	Conj. Base Formula	Conj. Base Name
(a)	HCN	CN^-	cyanide ion
(b)	HSO_4^-	SO_4^{2-}	sulfate ion
(c)	$H_2PO_3^-$	HPO_3^{2-}	hydrogen phosphite ion
(d)	HCO_3^-	CO_3^{2-}	carbonate ion

13.6 *See Section 13.1 and Example 13.1.*

Part	Reaction	Acid/Conj. Base	Base/Conj. Acid
(a)	$NH_3 + CH_3COOH \rightarrow NH_4CH_3COO$	CH_3COOH / CH_3COO^-	NH_3 / NH_4^+
(b)	$N_2H_5^+ + CO_3^{2-} \rightarrow N_2H_4 + HCO_3^-$	$N_2H_5^+ / N_2H_4$	CO_3^{2-} / HCO_3^-
(c)	$H_3O^+ + OH^- \rightarrow 2H_2O$	H_3O^+ / H_2O	OH^- / H_2O
(d)	$HSO_4^- + HCOO^- \rightarrow SO_4^{2-} + HCOOH$	HSO_4^- / SO_4^{2-}	$HCOO^- / HCOOH$

13.8 *See Section 13.2 and Examples 13.3 and 4.*

Using $pH = -\log[H_3O^+]$ and $[H_3O^+] = 10^{-pH} = \text{inv}\log(-pH)$ yields:

Part	pH	$[H_3O^+]$, *M*	Solution
(a)	2.34	**4.6×10^{-3}**	**acidic**
(b)	**12.98**	1.04×10^{-13}	**basic**
(c)	-1.09	**12.**	**acidic**
(d)	**10.67**	2.12×10^{-11}	**basic**
(e)	**1.13**	7.40×10^{-2}	**acidic**

Part	pH	$[H_3O^+]$, M	Solution
(f)	13.41	**3.9 x 10^{-14}**	**basic**
(g)	**4.15**	7.07 x 10^{-5}	**acidic**
(h)	9.80	**1.6 x 10^{-10}**	**basic**
(i)	**0.30**	0.50	**acidic**

13.10 ***See Section 13.2 and Examples 13.2 through 5.***

Using $pH = -\log[H_3O^+]$ $\qquad$ $[H_3O^+] = 10^{-pH} = \text{inv}\log(-pH)$

$pOH = -\log[OH^-]$ $\qquad$ $[OH^-] = 10^{-pOH} = \text{inv}\log(-pOH)$

$K_w = [H_3O^+][OH^-] = 1.0 \times 10^{-14}$ $\qquad$ $pH + pOH = 14.00$ $\qquad$ yields

Part	pH	$[H_3O^+]$, M	pOH	$[OH^-]$, M	Solution
(a)	-1.04	**11.**	**15.04**	**9.1 x 10^{-16}**	**acidic**
(b)	**13.66**	**2.2 x 10^{-14}**	0.34	**0.46**	**basic**
(c)	**6.70**	1.98 x 10^{-7}	**7.30**	**5.0 x 10^{-8}**	**acidic (very weak)**
(d)	**12.65**	**2.2 x 10^{-13}**	**1.35**	4.4 x 10^{-2}	**basic**

13.12 ***See Section 13.2, Table 13.4 and Examples 13.2, 4 and 5.***

Using $[H_3O^+] = 10^{-pH} = \text{inv}\log(-pH)$ and $[OH^-] = \dfrac{K_w}{[H_3O^+]} = \dfrac{1.0 \times 10^{-14}}{[H_3O^+]}$ yields

Part	Substance	pH	$[H_3O^+]$, *M*	$[OH^-]$, *M*
(a)	vinegar	2.9	**1 x 10^{-3}**	**1 x 10^{-11}**
(b)	stomach acid	1.7	**2 x 10^{-2}**	**5 x 10^{-13}**
(c)	coffee	5.0	**1 x 10^{-5}**	**1 x 10^{-9}**
(d)	milk	6.9	**1 x 10^{-7}**	**1 x 10^{-7}**

13.14 ***See Sections 13.2 and 3 and Examples 13.6 and 7.***

(a) $Ba(OH)_2 \rightarrow Ba^{2+} + 2OH^-$ $\qquad$ 0.0045 *M* $Ba(OH)_2 \rightarrow$ 0.0090 *M* OH^-

$\mathbf{pOH} = -\log[OH^-] = -\log(0.0090) = \mathbf{2.05}$ $\qquad$ $\mathbf{pH} = 14.00 - pOH = \mathbf{11.95}$

(b) $HI + H_2O \rightarrow H_3O^+ + I^-$ $\qquad$ 0.080 *M* HI $\rightarrow$ 0.080 *M* H_3O^+

$\mathbf{pH} = -\log[H_3O^+] = -\log(0.080) = \mathbf{1.10}$ $\qquad$ $\mathbf{pOH} = 14.00 - pH = \mathbf{12.90}$

(c) $Sr(OH)_2 \rightarrow Sr^{2+} + 2OH^-$

$0.030\ M\ Sr(OH)_2 \rightarrow 0.060\ M\ OH^-$

$\mathbf{pOH} = -\log[OH^-] = -\log(0.060) = \mathbf{1.22}$

$\mathbf{pH} = 14.00 - pOH = \mathbf{12.78}$

(d) $HNO_3 + H_2O \rightarrow H_3O^+ + NO_3^-$

$12.3\ M\ HNO_3 \rightarrow 12.3\ M\ H_3O^+$

$\mathbf{pH} = -\log[H_3O^+] = -\log(12.3) = \mathbf{-1.09}$

$\mathbf{pOH} = 14.00 - pH = \mathbf{15.09}$

13.16 ***See Sections 13.2 and 3 and Example 13.7.***

$$NaOH(s) \xrightarrow{H_2O} Na^+(aq) + OH^-(aq)$$

$$?\ M\ OH^- = \frac{15.0\text{ g NaOH}}{0.500\text{ L soln}} \times \frac{1\text{ mol NaOH}}{40.1\text{ g NaOH}} \times \frac{1\text{ mol OH}^-}{1\text{ mol NaOH}} = 0.748\ M\ OH^-$$

$\mathbf{pOH} = -\log[OH^-] = -\log(0.748) = \mathbf{0.13}$

$\mathbf{pH} = 14.00 - pOH = \mathbf{13.87}$

13.18 ***See Sections 13.1, 2 and 3.***

Unbalanced:	$HCl + Ba(OH)_2 \rightarrow H_2O + BaCl_2$	Start with $BaCl_2$.
Step 1:	$\underline{2}HCl + Ba(OH)_2 \rightarrow H_2O + BaCl_2$	Balances Cl.
Step 2:	$2HCl + Ba(OH)_2 \rightarrow \underline{2}H_2O + BaCl_2$	Balances H, O.

$$?\text{ mol HCl available} = 65\text{ mL} \times \frac{1\text{ L}}{10^3\text{ mL}} \times \frac{0.010\text{ mol}}{1\text{ L}} = 6.5 \times 10^{-4}\text{ mol HCl}$$

$$?\text{ mol Ba(OH)}_2\text{ available} = 35\text{ mL} \times \frac{1\text{ L}}{10^3\text{ mL}} \times \frac{0.020\text{ mol}}{1\text{ L}} = 7.0 \times 10^{-4}\text{ mol Ba(OH)}_2$$

? mol $Ba(OH)_2$ needed to react with mol HCl available

$$= 6.5 \times 10^{-4}\text{ mol HCl} \times \frac{1\text{ mol Ba(OH)}_2}{2\text{ mol HCl}} = 3.2 \times 10^{-4}\text{ mol Ba(OH)}_2$$

Since 7.0×10^{-4} mol $Ba(OH)_2$ are available and only 3.2×10^{-4} mol are needed, excess $Ba(OH)_2$ is present.

$$?\text{ mol Ba(OH)}_2\text{ excess} = 7.0 \times 10^{-4}\text{ mol available} - 3.2 \times 10^{-4}\text{ mol needed} = 3.8 \times 10^{-4}\text{ mol Ba(OH)}_2$$

The mixed soultion can now be treated as a solution containing only 3.8×10^{-4} mol $Ba(OH)_2$.
$Ba(OH)_2 \rightarrow Ba^{2+} + 2OH^-$

3.8×10^{-4} mol $Ba(OH)_2 \rightarrow 7.6 \times 10^{-4}$ mol OH^-

$$[OH^-]\text{ in mixed soln} = \frac{7.6 \times 10^{-4}\text{ mol OH}^-}{0.100\text{ L total soln}} = 7.6 \times 10^{-3}\ M\ OH^-$$

$\mathbf{pOH} = -\log[OH^-] = -\log(7.6 \times 10^{-3}) = \mathbf{2.12}$

$\mathbf{pH} = 14.00 - pOH = \mathbf{11.88}$

13.20 *See Sections 13.1, 2 and 3.*

$2HF \rightleftarrows H_2F^+ + F^-$ H_2F^+ is the conjugate acid of HF and F^- is the conjugate base of HF.

(a) KF is a source of F^- and therefore acts as a base.

(b) $HClO_4 + HF \rightarrow H_2F^+ + ClO_4^-$

(c) $NH_3 + HF \rightarrow NH_4^+ + F^-$

(d) $H^+ + NH_3 \rightarrow NH_4^+$

13.22 *See Sections 13.1 through 5 and Table 13.6.*

(a) $CH_3COOH + CH_3COOH \rightleftarrows CH_3COOH_2^+ + CH_3COO^-$

(b) According to Table 13.6, CH_3COOH is a stronger acid than H_2O. Hence, H_2O should act as the base.

$CH_3COOH + H_2O \rightleftarrows H_3O^+ + CH_3COO^-$

(c) The stongest acid and base that can exist in any protonic solvent is the acid and base that are formed in the autoionization reaction of the protonic solvent. Hence, $CH_3COOH_2^+$ is the strongest acid that can exist in CH_3COOH and CH_3COO^- is the strongest base that can exist in CH_3COOH. $NaCH_3COO$ or KCH_3COO could be added to add a strong base to pure CH_3COOH.

(d) The strongest acid that can exist in CH_3COOH is $CH_3COOH_2^+$; see solution for part (c). Strong acids such as $HClO_4$ react with CH_3COOH completely yielding $CH_3COOH_2^+$.

13.24 *See Section 13.4 and 5 and Tables 13.6 and 7.*

The extent of reaction with the solvent will depend on the acid strength of the solvent. In pure acetic acid, ammonia will act as a strong base and theobromine will act as a weak base. In liquid ammonia, neither ammonia nor theobromine will ionize appreciably. In water, ammonia will act as a weak base and theobromine will not ionize appreciably. Hence, **pure acetic acid** is the best solvent to use to differentiate between the base strengths of ammonia and theobromine, but **water** can also be used.

13.26 *See Section 13.5 and Example 13.9.*

$C_3H_7COOH + H_2O \rightleftarrows H_3O^+ + C_3H_7COO^-$

$$?\,M\,C_3H_7COOH = \frac{50.0\text{ mg }C_3H_7COOH}{1.00\text{ mL soln}} \times \frac{1\text{ mol }C_3H_7COOH}{88.0 \times 10^3\text{ mg }C_3H_7COOH} \times \frac{10^3\text{ mL soln}}{1\text{ L soln}} = 0.568\,M\,C_3H_7COOH$$

Conc, *M*	C_3H_7COOH	H_3O^+	$C_3H_7COO^-$
initial	0.568	0.	0.
Change	-3.0×10^{-3}	$+3.0 \times 10^{-3}$	$+3.0 \times 10^{-3}$
equilibrium	0.565	3.0×10^{-3}	3.0×10^{-3}

$$K_a = \frac{[H_3O^+][C_3H_7COO^-]}{[C_3H_7COOH]} \qquad K_a = \frac{(3.0 \times 10^{-3})(3.0 \times 10^{-3})}{0.565} = 1.6 \times 10^{-5}$$

$$pK_a = -\log K_a \qquad pK_a = -\log 1.6 \times 10^{-5} = \mathbf{4.80}$$

13.28 ***See Section 13.5 and Example 13.9.***

$$C_5H_{11}COOH + H_2O \rightleftarrows H_3O^+ + C_5H_{11}COO^-$$

$$[H_3O^+] = 10^{-pH} = \text{inv}\log(-pH) \qquad [H_3O^+] = \text{inv}\log(-3.43) = 3.7 \times 10^{-4}$$

Conc, *M*	$C_5H_{11}COOH$	H_3O^+	$C_5H_{11}COO^-$
initial	0.0100	0.	0.
Change	-3.7×10^{-4}	$+3.7 \times 10^{-4}$	$+3.7 \times 10^{-4}$
equilibrium	9.6×10^{-3}	3.7×10^{-4}	3.7×10^{-4}

$$K_a = \frac{[H_3O^+][C_5H_{11}COO^-]}{[C_5H_{11}COOH]} \qquad K_a = \frac{(3.7 \times 10^{-4})(3.7 \times 10^{-4})}{9.6 \times 10^{-3}} = \mathbf{1.4 \times 10^{-5}}$$

$$pK_a = -\log K_a \qquad pK_a = -\log 1.4 \times 10^{-5} = \mathbf{4.85}$$

13.30 ***See Section 13.5 and Table 13.6***

(a) $C_6H_5COOH + H_2O \rightleftarrows H_3O^+ + C_6H_5COO^-$ $\qquad K_a = 6.3 \times 10^{-5}$

Conc, *M*	C_6H_5COOH	H_3O^+	$C_6H_5COO^-$
initial	0.20	0.	0.
Change	$-y$	$+y$	$+y$
equilibrium	$0.20-y$	y	y

$$K_a = \frac{[H_3O^+][C_6H_5COO^-]}{[C_6H_5COOH]} \qquad K_a = \frac{(y)(y)}{(0.20-y)} \cong \frac{y^2}{0.20} = 6.3 \times 10^{-5}$$

$$[H_3O^+] = y = 3.5 \times 10^{-3} \qquad pH = -\log(3.5 \times 10^{-3}) = \mathbf{2.46}$$

Approximation check: $\left(\frac{3.5 \times 10^{-3}}{0.20}\right) \times 100\% = 1.8\%$, so the approximation is valid.

(b) $HCOOH + H_2O \rightleftarrows H_3O^+ + HCOO^-$ $K_a = 1.8 \times 10^{-4}$

Conc, *M*	HCOOH	H_3O^+	$HCOO^-$
initial	1.50	0.	0.
Change	−y	+y	+y
equilibrium	1.50 − y	y	y

$$K_a = \frac{[H_3O^+][HCOO^-]}{[HCOOH]} \qquad K_a = \frac{(y)(y)}{(1.50-y)} \cong \frac{y^2}{1.50} = 1.8 \times 10^{-4}$$

$$[H_3O^+] = y = 1.6 \times 10^{-2} \qquad pH = -\log(1.6 \times 10^{-2}) = \mathbf{1.80}$$

Approximation check: $\left(\frac{1.6 \times 10^{-2}}{1.50}\right) \times 100\% = 1.1\%$, so the approximation is valid.

(c) $HCN + H_2O \rightleftarrows H_3O^+ + CN^-$ $K_a = 7.2 \times 10^{-10}$

Conc, *M*	HCN	H_3O^+	CN^-
initial	0.0055	0.	0.
Change	−y	+y	+y
equilibrium	0.0055 − y	y	y

$$K_a = \frac{[H_3O^+][CN^-]}{[HCN]} \qquad K_a = \frac{(y)(y)}{(0.0055-y)} \cong \frac{y^2}{0.0055} = 7.2 \times 10^{-10}$$

$$[H_3O^+] = y = 2.0 \times 10^{-6} \qquad pH = -\log(2.0 \times 10^{-6}) = \mathbf{5.70}$$

Approximation check: $\left(\frac{2.0 \times 10^{-6}}{0.0055}\right) \times 100\% = 0.036\%$, so the approximation is valid.

(d) $HNO_2 + H_2O \rightleftarrows H_3O^+ + NO_2^-$ $K_a = 4.6 \times 10^{-4}$

Conc, *M*	HNO_2	H_3O^+	NO_2^-
initial	0.075	0.	0.
Change	−y	+y	+y
equilibrium	0.075 − y	y	y

$$K_a = \frac{[H_3O^+][NO_2^-]}{[HNO_2]} \qquad K_a = \frac{(y)(y)}{(0.075-y)} \cong \frac{y^2}{0.075} = 4.6 \times 10^{-4}$$

$$[H_3O^+] = y = 5.9 \times 10^{-3} \qquad pH = -\log(5.9 \times 10^{-3}) = 2.23$$

Approximation check: $\left(\frac{5.9 \times 10^{-3}}{0.075}\right) \times 100\% = 7.9\%$, so the approximation is not valid. The quadratic formula or succesive approximation method must be used to solve the equilibrium expression.

$$K_a = \frac{(y)(y)}{(0.075-y)} = 4.6 \times 10^{-4} \text{ yields } y^2 + 4.6 \times 10^{-4}\,y - 3.45 \times 10^{-5} = 0$$

Substituting into the quadratic formula gives $y = \frac{-4.6 \times 10^{-4} \pm \sqrt{(4.6 \times 10^{-4})^2 - 4(1)(-3.45 \times 10^{-5})}}{2(1)}$

and $y = \frac{-4.6 \times 10^{-4} \pm 1.2 \times 10^{-2}}{2} = 5.8 \times 10^{-3}$ or -6.2×10^{-3}.

The only reasonable value of y is 5.8×10^{-3}, since -6.2×10^{-3} predicts negative concentrations for H_3O^+ and NO_2^-. Hence, $[H_3O^+] = y = 5.8 \times 10^{-3}$ and $pH = -\log(5.8 \times 10^{-3}) = \mathbf{2.24}$.

13.32 ***See Section 13.5 and Table 13.6.***

(a) $HNO_2 + H_2O \rightleftarrows H_3O^+ + NO_2^-$ $K_a = 4.6 \times 10^{-4}$

Conc, *M*	HNO_2	H_3O^+	NO_2^-
initial	0.33	0.	0.
Change	$-y$	$+y$	$+y$
equilibrium	$0.33-y$	y	y

$$K_a = \frac{[H_3O^+][NO_2^-]}{[HNO_2]} \qquad K_a = \frac{(y)(y)}{(0.33-y)} \cong \frac{y^2}{0.33} = 4.6 \times 10^{-4}$$

$$[H_3O^+] = y = 1.2 \times 10^{-2} \qquad pH = -\log(1.2 \times 10^{-2}) = \mathbf{1.92}$$

Approximation check: $\left(\frac{1.2 \times 10^{-2}}{0.33}\right) \times 100\% = 3.6\%$, so the approximation is valid.

(b) $C_6H_5OH + H_2O \rightleftarrows H_3O^+ + C_6H_5O^-$ $K_a = 1.3 \times 10^{-10}$

Conc, *M*	C_6H_5OH	H_3O^+	$C_6H_5O^-$
initial	0.016	0.	0.
Change	−y	+y	+y
equilibrium	0.016 − y	y	y

$$K_a = \frac{[H_3O^+][C_6H_5O^-]}{[C_6H_5OH]} \qquad K_a = \frac{(y)(y)}{(0.016-y)} \cong \frac{y^2}{0.016} = 1.3 \times 10^{-10}$$

$$[H_3O^+] = y = 1.4 \times 10^{-6} \qquad pH = -\log(1.4 \times 10^{-6}) = \mathbf{5.85}$$

Approximation check: $\left(\frac{1.4 \times 10^{-6}}{0.016}\right) \times 100\% = 0.0088\%$, so the approximation is valid.

(c) $HF + H_2O \rightleftarrows H_3O^+ + F^-$ $K_a = 3.5 \times 10^{-4}$

Conc, *M*	HF	H_3O^+	F^-
initial	0.15	0.	0.
Change	−y	+y	+y
equilibrium	0.15 − y	y	y

$$K_a = \frac{[H_3O^+][F^-]}{[HF]} \qquad K_a = \frac{(y)(y)}{(0.15-y)} \cong \frac{y^2}{0.15} = 3.5 \times 10^{-4}$$

$$[H_3O^+] = y = 7.2 \times 10^{-3} \qquad pH = -\log(7.2 \times 10^{-3}) = \mathbf{2.14}$$

Approximation check: $\left(\frac{7.2 \times 10^{-3}}{0.15}\right) \times 100\% = 4.8\%$, so the approximation is valid.

(d) $HCOOH + H_2O \rightleftarrows H_3O^+ + HCOO^-$ $K_a = 1.8 \times 10^{-4}$

Conc, *M*	HCOOH	H_3O^+	$HCOO^-$
initial	0.010	0.	0.
Change	−y	+y	+y
equilibrium	0.010 − y	y	y

$$K_a = \frac{[H_3O^+][HCOO^-]}{[HCOOH]} \qquad K_a = \frac{(y)(y)}{(0.010-y)} \cong \frac{y^2}{0.010} = 1.8 \times 10^{-4}$$

$$[H_3O^+] = y = 1.3 \times 10^{-3} \qquad pH = -\log(1.3 \times 10^{-3}) = 2.89$$

Approximation check: $\left(\frac{1.3 \times 10^{-3}}{0.010}\right) \times 100\% = 13\%$, so the approximation is not valid. The quadaratic formula or successive approximation method must be used to solve the equilibrium expression.

$$K_a = \frac{(y)(y)}{(0.010-y)} = 1.8 \times 10^{-4} \text{ yields } y^2 + 1.8 \times 10^{-4} y - 1.8 \times 10^{-6} = 0$$

Substituting into the quadratic formula gives $y = \frac{-1.8 \times 10^{-4} \pm \sqrt{(1.8 \times 10^{-4})^2 - 4(1)(-1.8 \times 10^{-6})}}{2(1)}$

and $y = \frac{-1.8 \times 10^{-4} \pm 2.7 \times 10^{-3}}{2} = 1.3 \times 10^{-3}$ or -1.4×10^{-3}.

The only reasonable value of y is 1.3×10^{-3}, since -1.4×10^{-3} predicts negative concentrations for H_3O^+ and $HCOO^-$. Hence, $[H_3O^+] = y = 1.3 \times 10^{-3}$ and $pH = -\log(1.3 \times 10^{-3}) = \mathbf{2.89}$

13.34 ***See Sections 13.5 , Table 13.6 and Example 13.10.***

(a) $C_6H_5COOH + H_2O \rightleftarrows H_3O^+ + C_6H_5COO^- \quad K_a = 6.3 \times 10^{-5}$

Conc, *M*	C_6H_5COOH	H_3O^+	$C_6H_5COO^-$
initial	0.010	0.	0.
Change	$-y$	$+y$	$+y$
equilibrium	$0.010-y$	y	y

$$K_a = \frac{[H_3O^+][C_6H_5COO^-]}{[C_6H_5COOH]} \qquad K_a = \frac{(y)(y)}{(0.010-y)} \cong \frac{y^2}{0.010} = 6.3 \times 10^{-5}$$

Hence, $y = 7.9 \times 10^{-4}$ and $\left(\frac{7.9 \times 10^{-4}}{0.010}\right) \times 100\% = 7.9\%$, so the approximation is not valid. The quadratic formula or successive approximation method must be used to solve the equilibrium expression.

$$K_a = \frac{(y)(y)}{(0.010-y)} = 6.3 \times 10^{-5} \text{ yields } y^2 + 6.3 \times 10^{-5} y - 6.3 \times 10^{-7} = 0$$

Substituting into the quadratic formula gives $y = \dfrac{-6.3 \times 10^{-5} \pm \sqrt{(6.3 \times 10^{-5})^2 - 4(1)(-6.3 \times 10^{-7})}}{2(1)}$

and $y = \dfrac{-6.3 \times 10^{-5} \pm 1.6 \times 10^{-3}}{2} = 7.7 \times 10^{-4}$ or -8.3×10^{-4}.

The only reasonable value of y is 7.7×10^{-4}, since -8.3×10^{-4} predicts negative concentrations for H_3O^+ and

$C_6H_5COO^-$. Hence, $\alpha = \dfrac{[C_6H_5COO^-]}{C_{C_6H_5COOH}}$, $\alpha = \dfrac{7.7 \times 10^{-4}}{0.010} = \mathbf{0.077}$ and **% ionization = 7.7%.**

(b) $C_6H_5COOH + H_2O \rightleftarrows H_3O^+ + C_6H_5COO^-$ $K_a = 6.3 \times 10^{-5}$

Conc, *M*	C_6H_5COOH	H_3O^+	$C_6H_5COO^-$
initial	0.0010	0.	0.
Change	−y	+y	+y
equilibrium	0.0010−y	y	y

$K_a = \dfrac{[H_3O^+][C_6H_5COO^-]}{[C_6H_5COOH]}$ $K_a = \dfrac{(y)(y)}{(0.0010-y)} \cong \dfrac{y^2}{0.0010} = 6.3 \times 10^{-5}$

Hence, $y = 2.5 \times 10^{-4}$ and $\left(\dfrac{2.5 \times 10^{-4}}{0.010}\right) \times 100\% = 25\%$, so the approximation is not valid. The quadratic formula or successive approximation method must be used to solve the equilibrium expression.

$K_a = \dfrac{(y)(y)}{(0.0010-y)} = 6.3 \times 10^{-5}$ yields $y^2 + 6.3 \times 10^{-5} y - 6.3 \times 10^{-8} = 0$

Substituting into the quadratic formula gives $y = \dfrac{-6.3 \times 10^{-5} \pm \sqrt{(6.3 \times 10^{-5})^2 - 4(1)(-6.3 \times 10^{-8})}}{2(1)}$

and $y = \dfrac{-6.3 \times 10^{-5} \pm 5.1 \times 10^{-4}}{2} = 2.2 \times 10^{-4}$ or -2.9×10^{-4}.

The only reasonable value of y is 2.2×10^{-4}, since -2.9×10^{-4} predicts negative concentrations for H_3O^+ and

$C_6H_5COO^-$. Hence, $\alpha = \dfrac{[C_6H_5COO^-]}{C_{C_6H_5COOH}}$, $\alpha = \dfrac{2.2 \times 10^{-4}}{0.0010} = \mathbf{0.22}$ and **% ionization = 22%.**

13.36 *See Section 13.5.*

The percent ionization can be used to calculate the equilibrium concentrations of H_3O^+ and CN^- formed via

$HCN + H_2O \rightleftarrows H_3O^+ + CN^-$.

Using $\alpha = \frac{[CN^-]}{C_{HCN}}$ and recognizing that α is equal to the percent ionization expressed in decimal form gives

$$[CN^-] = \alpha C_{HCN} \qquad [CN^-] = (0.00038)(0.0050\ M) = 1.9 \times 10^{-6}\ M\ CN^-$$

Since this value is small compared to C_{HCN}, we obtain

$$K_a = \frac{[H_3O^+][CN^-]}{[HCN]} \qquad K_a = \frac{(1.9 \times 10^{-6})(1.9 \times 10^{-6})}{0.0050} = \mathbf{7.2 \times 10^{-10}}$$

13.38 *See Section 13.5.*

An increase in the concentration of a strong acid produces a directly proportional increase in the concentrations of ions and thus in conductivity whereas an increase in the concentration of a weak acid does not. This is due to the observation that strong acids ionize completely whereas weak acids do not.

We can determine whether conductivity is directly proportional to concentration by graphing conductivity versus concentration for each acid.

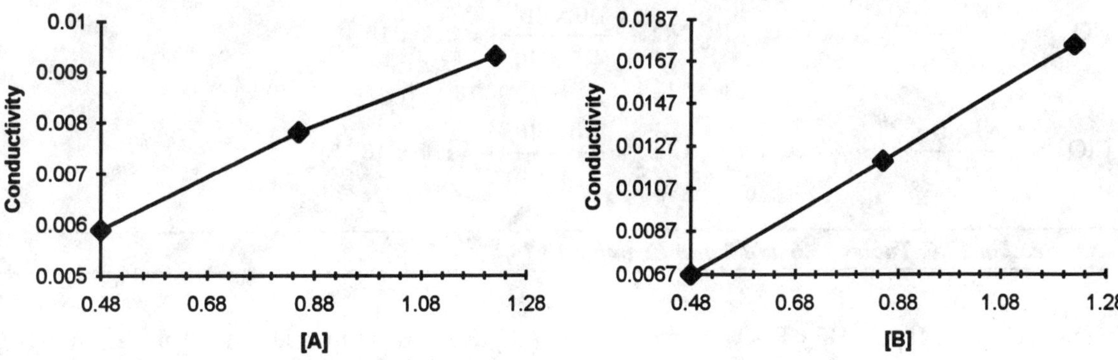

Since there is no directly proportional increase in conductivity with increase in concentration for Acid A, we must conclude Acid A is a weak acid. Since there is a directly proportional increase in conductivity with increase in concentration for Acid B, (within the limits of experimental error), we must conclude Acid B is a strong acid.

13.40 See Section 13.6 and Example 13.11.

$C_8H_{17}N + H_2O \rightleftarrows C_8H_{17}NH^+ + OH^-$

Conc, *M*	$C_8H_{17}N$	$C_8H_{17}NH^+$	OH^-
initial	0.50	0.	0.
Change	$-y$	$+y$	$+y$
equilibrium	$0.50-y$	y	y

$pK_b = 3.1$ $\qquad K_b = 10^{-pK_b} = \text{inv}\log(-pK_b) = 7.9 \text{ x } 10^{-4}$

$$K_b = \frac{[C_8H_{17}NH^+][OH^-]}{[C_8H_{17}N]} \qquad K_b = \frac{(y)(y)}{(0.50-y)} \cong \frac{y^2}{0.50} = 7.9 \text{ x } 10^{-4}$$

$[OH^-] = y = 2.0 \text{ x } 10^{-2}$ $\qquad pOH = -\log(2.0 \text{ x } 10^{-2}) = 1.70$ $\qquad pH = 14.00 - 1.70 = \mathbf{12.30}$

Approximation check: $\left(\frac{2.0 \text{ x } 10^{-2}}{0.50}\right) \times 100\% = 4\%$, so the approximation is valid.

13.42 See Section 13.6, Tables 13.6 and 9 and Example 13.12.

(a) $K_b\ HCOO^- = \frac{K_w}{K_a\ HCOOH}$ $\qquad K_b\ HCOO^- = \frac{1.0 \text{ x } 10^{-14}}{1.8 \text{ x } 10^{-4}} = \mathbf{5.6 \text{ x } 10^{-11}}$

(b) $K_b\ NO_2^- = \frac{K_w}{K_a\ HNO_2}$ $\qquad K_b\ NO_2^- = \frac{1.0 \text{ x } 10^{-14}}{4.6 \text{ x } 10^{-4}} = \mathbf{2.2 \text{ x } 10^{-11}}$

(c) $K_b\ ClO_2^- = \frac{K_w}{K_a\ HClO_2}$ $\qquad K_b\ ClO_2^- = \frac{1.0 \text{ x } 10^{-14}}{1..0 \text{ x } 10^{-2}} = \mathbf{1.0 \text{ x } 10^{-12}}$

13.44 See Section 13.6, Tables 13.6 and 7 and Example 13.13.

According to Table 13.6, $H_2O < HF < HSO_4^-$. According to Table 13.7, NH_3 is a weaker base than OH^-. Hence, NH_4^+ is a stronger acid than H_2O. To determine where NH_4^+ fits in this series, we must calculate $K_a\ NH_4^+$ using

$K_a\ NH_4^+ = \frac{K_w}{K_b\ NH_3}$ $\qquad K_a\ NH_4^+ = \frac{1.0 \text{ x } 10^{-14}}{1.8 \text{ x } 10^{-5}} = 5.6 \text{ x } 10^{-10}$

This gives $H_2O < NH_4^+ < HF < HSO_4^-$.

13.46 See Section 13.6 and Table 13.6.

According to Table 13.6, $H_2O < CH_3COOH < HCOOH$. F^- cannot act as Bronsted acid, since it doesn't contain partially positive (i.e. protonic) hydrogen atoms.

13.48 *See Section 13.6, Table 13.7 and Example 13.14.*

The pyridium ion, $C_5H_5NH^+$, is the conjugate acid of the weak base pyridine, C_5H_5N, and therefore acts as a weak acid. The iodide ion, I^-, is the conjugate base of the strong acid HI and therefore does not react appreciably with water. The equilibrium of interest is $C_5H_5NH^+ + H_2O \rightleftarrows H_3O^+ + C_5H_5N$.

Conc, *M*	$C_5H_5NH^+$	H_3O^+	C_5H_5N
initial	0.060	0.	0.
Change	−y	+y	+y
equilibrium	0.060 − y	y	y

$$K_a\ C_5H_5NH^+ = \frac{K_w}{K_b\ C_5H_5N} = \frac{[H_3O^+][C_5H_5N]}{[C_5H_5NH^+]}$$

$$K_a\ C_5H_5NH^+ = \frac{(y)(y)}{0.060 - y} \cong \frac{y^2}{0.060} = \frac{1.0 \times 10^{-14}}{1.8 \times 10^{-9}} = 5.6 \times 10^{-6}$$

$[H_3O^+] = y = 5.8 \times 10^{-4}$ and $pH = -\log(5.8 \times 10^{-4}) = \mathbf{3.24}$.

Approximation check: $\left(\frac{5.8 \times 10^{-4}}{0.060}\right) \times 100\% = 0.97\%$, so the approximation is valid.

13.50 *See Section 13.6, Table 13.6 and Example 13.14.*

The sodium ion, Na^+, does not react appreciably with water. The acetate ion, CH_3COO^-, is the conjugate base of the weak acid acetic acid, CH_3COOH, and therefore acts as a weak base. The equilibrium of interest is :

$CH_3COO^- + H_2O \rightleftarrows CH_3COOH + OH^-$.

Conc, *M*	CH_3COO^-	CH_3COOH	OH^-
initial	0.010	0.	0.
Change	−y	+y	+y
equilibrium	0.010 − y	y	y

$$K_b\ CH_3COO^- = \frac{K_w}{K_a\ CH_3COOH} = \frac{[CH_3COO^-][OH^-]}{[CH_3COOH]}$$

$$K_b\ CH_3COO^- = \frac{(y)(y)}{0.010 - y} \cong \frac{y^2}{0.010} = \frac{1.0 \times 10^{-14}}{1.8 \times 10^{-5}} = 5.6 \times 10^{-10}$$

$\left[OH^-\right] = y = 2.4 \times 10^{-6}$, $pOH = -\log\left(2.4 \times 10^{-6}\right) = 5.62$ and $pH = 14.00 - 5.62 = \mathbf{8.38}$

Approximation check: $\left(\frac{2.4 \times 10^{-6}}{0.010}\right) \times 100\% = 0.024\%$, so the approximtion is valid.

13.52 ***See Section 13.7, Table 13.6 and Example 13.6.***

HI is a strong acid whereas CH_3COOH is a weak acid. The pH will be determined by the strong acid HI:

$HI + H_2O \rightarrow H_3O^+ + I^-$.

$\left[H_3O^+\right] = C_{HI} = 0.050\ M$ and $pH = -\log(0.050) = \mathbf{1.30}$.

13.54 ***See Section 13.7, Table 13.6 and Example 13.6.***

Acetic acid, CH_3COOH, with $K_a = 1.8 \times 10^{-5}$ is a much stronger acid than hydrocyanic acid, HCN, with $K_a = 7.2 \times 10^{-10}$. We can therefore consider this to be just a solution of acetic acid:

$CH_3COOH + H_2O \rightleftarrows H_3O^+ + CH_3COO^-$.

Conc, M	CH_3COOH	H_3O^+	CH_3COO^-
initial	0.10	0.	0.
Change	$-y$	$+y$	$+y$
equilibrium	$0.10 - y$	y	y

$K_a = \frac{\left[H_3O^+\right]\left[CH_3COO^-\right]}{\left[CH_3COOH\right]}$ $\qquad K_a = \frac{(y)(y)}{0.10 - y} \cong \frac{y^2}{0.10} = 1.8 \times 10^{-5}$

$\left[H_3O^+\right] = y = 1.3 \times 10^{-3}$ and $pH = -\log\left(1.3 \times 10^{-3}\right) = \mathbf{2.89}$.

Approximation check: $\left(\frac{1.3 \times 10^{-3}}{0.10}\right) \times 100\% = 1.3\%$, so the approximation is valid.

13.56 ***See Section 13.7, Table 13.6 and Example 13.6.***

The benzoate ion, $C_6H_5COO^-$, is the conjugate base of the weak acid benzoic acid, C_6H_5COOH, and therefore acts as a weak base. KOH is a strong base, and the pH of the solution is determined by the KOH.

$\left[OH^-\right] = C_{OH^-} = 0.10\ M$, $pOH = -\log(0.10) = 1.00$ and $pH = 14.00 - 1.00 = \mathbf{13.00}$

13.58 *See Section 13.8.*

(a) $HClO_4$ is a stronger acid than $HClO_3$. The additional oxygen atom withdraws electron density from the Cl, which in turn withdraws additional electron density from the O-H bond, making it weaker and easier to ionize in water. The additional oxygen atom also helps stabilize the anion that is formed when the acid ionizes,

so ClO_4^- is more stable than ClO_3^-.

(b) HNO_3 is a stronger acid than HNO_2. The reasoning is similar to that given in the solution for part (a) above.

(c) AsH_3 is a stronger acid than GeH_4. Within any row of the periodic table the acidities of the binary hydrides increase as the central atom becomes more electronegative and can more easily accommodate negative charge. Note: This increase in acidity occurs in spite of increasing A-H bond strengths moving from left-to-right across a row or period; see text.

13.60 *See Sections 3.1, 13.1 and 9.*

System	Acid	Base
Arrhenius	Any substance that provides H^+ in water.	Any substance that provides OH^- in water.
Bronsted-Lowry	A proton donor.	A proton acceptor.
Lewis	An electron pair acceptor.	An electron pair donor.

(a) $HCl(aq) + NH_3(aq) \rightarrow NH_4Cl(aq)$ Arrhenius, Bronsted-Lowry and Lewis reaction.

(b) $SO_2(g) + NaOH(s) \rightarrow NaHSO_3(s)$ Lewis reaction.

(c) $LiH(s) + H_2O(l) \rightarrow LiOH(aq) + H_2(g)$ Arrhenius, Bronsted-Lowry and Lewis reaction.

(d) $HSO_4^-(aq) + F^-(aq) \rightleftarrows HF(aq) + SO_4^{2-}(aq)$ Arrhenius, Bronsted-Lowry and Lewis reaction.

Note: The acid-base theories become more general moving from Arrhenius to Bronsted-Lowry to Lewis. Hence, any Arrhenius reaction is also a Bronsted-Lowry reaction and a Lewis reaction. Any Bronsted-Lowry reaction is also a Lewis reaction. However, some Lewis reactions are neither Arrhenius nor Bronsted-Lowry reactions, since the Lewis theory is the most general and therefore includes additional reactions.

13.62 *See Section 13.4.*

The equilibrium will favor the formation of the weaker acid and the weaker base.

(a) $HCl(aq) + NH_3(aq) \rightleftarrows NH_4Cl(aq)$: The reaction favors **products**, since HCl(aq) is a stronger acid than $NH_4^+(aq)$ and $NH_3(aq)$ is a stronger base than $Cl^-(aq)$.

(b) $HNO_3(aq) + NaOH(aq) \rightleftarrows NaNO_3(aq) + H_2O(l)$: The reaction favors **products**, since HNO_3 is a strong aicd and NaOH is a strong base.

(c) $2KCl(aq) + Ba(OH)_2(aq) \rightleftarrows BaCl_2(aq) + 2KOH(aq)$: **No reaction** occurs.

(d) $HSO_4^-(aq) + NH_3(aq) \rightleftarrows NH_4^+(aq) + SO_4^{2-}(aq)$: The reaction favors **products**, since HSO_4^- is a stronger acid than NH_4^+ and NH_3 is a stronger base than SO_4^{2-}.

13.64 *See Section 13.5, Table 13.6 and Example 13.10.*

$HNO_2 + H_2O \rightleftarrows H_3O^+ + NO_2^-$ $\quad K_a = 4.6 \times 10^{-4}$

$[H_3O^+] = 10^{-pH} = \text{inv}\log(-pH)$ $\quad [H_3O^+] = \text{inv}\log(-4.75) = 1.8 \times 10^{-5}\ M$

Conc, M	HNO_2	H_3O^+	NO_2^-
initial	y	0.	0.
Change	-1.8×10^{-5}	$+1.8 \times 10^{-5}$	$+1.8 \times 10^{-5}$
equilibrium	$y - 1.8 \times 10^{-5}$	1.8×10^{-5}	1.8×10^{-5}

$$K_a = \frac{[H_3O^+][NO_2^-]}{[HNO_2]} \qquad K_a = \frac{(1.8 \times 10^{-5})(1.8 \times 10^{-5})}{(y - 1.8 \times 10^{-5})} = 4.6 \times 10^{-4}$$

Rearranging and solving for y gives

$3.2 \times 10^{-10} = 4.6 \times 10^{-4}\, y - 8.3 \times 10^{-9}$ and $[HNO_2]$ initial $= y = 1.9 \times 10^{-5}\ M$.

Solving $M(\text{con}) \times V(\text{con}) = M(\text{dil}) \times V(\text{dil})$ for $V(\text{con})$ gives $V(\text{con}) = \dfrac{M(\text{dil}) \times V(\text{dil})}{M(\text{con})}$

$$V(\text{con}) = \frac{1.9 \times 10^{-5}\ M \times 1.00\ \text{L}}{0.083\ M} = 2.3 \times 10^{-4}\ \text{L or } \mathbf{0.23\ mL}.$$

13.66 *See Section 13.5, Table 13.6 and Example 13.10.*

$C_6H_5COOH + H_2O \rightleftarrows H_3O^+ + C_6H_5COO^-$ $\quad K_a = 6.3 \times 10^{-5}$

$[H_3O^+] = 10^{-pH} = \text{inv}\log(-pH)$ $\quad [H_3O^+] = \text{inv}\log(-3.50) = 3.2 \times 10^{-4}\ M$

Conc, M	C_6H_5COOH	H_3O^+	$C_6H_5COO^-$
initial	y	0.	0.
Change	-3.2×10^{-4}	$+3.2 \times 10^{-4}$	$+3.2 \times 10^{-4}$
equilibrium	$y - 3.2 \times 10^{-4}$	3.2×10^{-4}	3.2×10^{-4}

$$K_a = \frac{[H_3O^+][C_6H_5COO^-]}{[C_6H_5COOH]} \qquad K_a = \frac{(3.2 \times 10^{-4})(3.2 \times 10^{-4})}{(y - 3.2 \times 10^{-4})} = 6.3 \times 10^{-5}$$

Rearranging and solving for y gives

$1.0 \times 10^{-7} = 6.3 \times 10^{-5}\,y - 2.0 \times 10^{-8}$, and $[C_6H_5COOH]$ initial $= y = 1.9 \times 10^{-3}\ M$.

Hence, $? \text{ g } C_6H_5COOH = 1.00 \text{ L } C_6H_5COOH \text{ soln} \times \dfrac{1.9 \times 10^{-3} \text{ mol } C_6H_5COOH}{1 \text{ L } C_6H_5COOH \text{ soln}} \times \dfrac{122.0 \text{ g } C_6H_5COOH}{1 \text{ mol } C_6H_5COOH}$

$= \mathbf{0.23\ g\ C_6H_5COOH}$

13.68 ***See Section 13.6 and Example 13.11.***

$NH_3 + H_2O \rightleftarrows NH_4^+ + OH^-$ $K_b = 1.8 \times 10^{-5}$

$$? \ M\,NH_3 = \frac{10.0 \text{ mL conc } NH_3 \text{ soln}}{1.00 \text{ L}} \times \frac{0.90 \text{ g conc } NH_3 \text{ soln}}{1 \text{ mL conc } NH_3 \text{ soln}} \times \frac{28 \text{ g } NH_3}{100 \text{ g conc } NH_3 \text{ soln}} \times \frac{1 \text{ mol } NH_3}{17.0 \text{ g } NH_3} = 0.15\ M\,NH_3$$

Conc, M	NH_3	NH_4^+	OH^-
initial	0.15	0.	0.
Change	$-y$	$+y$	$+y$
equilibrium	$0.15-y$	y	y

$$K_b = \frac{[NH_4^+][OH^-]}{[NH_3]} \qquad K_b = \frac{(y)(y)}{0.15-y} \cong \frac{y^2}{0.15} = 1.8 \times 10^{-5}$$

$[OH^-] = y = 1.6 \times 10^{-3}$, $pOH = -\log(1.6 \times 10^{-3}) = 2.80$ and $pH = 14.00 - 2.80 = \mathbf{11.20}$

Approximation check: $\left(\dfrac{1.6 \times 10^{-3}}{0.15}\right) \times 100\% = 1.1\%$, so the approximtion is valid.

13.70 ***See Section 13.8.***

$HClO_4$ $[ClO_3OH]$ has three oxygen atoms that are not bonded to hydrogen whereas H_2SeO_4 $[SeO_2(OH)_2]$ has just two. Since oxygen is more electronegative than the central atoms in these acids, each oxygen atom that is not bonded to hydrogen withdraws electron density from the central atom, and it in turn withdraws electron density from the O-H bonds. The higher the number of oxygen atoms not bonded to hydrogen, the weaker the O-H bonds and the easier they are to ionize in water.

Chapter 14: Acid/Base Reactions

14.2 ***See Section 14.1 and Example 14.1.***

$H_2SO_4(aq) + 2NaOH(aq) \rightarrow Na_2SO_4(aq) + 2H_2O(l)$

Strategy: L NaOH soln → mol NaOH → mol H_2SO_4 → M H_2SO_4

$$? \text{ mol } H_2SO_4 = 0.03177 \text{ L NaOH soln} \times \frac{0.102 \text{ mol NaOH}}{1 \text{ L NaOH soln}} \times \frac{1 \text{ mol } H_2SO_4}{2 \text{ mol NaOH}} = 0.00162 \text{ mol } H_2SO_4$$

$$? M \; H_2SO_4 = \frac{0.00162 \text{ mol } H_2SO_4}{0.01000 \text{ L } H_2SO_4 \text{ soln}} = \mathbf{0.162} \; \boldsymbol{M}$$

14.4 ***See Section 14.1 and Example 14.1.***

$H_2A(aq) + 2KOH(aq) \rightarrow K_2A(aq) + 2H_2O(l)$

Strategy: g H_2A → mol H_2A → mol KOH → L KOH soln

$$? \text{L KOH soln} = 24.93 \text{ g } H_2A \times \frac{1 \text{ mol } H_2A}{124.0 \text{ g } H_2A} \times \frac{2 \text{ mol KOH}}{1 \text{ mol } H_2A} \times \frac{1 \text{ L KOH soln}}{0.221 \text{ mol KOH}} = \mathbf{1.82 \text{ L KOH soln}}$$

14.6 ***See Section 14.1 and Example 14.1.***

(a) $HCl(aq) + KOH(aq) \rightarrow KCl(aq) + H_2O(l)$

Strategy: L KOH soln → mol KOH → mol HCl → L HCl soln → mL HCl soln

$$? \text{ mL HCl soln} = 0.0100 \text{ L KOH soln} \times \frac{0.150 \text{ mol KOH}}{1 \text{ L KOH soln}} \times \frac{1 \text{ mol HCl}}{1 \text{ mol KOH}} \times \frac{1 \text{ L HCl soln}}{0.100 \text{ mol HCl}} \times \frac{10^3 \text{ mL HCl soln}}{1 \text{ L HCl soln}}$$
$$= \mathbf{15.0 \text{ mL HCl soln}}$$

(b) $2HCl(aq) + Ba(OH)_2(aq) \rightarrow BaCl_2(aq) + 2H_2O(l)$

Strategy: L $Ba(OH)_2$ soln → mol $Ba(OH)_2$ → mol HCl → L HCl soln → mL HCl soln

$$? \text{ L HCl soln} = 0.2500 \text{ L } Ba(OH)_2 \text{ soln} \times \frac{0.00520 \text{ mol } Ba(OH)_2}{1 \text{ L } Ba(OH)_2 \text{ soln}} \times \frac{2 \text{ mol HCl}}{1 \text{ mol } Ba(OH)_2} \times \frac{1 \text{ L HCl soln}}{0.100 \text{ mol HCl}}$$

$$\times \frac{10^3 \text{ mL HCl soln}}{1 \text{ L HCl soln}} = \mathbf{26.0 \text{ mL HCl soln}}$$

(c) $HCl(aq) + NH_3(aq) \rightarrow NH_4Cl(aq)$

Strategy: L NH_3 soln → mol NH_3 → mol HCl → L HCl soln → mL HCl soln

$$? \text{ L HCl soln} = 0.1000 \text{ L NH}_3 \text{ soln} \times \frac{0.100 \text{ mol NH}_3}{1 \text{ L NH}_3 \text{ soln}} \times \frac{1 \text{ mol HCl}}{1 \text{ mol NH}_3} \times \frac{1 \text{ L HCl soln}}{0.100 \text{ mol HCl}} \times \frac{10^3 \text{ mL HCl soln}}{1 \text{ L HCl soln}}$$

$$= \mathbf{100 \text{ mL HCl soln}}$$

14.8	***See Section 14.1 and Example 14.1.***

The key to solving this problem is recognizing that $\%N = \frac{g\ N}{g \text{ plant material}} \times 100\%$.
To calculate g N, we need to know moles NH_3 released by the plant material.
Since $NH_3(g) + HCl(aq) \rightarrow NH_4Cl(aq)$, mol NH_3 released = mol HCl reacted with NH_3.
Furthermore, since $HCl(aq) + NaOH(aq) \rightarrow NaCl(aq) + H_2O(l)$,
mol HCl remaining after reaction with NH_3 = mol NaOH required for titration.
Hence, mol NH_3 released = mol HCl reacted with NH_3
= (mol HCl initial - mol HCl remaining)
= (mol HCl initial - mol NaOH required).
This gives

$$? \text{ mol HCl initial} = 0.1000 \text{ L HCl soln} \times \frac{0.121 \text{ mol HCl}}{1 \text{ L HCl soln}} = 0.0121 \text{ mol HCl}$$

$$? \text{ mol NaOH required} = 0.03422 \text{ L NaOH soln} \times \frac{0.118 \text{ mol NaOH}}{1 \text{ L NaOH soln}} = 0.00404 \text{ mol NaOH}$$

$$? \text{ mol HCl reacted with NH}_3 = 0.0121 \text{ mol HCl initial} - 0.00404 \text{ mol HCl remaining} = 0.0081 \text{ mol HCl}$$

$$? \text{ mol NH}_3 \text{ released} = \text{mol HCl reacted with NH}_3 = 0.0081 \text{ mol NH}_3$$

$$? \text{ g NH}_3 = 0.0081 \text{ mol NH}_3 \times \frac{14.0 \text{ g N}}{1 \text{ mol NH}_3} = 0.11 \text{ g N}$$

$$\%N = \frac{g\ N}{g \text{ sample}} \times 100\% \qquad \%N = \frac{0.11 \text{ g N}}{21.34 \text{ g plant material}} \times 100\% = \mathbf{0.52\ \%N}$$

14.10	***See Section 14.1 and Example 14.2.***

Overall: $HNO_3(aq) + KOH(aq) \rightarrow KNO_3(aq) + H_2O(l)$
Net ionic: $H_3O^+(aq) + OH^-(aq) \rightarrow 2H_2O(l)$

1. **0. mL KOH soln added**:

$$? \text{ mmol H}_3\text{O}^+ = 50.00 \text{ mL HNO}_3 \text{ soln} \times \frac{0.250 \text{ mmol H}_3\text{O}^+}{1 \text{ ml HNO}_3 \text{ soln}} = 12.5 \text{ mmol H}_3\text{O}^+$$

$$\left[\text{H}_3\text{O}^+\right] = \frac{12.5 \text{ mmol H}_3\text{O}^+}{50.00 \text{ mL}} = 0.250\ M, \text{ and pH} = -\log(0.250) = \mathbf{0.60}.$$

2. **12.50 mL KOH soln added**:

$$? \text{ mmol OH}^- \text{ added} = 12.50 \text{ mL KOH soln} \times \frac{0.500 \text{ mmol OH}^-}{1 \text{ mL KOH soln}} = 6.25 \text{ mmol OH}^-$$

$V_T = 50.00 \text{ mL} + 12.50 \text{ mL} = 62.50 \text{ mL}$

	H_3O^+	OH^-	H_2O
s, mmol	12.5	6.25	excess
R, mmol	-6.25	−6.25	+12.5
f, mmol	6.25	0.	excess
c, *M*	0.100	0.	excess

pH = -log(0.100) = **1.00**

3. **25.00 mL KOH soln added**:

$$? \text{ mmol OH}^- \text{ added} = 25.00 \text{ mL KOH soln} \times \frac{0.500 \text{ mmol OH}^-}{1 \text{ mL KOH soln}} = 12.5 \text{ mmol OH}^-$$

$V_T = 50.00 \text{ mL} + 25.00 \text{ mL} = 75.00 \text{ mL}$

	H_3O^+	OH^-	H_2O
s, mmol	12.5	12.5	excess
R, mmol	-12.5	−12.5	+25.0
f, mmol	0.	0.	excess
c, *M*	1.0×10^{-7} *	1.0×10^{-7} *	excess

* Concentrations due to autoionization of water at 25° C

pH = -log(1.0×10^{-7}) = **7.00**

4. **40.00 mL KOH soln added**:

$$? \text{ mmol OH}^- \text{ added} = 40.00 \text{ mL KOH soln} \times \frac{0.500 \text{ mmol OH}^-}{1 \text{ mL KOH soln}} = 20.0 \text{ mmol OH}^-$$

$V_T = 50.00 \text{ mL} + 40.00 \text{ mL} = 90.00 \text{ mL}$

	H_3O^+	OH^-	H_2O
s, mmol	12.5	20.0	excess
R, mmol	-12.5	−12.5	+25.0
f, mmol	0.0	7.5	excess
c, *M*	0.	0.083	excess

pOH = -log(0.083) = 1.08 pH = 14.00 - 1.08 = **12.92**

Summary:

mL KOH soln	pH
0.	0.60
12.50	1.00
25.00	7.00
40.00	12.92

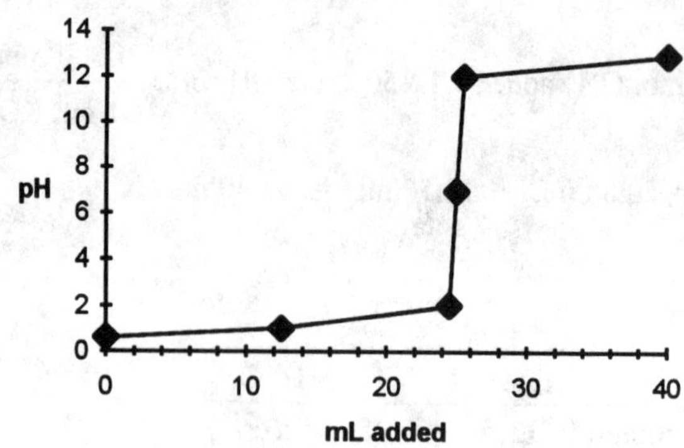

14.12 ***See Section 14.1 and Example 14.2.***

Overall: $HNO_3(aq) + LiOH(aq) \rightarrow LiNO_3(aq) + H_2O(l)$
Net ionic: $H_3O^+(aq) + OH^-(aq) \rightarrow 2H_2O(l)$

1. **0. mL HNO_3 soln added**:

$$? \text{ mmol } OH^- = 1.00 \text{ mL LiOH soln} \times \frac{0.240 \text{ mmol } OH^-}{1 \text{ mL LiOH soln}} = 0.240 \text{ mmol } OH^-$$

$$\left[OH^-\right] = \frac{0.240 \text{ mmol } OH^-}{1.00 \text{ mL}} = 0.240\ M$$

pOH = -log(0.240) = 0.620 pH = 14.00 - 0.620 = **13.38**

2. **0.25 mL HNO_3 soln added**:

$$? \text{ mmol } H_3O^+ \text{ added} = 0.25 \text{ mL } HNO_3 \text{ soln} \times \frac{0.200 \text{ mmol } H_3O^+}{1 \text{ mL } HNO_3 \text{ soln}} = 5.0 \times 10^{-2} \text{ mmol } H_3O^+$$

V_T = 1.00 mL + 0.25 mL = 1.25 mL

	H_3O^+	OH^-	H_2O
s, mmol	0.050	0.240	excess
R, mmol	-0.050	-0.050	+0.100
f, mmol	0.	0.190	excess
c, *M*	0.	0.152	excess

pOH = -log(0.152) = 0.818 pH = 14.00 - 0.818 = **13.18**

3. **0.50 mL HNO_3 soln added**:

$$? \text{ mmol } H_3O^+ \text{ added} = 0.50 \text{ mL } HNO_3 \text{ soln} \times \frac{0.200 \text{ mmol } H_3O^+}{1 \text{ mL } HNO_3 \text{ soln}} = 0.10 \text{ mmol } H_3O^+$$

V_T = 1.00 mL + 0.50 mL = 1.50 mL

	H_3O^+	OH^-	H_2O
s, mmol	0.10	0.240	excess
R, mmol	-0.10	-0.10	+0.20
f, mmol	0.	0.14	excess
C, *M*	0.	0.093	excess

pOH = -log(0.093) = 1.03   pH = 14.00 - 1.03 = **12.97**

4. **1.20 mL HNO_3 soln added**:

$$? \text{ mmol } H_3O^+ \text{ added} = 1.20 \text{ mL } HNO_3 \text{ soln} \times \frac{0.200 \text{ mmol } H_3O^+}{1 \text{ mL } HNO_3 \text{ soln}} = 0.240 \text{ mmol } H_3O^+$$

V_T = 1.00 mL + 1.20 mL = 2.20 mL

	H_3O^+	OH^-	H_2O
s, mmol	0.240	0.240	excess
R, mmol	-0.240	-0.240	+0.480
f, mmol	0.	0.	excess
c, *M*	1.0×10^{-7} *	1.0×10^{-7} *	excess

* Concentrations due to autoionization of water at 25° C.

pH = -log(1.0 x 10^{-7}) = **7.00**

5. **1.50 mL HNO_3 soln added**:

$$? \text{ mmol } H_3O^+ \text{ added} = 1.50 \text{ mL } HNO_3 \text{ soln} \times \frac{0.200 \text{ mmol } H_3O^+}{1 \text{ mL } HNO_3 \text{ soln}} = 0.300 \text{ mmol } H_3O^+$$

V_T = 1.00 mL + 1.50 mL = 2.50 mL

	H_3O^+	OH^-	H_2O
s, mmol	0.300	0.240	excess
R, mmol	-0.240	-0.240	+0.480
f, mmol	0.060	0.	excess
c, *M*	0.024	0.	excess

pH = -log(0.024) = **1.62**

Summary:

mL HNO_3 soln	pH
0.	13.38
0.25	13.18
0.50	12.97
1.20	7.00
1.50	1.62

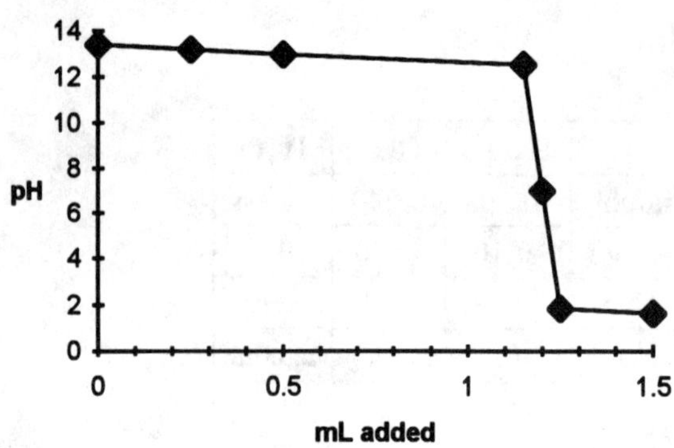

14.14 ***See Tables 13.6 and 7, Section 14.2 and Examples 14.4 and 5.***

(a) $pH = pK_a + \log\left(\frac{C_b}{C_a}\right)$ The K_a for HF is 3.5 x 10^{-4}, and its pK_a is $-\log(3.5 \times 10^{-4}) = 3.46$.

$? C_b = \frac{15.45 \text{ g KF}}{0.1000 \text{ L soln}} \times \frac{1 \text{ mol KF}}{58.10 \text{ g KF}} = 2.66\ M$ $pH = 3.46 + \log\left(\frac{2.66}{0.850}\right) = \mathbf{3.96}$

(b) $pH = pK_a + \log\left(\frac{C_b}{C_a}\right)$ The K_b for NH_3 is 1.8 x 10^{-5}.

Hence, $K_a NH_4^+ = \frac{K_w}{K_b NH_3}$ $K_a NH_4^+ = \frac{1.0 \times 10^{-14}}{1.8 \times 10^{-5}} = 5.6 \times 10^{-10}$ $pK_a NH_4^+ = -\log(5.6 \times 10^{-10}) = 9.26$

$? C_a = \frac{45.00 \text{ g } NH_4Cl}{0.2500 \text{ L soln}} \times \frac{1 \text{ mol } NH_4Cl}{53.49 \text{ g } NH_4Cl} = 3.36\ M$ $pH = 9.26 + \log\left(\frac{0.455}{3.37}\right) = \mathbf{8.39}$

14.16 ***See Tables 13.6 and 7, Section 14.2 and Examples 14.4 and 5.***

(a) $pH = pK_a + \log\left(\frac{C_b}{C_a}\right)$ The K_a for HCOOH is 1.8 x 10^{-4}, and its pK_a is $-\log(1.8 \times 10^{-4}) = 3.74$.

V_T = 100.0 mL + 200.0 mL = 300.0 mL

Solving $M(\text{con}) \times V(\text{con}) = M(\text{dil}) \times V(\text{dil})$ for M(dil) gives $M(\text{dil}) = M(\text{con}) \times \frac{V(\text{con})}{V(\text{dil})}$. Hence,

$C_b = 0.100\ M \times \frac{200.0 \text{ mL}}{300.0 \text{ mL}} = 0.0667\ M$ and $C_a = 0.800\ M \times \frac{100.0 \text{ mL}}{300.0 \text{ mL}} = 0.267\ M$

$pH = 3.74 + \log\left(\frac{0.0667}{0.267}\right) = \mathbf{3.14}$

(b) $pH = pK_a + \log\left(\dfrac{C_b}{C_a}\right)$ The K_b for NH_3 is 1.8 x 10^{-5}.

Hence, $K_a NH_4^+ = \dfrac{K_w}{K_b\ NH_3}$ $K_a NH_4^+ = \dfrac{1.0 \times 10^{-14}}{1.8 \times 10^{-5}} = 5.6 \times 10^{-10}$ $pK_a NH_4^+ = -\log(5.6 \times 10^{-10}) = 9.25$

$V_T = 300.0\text{ mL} + 200.0\text{ mL} = 500.0\text{ mL}$

Solving $M(\text{con}) \times V(\text{con}) = M(\text{dil}) \times V(\text{dil})$ for M(dil) gives $M(\text{dil}) = M(\text{con}) \times \dfrac{V(\text{con})}{V(\text{dil})}$. Hence,

$C_b = 0.350\ M \times \dfrac{300.0\text{ mL}}{500.0\text{ mL}} = 0.210\ M$ and $C_a = 0.150\ M \times \dfrac{200.0\text{ mL}}{500.0\text{ mL}} = 0.0600\ M$

$pH = 9.25 + \log\left(\dfrac{0.210}{0.0600}\right) = \mathbf{9.79}$

14.18 ***See Table 13.6, Section 14.2 and Example 14.5.***

$pH = pK_a + \log\left(\dfrac{n_b}{n_a}\right)$

The K_a for HF is 3.5 x 10^{-4}, and its pK_a is $-\log(3.5 \times 10^{-4}) = 3.46$.

$?\ n_b = 0.750\text{ L NaF soln} \times \dfrac{0.200\text{ mol NaF}}{1\text{ L NaF soln}} = 0.150\text{ mol NaF}$

Substituting into the Henderson-Hasselbach equation gives $3.95 = 3.46 + \log\left(\dfrac{0.150}{n_a}\right)$

$\log\left(\dfrac{0.150}{n_a}\right) = 3.95 - 3.46 = 0.49$ $\dfrac{0.150}{n_a} = 10^{0.49} = \text{inv}\log(0.49) = 3.09$

$n_a = \dfrac{0.150}{3.09} = 0.0485\text{ mol}$ $?\text{ L HF soln} = 0.0485\text{ mol HF} \times \dfrac{1\text{ L HF soln}}{0.500\text{ mol HF}} = \mathbf{0.0970\ L\ HF\ soln}$

14.20 ***See Table 13.6, Section 14.2 and Examples 14.4 and 6.***

(a) $pH = pK_a + \log\left(\dfrac{C_b}{C_a}\right) = pK_a + \log\left(\dfrac{n_b}{n_a}\right)$

The K_a for CH_3COOH is 1.8 x 10^{-5}, and its pK_a is $-\log(1.8 \times 10^{-5}) = 4.74$.

Initial $pH = 4.74 + \log\left(\dfrac{1.00}{0.500}\right) = \mathbf{5.04}$

When NaOH is added, OH^- reacts with CH_3COOH: $OH^- + CH_3COOH \rightarrow H_2O + CH_3COO^-$

$$? \text{ mmol } OH^- \text{ added} = 1.00 \text{ mL} \times \frac{0.100 \text{ mmol}}{1 \text{ mL}} = 0.100 \text{ mmol}$$

$$? \text{ starting mmol } CH_3COOH = 100.0 \text{ mL} \times \frac{0.500 \text{ mmol}}{1 \text{ mL}} = 50.0 \text{ mmol}$$

$$? \text{ starting mmol } CH_3COO^- = 100.0 \text{ mL} \times \frac{1.00 \text{ mmol}}{1 \text{ mL}} = 100.0 \text{ mmol}$$

$V_T = 100.0 \text{ mL} + 1.00 \text{ mL} = 101.0 \text{ mL}$

	OH^-	CH_3COOH	CH_3COO^-	H_2O
s, mmol	0.100	50.00	100.0	excess
R, mmol	-0.100	-0.100	0.100	+0.100
f, mmol	0.	49.90	100.1	excess

$$\text{Final pH} = 4.74 + \log\left(\frac{100.1}{49.90}\right) = \mathbf{5.04}$$

Change in pH = 5.04 - 5.04 = **0.00**

(b) Initial pH = **7.00**, assuming we are using pure water at 25°C.

$$\text{Final } \left[OH^-\right] = \frac{0.100 \text{ mmol}}{101.00 \text{ mL}} = 9.90 \times 10^{-4} \, M$$

pOH = -log(9.90×10^{-4}) = 3.00 and pH = 14.00 - 3.00 = **11.00**

Change in pH = 11.00 - 7.00 = **4.00**

14.22 ***See Table 13.6 and Section 14.2.***

$$pH = pK_a + \log\left(\frac{C_b}{C_a}\right) = pK_a + \log\left(\frac{n_b}{n_a}\right)$$

The K_a for HCOOH is 1.8×10^{-4}, and its pK_a is -log(1.8×10^{-4}) = 3.74.

$$? \text{ mmol strong acid or strong base to be added} = 1.00 \text{ mL} \times \frac{0.100 \text{ mmol}}{1 \text{ mL}} = 0.100 \text{ mmol}$$

For pH = 3.80: $3.80 = 3.74 + \log\left(\frac{n_b}{n_a}\right)$

$$\log\left(\frac{n_b}{n_a}\right) = 3.80 - 3.74 = 0.06 \qquad \frac{n_b}{n_a} = 10^{0.06} = \text{invlog}(0.06) = 1.15$$

To achieve pH = 3.90, n_b must increase by 0.100 and n_a must decrease by 0.100 mmol due to addition of strong base and its reaction with the acid component of the buffer. This gives:

$$3.90 = 3.74 + \log\left(\frac{n_b + 0.100}{n_a - 0.100}\right)$$

$$\log\left(\frac{n_b + 0.100}{n_a - 0.100}\right) = 3.90 - 3.74 = 0.16 \qquad \frac{n_b + 0.100}{n_a - 0.100} = 10^{0.16} = \text{inv}\log(0.16) = 1.45$$

Thus, we have $\frac{n_b}{n_a} = 1.15$ and $\frac{n_b + 0.100\text{ mmol}}{n_a - 0.100\text{ mmol}} = 1.45$

Solving the first expression for n_b and substituting into the second expression gives

$$\frac{1.15n_a + 0.100\text{ mmol}}{n_a - 0.100\text{ mmol}} = 1.45 \qquad 1.15n_a + 0.100\text{ mmol} = 1.45n_a - 0.145\text{ mmol}$$

$$0.30n_a = 0.245\text{ mmol} \qquad n_a = 0.82\text{ mmol}$$

Thus, $n_b = 1.15n_a = (1.15)(0.82) = .94$ mmol

The minimum concentration of formic acid is therefore $\frac{0.82 \times 10^{-3}\text{ mol}}{0.5010\text{ L}} = \mathbf{1.64 \times 10^{-3}}$ ***M*** and that of sodium formate is therefore $\frac{0.94 \times 10^{-3}\text{ mol}}{0.5010\text{ L}} = \mathbf{1.88 \times 10^{-3}}$ ***M***.

14.24 ***See Table 13.6, Section 14.2 and Examples 14.4 and 5.***

The objective is to calculate the ratio n_b / n_a that is needed to give a pH of 5.00 using the CH_3COOH / CH_3COO^- buffer system and then use these numbers of moles to calculate the concentrations of acetic acid and sodium acetate in solution. The K_a for CH_3COOH is 1.8×10^{-5}, and its pK_a is 4.74. Substituting into the Henderson-Hasselbach equation yields

$$5.00 = 4.74 + \log\left(\frac{n_b}{n_a}\right), \quad \log\left(\frac{n_b}{n_a}\right) = 5.00 - 4.74 \text{ and } \frac{n_b}{n_a} = 10^{0.26} = \text{inv}\log(0.26) = 1.8$$

This means the initial n_b / n_a ratio of 0.0500 mol/ 0.0500 mol or 1.00 must be adjusted to 1.8 to give a pH of 5.00. Hence,

$$\frac{0.0500 + y}{0.0500 - y} = 1.8, \quad 2.8y = 0.0400 \quad \text{and} \quad y = 0.014.$$

This gives $[\mathbf{CH_3COO^-}] = \frac{(0.0500 + 0.014)\text{ mol}}{1.00\text{ L}} = \mathbf{0.064}$ ***M***

and $[\mathbf{CH_3COOH}] = \frac{(0.0500 - 0.014)\text{ mol}}{1.00\text{ L}} = \mathbf{0.036}$ ***M***.

14.26 ***See Table 13.6 and Section 14.3.***

Overall: $HCOOH(aq) + NaOH(aq) \rightarrow H_2O(l) + NaHCOO(aq)$
Net ionic: $HCOOH(aq) + OH^-(aq) \rightarrow H_2O(l) + HCOO^-(aq)$

1. **0. NaOH soln added**:

$$HCOOH(aq) + H_2O(l) \rightleftarrows H_3O^+(aq) + HCOO^-(aq)$$

The K_a for HCOOH is 1.8×10^{-4}, and it pK_a is $-\log(1.8 \times 10^{-4}) = 3.74$

Conc, *M*	HCOOH	H_3O^+	$HCOO^-$
initial	0.250	0.	0.
Change	−y	+y	+y
equilibrium	0.250 − y	y	y

$$K_a = \frac{[H_3O^+][HCOO^-]}{[HCOOH]} \qquad K_a = \frac{(y)(y)}{(0.250-y)} \cong \frac{(y)(y)}{0.250} = 1.8 \times 10^{-4}$$

$[H_3O^+] = y = 6.7 \times 10^{-3}$ and pH = $-\log(6.7 \times 10^{-3})$ = **2.17**.

Approximation check: $\left(\frac{6.7 \times 10^{-3}}{0.250}\right) \times 100\% = 2.68\%$, so the approximation is valid.

2. **5.00 mL NaOH soln added**:

$$? \text{ starting mmol HCOOH} = 20.00 \text{ mL HCOOH soln} \times \frac{0.250 \text{ mmol HCOOH}}{1 \text{ mL HCOOH soln}} = 5.00 \text{ mmol HCOOH}$$

$$? \text{ mmol OH}^- \text{ added} = 5.00 \text{ mL NaOH soln} \times \frac{0.500 \text{ mmol OH}^-}{1 \text{ mL NaOH soln}} = 2.50 \text{ mmol OH}^-$$

V_T = 20.00 mL + 5.00 mL = 25.00 mL

	HCOOH	OH^-	$HCOO^-$
s, mmol	5.00	2.50	0.
R, mmol	-2.50	−2.50	+2.50
f, mmol	2.50	0.	2.50
C, *M*	0.100	0.	0.100

$$pH = pK_a + \log\left(\frac{C_b}{C_a}\right) \qquad pH = 3.74 + \log\left(\frac{0.100}{0.100}\right) = \mathbf{3.74}$$

3. **10.00 mL NaOH added**:

$$? \text{ mmol OH}^- \text{ added} = 10.00 \text{ mL NaOH soln} \times \frac{0.500 \text{ mmol OH}^-}{1 \text{ mL NaOH soln}} = 5.00 \text{ mmol OH}^-$$

$V_T = 20.00 \text{ mL} + 10.00 \text{ mL} = 30.00 \text{ mL}$

	HCOOH	OH^-	$HCOO^-$
s, mmol	5.00	5.00	0.
R, mmol	−5.00	−5.00	+5.00
f, mmol	0.	0.	5.00
C, *M*	0.	0.	0.167

The pH at the equivalence point is due to the reaction of $HCOO^-$ with H_2O:

$$HCOO^- + H_2O \rightleftarrows HCOOH + OH^-$$

$$K_b \text{ HCOO}^- = \frac{K_w}{K_a \text{ HCOOH}} \qquad K_b \text{ HCOO}^- = \frac{1.0 \times 10^{-14}}{1.8 \times 10^{-4}} = 5.6 \times 10^{-11}$$

Conc, *M*	$HCOO^-$	HCOOH	OH^-
initial	0.167	0.	0.
Change	−y	+y	+y
equilibrium	0.167 − y	y	y

$$K_b \text{ HCOO}^- = \frac{[\text{HCOOH}][\text{OH}^-]}{[\text{HCOO}^-]} \qquad K_b \text{ HCOO}^- = \frac{(y)(y)}{(0.167-y)} \cong \frac{(y)(y)}{0.167} = 5.6 \times 10^{-11}$$

$[OH^-] = y = 3.1 \times 10^{-6}$, $pOH = -\log(3.1 \times 10^{-6}) = 5.51$ and $pH = 14.00 - 5.51 =$ **8.49**

4. **15.00 mL NaOH added**:

$$? \text{ mmol OH}^- \text{ added} = 15.00 \text{ mL NaOH soln} \times \frac{0.500 \text{ mmol OH}^-}{1 \text{ mL NaOH soln}} = 7.50 \text{ mmol OH}^-$$

$V_T = 20.00 \text{ mL} + 15.00 \text{ mL} = 35.00 \text{ mL}$

	HCOOH	OH^-	$HCOO^-$
s, mmol	5.00	7.50	0.
R, mmol	−5.00	−5.00	+5.00
f, mmol	0.	2.50	5.00
C, *M*	0.	0.071	0.143

The excess of strong base determines the pH.

$[OH^-] = 0.071\ M$, pOH = -log(0.071) = 1.15 and pH = 14.00 - 1.15 = **12.85**

Summary:

mL NaOH soln	pH
0.	2.17
5.00	3.74
10.00	8.49
15.00	12.85

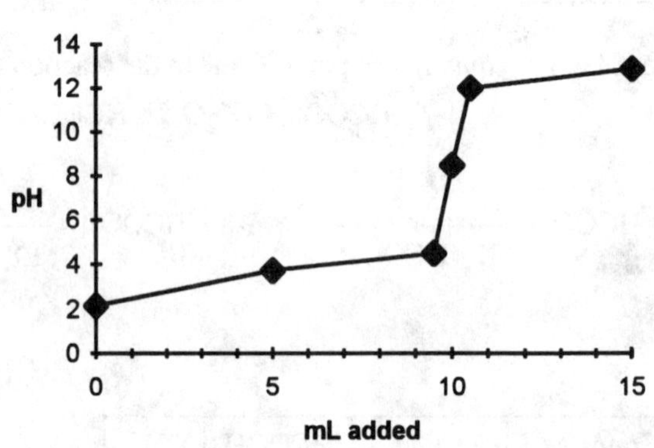

14.28 ***See Table 13.6 and Section 14.3.***

Overall: $HF(aq) + NaOH(aq) \rightarrow H_2O(l) + NaF(aq)$

Net ionic: $HF(aq) + OH^-(aq) \rightarrow H_2O(l) + F^-(aq)$

1. **0. mL NaOH soln added**:

$$HF(aq) + H_2O(l) \rightleftarrows H_3O^+(aq) + F^-(aq)$$

The K_a for HF is 3.5×10^{-4}, and its pK_a is $-\log(3.5 \times 10^{-4}) = 3.46$

Conc, *M*	HF	H_3O^+	F^-
initial	0.230	0.	0.
Change	−y	+y	+y
equilibrium	0.230 − y	y	y

$$K_a = \frac{[H_3O^+][F^-]}{[HF]} \qquad K_a = \frac{(y)(y)}{(0.230 - y)} \cong \frac{(y)(y)}{0.230} = 3.5 \times 10^{-4}$$

$[H_3O^+] = y = 9.0 \times 10^{-3}$ pH = **2.05**

Approximation check: $\left(\frac{9.0 \times 10^{-3}}{0.230}\right) \times 100\% = 3.9\%$, so the approximation is valid.

2. **50% of amount of base needed to reach the equivalence point:**

$$? \text{ starting mmol HF} = 100.0 \text{ mL HF soln} \times \frac{0.230 \text{ mmol HF}}{1 \text{ mL HF soln}} = 23.0 \text{ mmol HF}$$

$$? \text{ mmol HF at } 50\% = \text{mmol F}^- = (0.50)(23.0 \text{ mmol}) = 11.5 \text{ mmol}$$

$$\text{pH} = \text{pK}_a + \log\left(\frac{\text{mmol b}}{\text{mmol a}}\right) \qquad \text{pH} = 3.46 + \log\left(\frac{11.5}{11.5}\right) = \mathbf{3.46}$$

3. **95% of amount of base needed to reach the equivalence point:**

$$? \text{ mmol HF} = (0.05)(23.0 \text{ mmol}) = 1.15 \text{ mmol HF}$$

$$? \text{ mmol F}^- = (0.95)(23.0 \text{ mmol}) = 21.9 \text{ mmol F}^-$$

$$\text{pH} = \text{pK}_a + \log\left(\frac{\text{mmol b}}{\text{mmol a}}\right) \qquad \text{pH} = 3.46 + \log\left(\frac{21.9}{1.15}\right) = \mathbf{4.74}$$

4. **100% of amount of base needed to reach the equivalence point:**

$$? \text{ mmol HF} = (0.00)(23.0 \text{ mmol}) = 0. \text{ mmol HF}$$

$$? \text{ mmol F}^- = (1.00)(23.0 \text{ mmol}) = 23.0 \text{ mmol F}^-$$

The pH at the equivalence point is due to the reaction of F^- with H_2O:

$$F^-(aq) + H_2O(l) \rightleftarrows HF(aq) + OH^-(aq)$$

$$K_b\ F^- = \frac{K_w}{K_a\ HF} \qquad K_b\ F^- = \frac{1.0 \times 10^{-14}}{3.5 \times 10^{-4}} = 2.9 \times 10^{-11}$$

We need to know $[F^-]$ at the equivalence point and therefore need to know the total volume of the solution at the equivalence point. Hence, we need to know the volume of NaOH containing 23.0 mmol OH^-, since HF and OH^- react in a 1:1 ratio.

$$? \text{ L NaOH soln} = 23.0 \text{ mmol OH}^- \times \frac{1 \text{ mmol NaOH}}{1 \text{ mmol OH}^-} \times \frac{1 \text{ mL NaOH soln}}{0.500 \text{ mmol NaOH}} = 46.0 \text{ mL NaOH soln}$$

$V_T = 100.0 \text{ mL} + 46.0 \text{ mL} = 146.0 \text{ mL}$

$$[F^-] \text{ at the equivalence point} = \frac{23.0 \text{ mmol F}^-}{146.0 \text{ mL soln}} = 0.158\ M$$

Conc, M	F^-	HF	OH^-
initial	0.158	0.	0.
Change	$-y$	$+y$	$+y$
equilibrium	$0.158 - y$	y	y

$$K_b = \frac{[HF][OH^-]}{[F^-]} \qquad K_b = \frac{(y)(y)}{(0.158-y)} \cong \frac{(y)(y)}{0.158} = 2.9 \times 10^{-11}$$

$[OH^-] = y = 2.1 \times 10^{-6}$, pOH = -log(2.1 x 10^{-6}) = 5.68 and pH = 14.00 - 5.68 = **8.32**

5. **105% of amount of base needed to reach equivalence point**:

? mmol excess OH^- = (0.05)(23.0 mmol) = 1.15 mmol

? mmol total OH^- = (1.05)(23.0 mmol) = 24.2 mmol

$$? \text{ mL NaOH soln added} = 24.2 \text{ mmol } OH^- \times \frac{1 \text{ mmol NaOH}}{1 \text{ mmol } OH^-} \times \frac{1 \text{ mL NaOH soln}}{0.500 \text{ mmol NaOH}} = 48.4 \text{ mL NaOH soln}$$

V_T = 100.0 mL + 48.4 mL = 148.4 mL

$$[OH^-] = \frac{1.15 \text{ mmol}}{148.4 \text{ mL soln}} = 7.75 \times 10^{-3}\ M$$

pOH = -log(7.75 x 10^{-3}) = 2.11 and pH = 14.00 - 2.11 = **11.89**

Summary:

%OH added	pH
0.	2.05
50	3.46
95	4.74
100	8.32
105	11.89

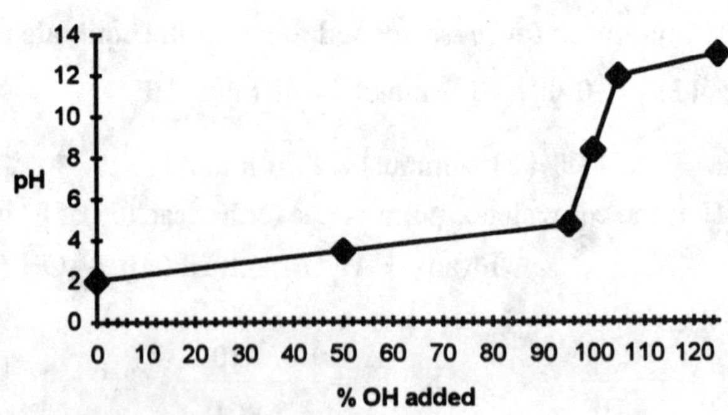

14.30 ***See Table 13.7 and Section 4.3.***

Overall: $C_5H_5N(aq) + HCl(aq) \rightarrow C_5H_5NH^+(aq) + Cl^-(aq)$
Net ionic: $C_5H_5N(aq) + H_3O^+(aq) \rightarrow C_5H_5NH^+(aq) + H_2O(l)$

1. **0. mL HCl added**:

$C_5H_5N(aq) + H_2O(l) \rightleftarrows C_5H_5NH^+(aq) + OH^-(aq)$ $\qquad K_b = 1.8 \times 10^{-9}$

Conc, M	C_5H_5N	$C_5H_5NH^+$	OH^-
initial	0.200	0.	0.
Change	−y	+y	+y
equilibrium	0.200 − y	y	y

$$K_b = \frac{[C_5H_5NH^+][OH^-]}{[C_5H_5N]} \qquad K_b = \frac{(y)(y)}{(0.200 - y)} \cong \frac{(y)(y)}{0.200} = 1.8 \text{ x } 10^{-9}$$

$[OH^-] = y = 1.9 \text{ x } 10^{-5}$, pOH = -log(1.9 x 10^{-5}) = 4.72 and pH = 14.00 - 4.72 = **9.28**

Approximation check: $\left(\frac{1.9 \text{ x } 10^{-5}}{0.200}\right) \times 100\% = 9.5 \text{ x } 10^{-3}\ \%$, so the approximation is valid.

2. **15.00 mL HCl added**:

$$? \text{ starting mmol } C_5H_5N = 30.00 \text{ mL } C_5H_5N \text{ soln} \times \frac{0.200 \text{ mmol } C_5H_5N}{1 \text{ mL } C_5H_5N \text{ soln}} = 6.00 \text{ mmol } C_5H_5N$$

$$? \text{ mmol } H_3O^+ \text{ added} = 15.00 \text{ mL HCl soln} \times \frac{0.200 \text{ mmol } H_3O^+}{1 \text{ mL HCl soln}} = 3.00 \text{ mmol } H_3O^+$$

V_T = 30.00 mL + 15.00 mL = 45.00 mL

	C_5H_5N	H_3O^+	$C_5H_5NH^+$
s. mmol	6.00	3.00	0.
R, mmol	–3.00	–3.00	+3.00
f, mmol	3.00	0.	3.00
C, *M*	0.0667	0.	0.0667

After HCl has been added, the system can be treated as an $C_5H_5NH^+$/ C_5H_5NH buffer system until the equivalence point is reached.

$$C_5H_5NH^+(aq) + H_2O(l) \rightleftarrows C_5H_5N(aq) + H_3O^+(aq)$$

$$K_a\ C_5H_5NH^+ = \frac{K_w}{K_b\ C_5H_5N} \qquad K_a\ C_5H_5NH^+ = \frac{1.0 \text{ x } 10^{-14}}{1.8 \text{ x } 10^{-9}} = 5.6 \text{ x } 10^{-6} \quad pK_a = -\log(5.6 \text{ x } 10^{-6}) = 5.25$$

$$pH = pK_a + \log\left(\frac{C_b}{C_a}\right) \qquad pH = 5.25 + \log\left(\frac{0.0667}{0.0667}\right) = \mathbf{5.25}$$

3. **30.00 mL HCl added**:

$$? \text{ mmol } H_3O^+ \text{ added} = 30.00 \text{ mL HCl soln} \times \frac{0.200 \text{ mmol } H_3O^+}{1 \text{ mL HCl soln}} = 6.00 \text{ mmol } H_3O^+$$

V_T = 30.00 mL + 30.00 mL = 60.00 mL

	C_5H_5N	H_3O^+	$C_5H_5NH^+$
s. mmol	6.00	6.00	0.
R, mmol	−6.00	−6.00	+6.00
f, mmol	0.	0.	6.00
C, *M*	0.	0.	0.100

The pH at the equivalence point is due to the reaction of $C_5H_5NH^+$ with H_2O:

$$C_5H_5NH^+(aq) + H_2O(l) \rightleftarrows C_5H_5N(aq) + H_3O^+(aq) \qquad K_a = 5.6 \times 10^{-6}$$

Conc, *M*	$C_5H_5NH^+$	C_5H_5N	H_3O^+
initial	0.100	0.	0.
Change	−y	+y	+y
equilibrium	0.100 − y	y	y

$$K_a\ C_5H_5NH^+ = \frac{[C_5H_5N][H_3O^+]}{[C_5H_5NH^+]} \qquad K_a\ C_5H_5NH^+ = \frac{(y)(y)}{(0.100-y)} \cong \frac{(y)(y)}{0.100} = 5.6 \times 10^{-6}$$

$[H_3O^+] = y = 7.5 \times 10^{-4}$ and $pH = -\log(7.5 \times 10^{-4}) = \mathbf{3.12}$

Approximation check: $\left(\frac{7.5 \times 10^{-4}}{0.100}\right) \times 100\% = 0.75\%$, so the approximation is valid.

4. **40.00 mL HCl added**:

$$? \text{ mmol } H_3O^+ \text{ added} = 40.00 \text{ mL HCl soln} \times \frac{0.200 \text{ mol } H_3O^+}{1 \text{ mL HCl soln}} = 8.00 \text{ mmol } H_3O^+$$

$V_T = 30.00 \text{ mL} + 40.00 \text{ mL} = 70.00 \text{ mL}$

	C_5H_5N	H_3O^+	$C_5H_5NH^+$
s. mmol	6.00	6.00	0.
R, mmol	−6.00	−6.00	+6.00
f, mmol	0.	2.00	6.00
C, *M*	0.	0.0286	0.0857

The excess of strong acid determines the pH: $[H_3O^+] = 0.0286$ and $pH = -\log(0.0286) = \mathbf{1.54}$

Summary:

mL HCl soln	pH
0.	9.28
15.	5.25
30.00	3.12
40.00	1.54

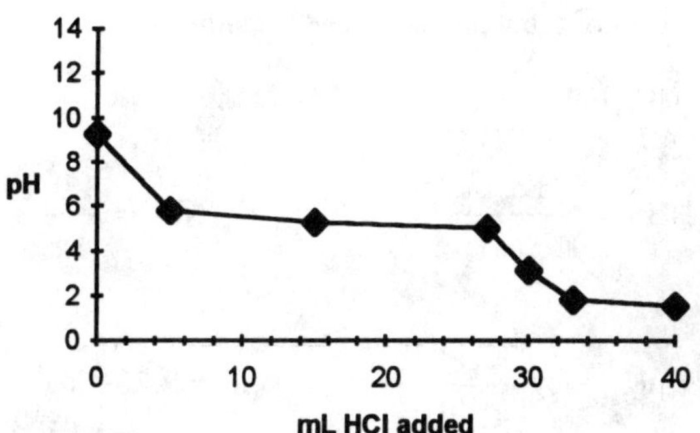

Note: Pyridine is too weak a base to show a titratrion curve with a clear inflection point.

14.32 ***See Table 13.6 and 7 and Section 14.3.***

(a)

Overall: $HCOOH(aq) + NaOH(aq) \rightarrow NaHCOO(aq) + H_2O(l)$

Net ionic: $HCOOH(aq) + OH^-(aq) \rightarrow HCOO^-(aq) + H_2O(l)$

$$? \text{ starting mmol HCOOH} = 1.00 \text{ mL HCOOH soln} \times \frac{0.150 \text{ mmol HCOOH}}{1 \text{ mL HCOOH soln}} = 0.150 \text{ mmol HCOOH}$$

$$? \text{ mmol OH}^- \text{ added} = 2.00 \text{ mL NaOH soln} \times \frac{0.100 \text{ mmol OH}^-}{1 \text{ mL NaOH soln}} = 0.200 \text{ mmol OH}^-$$

$V_T = 1.00 \text{ mL} + 2.00 \text{ mL} = 3.00 \text{ mL}$

An excess of 0.050 mmol of OH^- has been added, and this excess will determine the pH of the solution.

$$\left[OH^-\right] = \frac{0.050 \text{ mmol OH}^-}{3.00 \text{ mL}} = 0.017, \qquad pOH = -\log(0.017) = 1.77 \quad \text{and} \quad pH = 14.00 - 1.77 = \mathbf{12.23}$$

(b)

Overall: $NH_3(aq) + HI(aq) \rightarrow NH_4I(aq)$

Net ionic: $NH_3(aq) + H_3O^+(aq) \rightarrow NH_4^+(aq)$

$$? \text{ starting mmol NH}_3 = 25.00 \text{ mL NH}_3 \text{ soln} \times \frac{0.250 \text{ mmol NH}_3}{1 \text{ mL NH}_3 \text{ soln}} = 6.25 \text{ mmol NH}_3$$

$$? \text{ mmol H}_3\text{O}^+ \text{ added} = 5.00 \text{ mL HI soln} \times \frac{0.100 \text{ mmol H}_3\text{O}^+}{1 \text{ mL HI soln}} = 0.500 \text{ mmol H}_3\text{O}^+$$

$V_T = 25.00 \text{ mL} + 5.00 \text{ mL} = 30.00 \text{ mL}$

	NH_3	H_3O^+	NH_4^+
s, mmol	6.25	0.500	0.
R, mmol	–0.500	–0.500	+0.500
f, mmol	5.75	0.	0.500
C, *M*	0.192	0.	0.0167

An excess of the weak base NH_3 remains, and the system can be treated as a NH_4^+/NH_3 buffer system.

For $NH_4^+(aq) + H_2O(l) \rightleftarrows NH_3(aq)$,

$$K_a\ NH_4^+ = \frac{K_w}{K_b\ NH_3} \qquad K_a\ NH_4^+ = \frac{1.0 \times 10^{-14}}{1.8 \times 10^{-5}} = 5.6 \times 10^{-10} \quad \text{and} \quad pK_a = -\log(5.6 \times 10^{-10}) = 9.25$$

$$pH = pK_a + \log\left(\frac{C_b}{C_a}\right) \qquad pH = 9.25 + \log\left(\frac{0.192}{0.0167}\right) = \mathbf{10.31}$$

(c) Overall: $Ba(OH)_2(aq) + 2HBr(aq) \rightarrow BaBr_2(aq) + 2H_2O(l)$
Net ionic: $2OH^-(aq) + 2H_3O^+(aq) \rightarrow 4H_2O(l)$

$$?\ \text{starting mmol OH}^- = 5.00\ \text{mL Ba(OH)}_2\ \text{soln} \times \frac{2 \times (0.200\ \text{mmol OH}^-)}{1\ \text{mL Ba(OH)}_2\ \text{soln}} = 2.00\ \text{mmol OH}^-$$

$$?\ \text{mmol H}_3\text{O}^+\ \text{added} = 50.00\ \text{mL HBr soln} \times \frac{0.400\ \text{mmol H}_3\text{O}^+}{1\ \text{mL HBr soln}} = 20.0\ \text{mmol H}_3\text{O}^+$$

$V_T = 5.00\ \text{mL} + 50.00\ \text{mL} = 55.00\ \text{mL}$

An excess of 18.0 mmol of H_3O^+ has been added, and this excess will determine the pH of the solution.

$$[H_3O^+] = \frac{18.00\ \text{mmol H}_3\text{O}^+}{55.00\ \text{mL}} = 0.327 \quad \text{and} \quad pH = -\log(0.327) = \mathbf{0.48}$$

14.34 ***See Sections 14.3 and 4 and Table 14.2.***

Overall: $HLac(aq) + NaOH(aq) \rightarrow NaLac(aq) + H_2O(l)$
Net ionic: $HLac(aq) + OH^-(aq) \rightarrow Lac^-(aq) + H_2O(l)$

$$?\ \text{starting mmol HLac} = 100\ \text{mL HLac soln} \times \frac{0.100\ \text{mmol HLac}}{1\ \text{mL HLac soln}} = 10.0\ \text{mmol HLac}$$

$$?\ \text{mmol OH}^-\ \text{required to reach equivalence point} = 10.0\ \text{mmol HLac} \times \frac{1\ \text{mmol OH}^-}{1\ \text{mmol HLac}} = 10.0\ \text{mmol OH}^-$$

$$?\ \text{mL NaOH required to reach equivalence point} = 10.0\ \text{mmol OH}^- \times \frac{1\ \text{mL NaOH soln}}{0.500\ \text{mmol OH}^-} = 20.0\ \text{mL NaOH soln}$$

$V_T = 100\ \text{mL} + 20.0\ \text{mL} = 120\ \text{mL}$

	HLac	OH^-	Lac^-
s, mmol	10.0	10.0	0.
R, mmol	−10.0	−10.0	+10.0
f, mmol	0.	0.	10.0
C, *M*	0.	0.	0.0833

The pH at the equivalence point is due to the reaction of Lac^- with H_2O:

$$Lac^-(aq) + H_2O(l) \rightleftarrows HLac(aq) + OH^-(aq)$$

$$K_b\ Lac^- = \frac{K_w}{K_a\ HLac} \qquad K_b\ Lac^- = \frac{1.0 \times 10^{-14}}{8.4 \times 10^{-4}} = 1.2 \times 10^{-11}$$

Conc, *M*	Lac^-	HLac	OH^-
initial	0.0833	0.	0.
Change	−y	+y	+y
equilibrium	0.0833 − y	y	y

$$K_b\ Lac^- = \frac{[HLac][OH^-]}{[Lac^-]} \qquad K_b\ Lac^- = \frac{(y)(y)}{(0.0833-y)} \cong \frac{(y)(y)}{0.0833} = 1.2 \times 10^{-11}$$

$[OH^-] = y = 1.0 \times 10^{-6}$, pOH = $-\log(1.0 \times 10^{-6}) = 6.00$ and **pH** = 14.00 − 6.00 = **8.00**

Approximation check: $\left(\frac{1.0 \times 10^{-6}}{0.0833}\right) \times 100\% = 1.2 \times 10^{-3}$ %, so the approximation is valid.

According to Table 14.2, phenolphthalein has a pK_{In} of 8.7 and pH range of 8.3 - 10.0. Hence, **phenolphthalein** can be used for titrating lactic acid with sodium hydroxide.

14.36 ***See Sections 14.3 and 4 and Table 14.2.***

Overall: $NH_3(aq) + HCl(aq) \rightarrow NH_4Cl(aq)$

Net ionic: $NH_3(aq) + H_3O^+(aq) \rightarrow NH_4^+(aq) + H_2O(l)$

$$?\ \text{starting mmol } NH_3 = 25.0\ \text{mL } NH_3\ \text{soln} \times \frac{1.44\ \text{mmol } NH_3}{1\ \text{mL } NH_3\ \text{soln}} = 36.0\ \text{mmol } NH_3$$

$$?\ \text{mmol } H_3O^+ \text{ required to reach equivalence point} = 36.0\ \text{mmol } NH_3 \times \frac{1\ \text{mmol } H_3O^+}{1\ \text{mmol } NH_3} = 36.0\ \text{mmol } H_3O^+$$

$$?\ \text{mL HCl required to reach equivalence point} = 36.0\ \text{mmol } H_3O^+ \times \frac{1\ \text{mL HCl soln}}{1.50\ \text{mmol } H_3O^+} = 24.0\ \text{mL HCl soln}$$

V_T = 25.0 ml + 24.0 ml = 49.0 mL

	NH_3	H_3O^+	NH_4^+
s, mmol	36.0	36.0	0.
R, mmol	−36.0	−36.0	+36.0
f, mmol	0.	0.	36.0
C, *M*	0.	0.	0.735

The pH at the equivalence point is due to the reaction of NH_4^+ with H_2O:

$$NH_4^+(aq) + H_2O(l) \rightleftarrows NH_3(aq) + H_3O^+(aq)$$

$$K_a\ NH_4^+ = \frac{K_w}{K_b\ NH_3} \qquad K_a\ NH_4^+ = \frac{1.0 \times 10^{-14}}{1.8 \times 10^{-5}} = 5.6 \times 10^{-10}$$

Conc, *M*	NH_4^+	NH_3	H_3O^+
initial	0.735	0.	0.
Change	−y	+y	+y
equilibrium	0.735−y	y	y

$$K_a\ NH_4^+ = \frac{[NH_3][H_3O^+]}{[NH_4^+]} \qquad K_a\ NH_4^+ = \frac{(y)(y)}{(0.735-y)} \cong \frac{(y)(y)}{0.735} = 5.6 \times 10^{-10}$$

$[H_3O^+] = y = 2.0 \times 10^{-5}$ and **pH** $= -\log(2.0 \times 10^{-5}) =$ **4.69**

Approximation check: $\left(\frac{2.0 \times 10^{-5}}{0.735}\right) \times 100\% = 2.7 \times 10^{-3}$ %, so the approximation is valid.

According to Table 14.2, methyl red has $pK_{In} = 5.00$ and a pH range of 4.2 - 6.3. Hence, **methyl red** can be used for titrating NH_3 with HCl.

14.38 ***See Section 14.5.***

(a) $$H_2C_2O_4(aq) + H_2O(l) \rightleftarrows H_3O^+(aq) + HC_2O_4^-(aq) \qquad K_{a1} = \frac{[H_3O^+][HC_2O_4^-]}{[H_2C_2O_4]}$$

$$HC_2O_4^-(aq) + H_2O(l) \rightleftarrows H_3O^+(aq) + C_2O_4^{2-}(aq) \qquad K_{a2} = \frac{[H_3O^+][C_2O_4^{2-}]}{[HC_2O_4^-]}$$

(b) $H_2SO_3(aq) + H_2O(l) \rightleftarrows H_3O^+(aq) + HSO_3^-(aq)$ $\quad K_{a1} = \frac{[H_3O^+][HSO_3^-]}{[H_2SO_3]}$

$HSO_3^-(aq) + H_2O(l) \rightleftarrows H_3O^+(aq) + SO_3^{2-}(aq)$ $\quad K_{a2} = \frac{[H_3O^+][SO_3^{2-}]}{[HSO_3^-]}$

14.40 ***See Section 14.5 and Table 14.3.***

$$SO_4^{2-}(aq) + H_2O(l) \rightleftarrows HSO_4^-(aq) + OH^-(aq)$$

$$K_b SO_4^{2-} = \frac{[HSO_4^-][OH^-]}{[SO_4^{2-}]} = \frac{K_w}{K_{a2}\ H_2SO_4} \qquad K_b SO_4^{2-} = \frac{1.0 \times 10^{-14}}{1.2 \times 10^{-2}} = \mathbf{8.3 \times 10^{-13}}$$

14.42 ***See Section 14.5 and Table 14.3.***

$$C_6H_5O_7^{3-}(aq) + H_2O(l) \rightleftarrows HC_6H_5O_7^{2-}(aq) + OH^-(aq)$$

$$K_b C_6H_5O_7^{3-} = \frac{[HC_6H_5O_7^{2-}][OH^-]}{[C_6H_5O_7^{3-}]} = \frac{K_w}{K_{a3}\ H_3C_6H_5O_7^{2-}} \qquad K_b C_6H_5O_7^{3-} = \frac{1.0 \times 10^{-14}}{4.0 \times 10^{-7}} = \mathbf{2.5 \times 10^{-8}}$$

14.44 ***See Section 14.5.***

The equivalence point occurs when three moles of NaOH have been added per one mole of acid. The only triprotic acid in the list is **citric acid**. The pK_a values for the three ionizable hydrogen atoms are too close to give separate equivalence points.

14.46 ***See Section 14.5.***

The shape of the titration curve indicates the acid is a diprotic acid. The separation of the equivalence points and the pH values at the half-way points for the two equivalence points indicate the acid being titrated is **oxalic acid**.

14.48 ***See Section 14.3.***

Overall: $HA(aq) + NaOH(aq) \rightarrow H_2O(l) + NaA(aq)$
Net ionic: $HA(aq) + OH^-(aq) \rightarrow H_2O(l) + A^-(aq)$

Since M = g/ mol, our first objective is to determine the number of moles of acid titrated.

$$? \text{ mol HA titrated} = 0.04135 \text{ L NaOH soln} \times \frac{0.0982 \text{ mol OH}^-}{1 \text{ L NaOH soln}} \times \frac{1 \text{ mol HA}}{1 \text{ mol OH}^-} = 4.06 \times 10^{-3} \text{ mol HA}$$

$$M = \frac{0.3419 \text{ g HA}}{4.06 \times 10^{-3} \text{ mol HA}} = \mathbf{84.2\ g/mol}$$

Since $pH = pK_a + \log\left(\frac{n_b}{n_a}\right)$, it is necessary to determine n_{A^-} and n_{HA} after 24.35 mL of NaOH solution have been added to be able to calculate pK_a.

After 24.35 mL NaOH soln have been added,

$$? \text{ mol HA reacted} = 0.02435 \text{ L NaOH soln} \times \frac{0.0982 \text{ mol OH}^-}{1 \text{ L NaOH soln}} \times \frac{1 \text{ mol HA}}{1 \text{ mol OH}^-} = 2.39 \times 10^{-3} \text{ mol HA}$$

Hence, moles A^- formed = moles HA reacted = 2.39×10^3 mol and
moles HA remaining = 4.06×10^{-3} mol HA initial - 2.39×10^{-3} mol HA reacted = 1.67×10^{-3} mol HA.

This yields $pH = pK_a + \log\left(\frac{n_b}{n_a}\right)$ $\quad pK_a = pH - \log\left(\frac{n_b}{n_a}\right)$ $\quad \mathbf{pK_a} = 4.56 - \log\left(\frac{2.39 \times 10^{-3}}{1.67 \times 10^{-3}}\right) = \mathbf{4.40}$

14.50 ***See Sections 14.5 and 6, Example 14.9 and Table 14.3.***

(a) The reaction of $HC_2O_4^-$ as an acid is:

$$HC_2O_4^-(aq) + H_2O(l) \rightleftarrows C_2O_4^{2-}(aq) + H_3O^+(aq)$$

$K_a = K_{a2}\ H_2C_2O_4 = 6.4 \times 10^{-5}$

The reaction of $HC_2O_4^-$ as a base is:

$$HC_2O_4^-(aq) + H_2O(l) \rightleftarrows H_2C_2O_4(aq) + OH^-(aq)$$

$$K_{b2} = \frac{K_w}{K_{a1}\ H_2C_2O_4} = \frac{1.0 \times 10^{-14}}{5.9 \times 10^{-2}} = 1.7 \times 10^{-13}$$

Since $K_a\ HC_2O_4^- > K_b\ HC_2O_4^-$, the solution will be **acidic**.

(b) The reaction of $HC_3H_2O_4^-$ as an acid is:

$$HC_3H_2O_4^-(aq) + H_2O(l) \rightleftarrows C_3H_2O_4^{2-}(aq) + H_3O^+(aq)$$

$K_a = K_{a2}\ H_2C_3H_2O_4 = 2.1 \times 10^{-6}$

The reaction of $HC_3H_2O_4^-$ as a base is:

$$HC_3H_2O_4^-(aq) + H_2O(l) \rightleftarrows H_2C_3H_2O_4(aq) + OH^-(aq)$$

$$K_{b2} = \frac{K_w}{K_{a1}\ H_2C_3H_2O_4} = \frac{1.0 \times 10^{-14}}{1.6 \times 10^{-2}} = 6.2 \times 10^{-13}$$

Since $K_a\ HC_3H_2O_4^- > K_b\ HC_3H_2O_4^-$, the solution will be **acidic**.

14.52 *See Sections 14.5 and 6, Example 14.9 and Table 14.3.*

(a) The reaction of $H_2PO_4^-$ as an acid is:

$$H_2PO_4^-(aq) + H_2O(l) \rightleftarrows HPO_4^{2-}(aq) + H_3O^+(aq) \qquad K_a = K_{a2}\ H_3PO_4 = 6.2 \times 10^{-8}$$

The reaction of $H_2PO_4^-$ as a base is:

$$H_2PO_4^-(aq) + H_2O(l) \rightleftarrows H_3PO_4(aq) + OH^-(aq)$$

$$K_{b3} = \frac{K_w}{K_{a1}\ H_3PO_4} = \frac{1.0 \times 10^{-14}}{7.5 \times 10^{-3}} = 1.3 \times 10^{-12}$$

Since $K_a\ H_2PO_4^-$ is greater than $K_b\ H_2PO_4^-$, the solution will be **acidic**.

(b) The reaction for CO_3^{2-} as a base is:

$$CO_3^{2-}(aq) + H_2O(l) \rightleftarrows HCO_3^-(aq) + OH^-(aq)$$

$$K_{b1} = \frac{K_w}{K_{a2}\ H_2CO_3} \qquad K_{b1} = \frac{1.0 \times 10^{-14}}{5.6 \times 10^{-11}} = 1.8 \times 10^{-4}$$

Since $CO_3^{2-}(aq)$ cannot act as an acid and K_b is appreciable, the solution will be **basic**.

14.54 *See Section 14.7.*

(a) The acetate ion, CH_3COO^-, is the conjugate base of the weak acid acetic acid, CH_3COOH, and therefore acts as a weak base. The acetate ion will therefore react with the H_3O^+ from HCl. The pertinent reactions are:

$$Ca(CH_3COO)_2(s) \rightleftarrows Ca^{2+}(aq) + 2CH_3COO^-(aq)$$

$$2CH_3COO^-(aq) + 2H_3O^+(aq) \rightleftarrows 2CH_3COOH(aq) + 2H_2O(l)$$

$$Ca(CH_3COO)_2(s) + 2H_3O^+(aq) \rightleftarrows Ca^{2+}(aq) + 2CH_3COOH(aq) + 2H_2O(l)$$

Hence, HCl **increases** the solubility of calcium acetate.

(b) The fluoride ion, F^-, is the conjugate base of the weak acid hydrofluoric acid, HF, and therfore acts as a weak base. The fluoride ion will therefore react with the H_3O^+ from HCl. The pertinent reactions are:

$$MgF_2(s) \rightleftarrows Mg^{2+}(aq) + 2F^-(aq)$$

$$2F^-(aq) + 2H_3O^+(aq) \rightleftarrows 2HF(aq) + 2H_2O(l)$$

$$MgF_2(s) + 2H_3O^+(aq) \rightleftarrows Mg^{2+}(aq) + 2HF(aq) + 2H_2O(l)$$

Hence, HCl **increases** the solubility of MgF_2.

14.56 See Section 14.7.

(a) The oxalate ion is related to the weak acid oxalic acid, $H_2C_2O_4$, and therefore acts as a weak base. The oxalate ion therefore reacts with the H_3O^+ from HCl. The pertinent reactions are:

$$CaC_2CO_4(s) \rightleftarrows Ca^{2+}(aq) + C_2O_4^{2-}(aq)$$

$$C_2O_4^{2-}(aq) + 2H_3O^+(aq) \rightleftarrows H_2C_2O_4(aq) + 2H_2O(l)$$

$$CaC_2O_4(s) + 2H_3O^+(aq) \rightleftarrows Ca^{2+}(aq) + H_2C_2O_4(aq) + 2H_2O(l)$$

Hence, HCl **increases** the solubility of CaC_2O_4.

(b) NH_3 forms a complex ion with Cu^{2+}, $Cu(NH_3)_4^{2+}$. The pertinent reactions are:

$$CuSO_4(s) \rightleftarrows Cu^{2+}(aq) + SO_4^{2-}(aq)$$

$$Cu^{2+}(aq) + 4NH_3(aq) \rightleftarrows Cu(NH_3)_4^{2+}(aq)$$

$$CuSO_4(s) + 4NH_3(aq) \rightleftarrows Cu(NH_3)_4^{2+}(aq) + SO_4^{2-}(aq)$$

Hence, NH_3 **increases** the solubility of $CuSO_4$.

14.58 See Sections 12.6 and 13.2 and Appendix F.

$$Mg(OH)_2(s) \rightleftarrows Mg^{2+}(aq) + 2OH^-(aq)$$

Conc, M	$Mg(OH)_2$	Mg^{2+}	OH^-
initial	excess	0.	0.
Change	$-s$	$+s$	$+2s$
equilibrium	excess	s	2s

$K_{sp} = [Mg^{2+}][OH^-]^2$ $K_{sp} = (s)(2s)^2 = 8.9 \times 10^{-12}$ Hence, $4s^3 = 8.9 \times 10^{-12}$ and $s = 1.3 \times 10^{-4}$.

$[OH^-] = 2s = (2)(1.3 \times 10^{-4}) = 2.6 \times 10^{-4}$, $pOH = -\log(2.6 \times 10^{-4}) = 3.58$, **pH** $= 14.00 - 3.58 =$ **10.42**

14.60 See Sections 14.1 and 4 and Table 14.2.

Overall: $HI(aq) + NaOH(aq) \rightarrow NaI(s) + H_2O(l)$
Net ionic: $H_3O^+(aq) + OH^-(aq) \rightarrow H_2O(l)$

$$? \text{ mol } H_3O^+ \text{ present} = 10.0 \text{ mol HI soln} \times \frac{1.70 \text{ g HI soln}}{1 \text{ mL HI soln}} \times \frac{57 \text{ g HI}}{100 \text{ g HI soln}} \times \frac{1 \text{ mol HI}}{127.9 \text{ g HI}} \times \frac{1 \text{ mol } H_3O^+}{1 \text{ mol HI}}$$

$$= 7.6 \times 10^{-2} \text{ mol } H_3O^+$$

? mol OH^- needed to reach equivalence point= 7.6×10^{-2} mol $H_3O^+ \times \frac{1 \text{ mol } OH^-}{1 \text{ mol } H_3O^+}$

$$= 7.6 \times 10^{-2} \text{ mol } OH^-$$

? mL NaOH soln required to reach equivalence point= 7.6×10^{-2} mol $OH^- \times \frac{1 \text{ mol NaOH}}{1 \text{ mol } OH^-}$

$$\times \frac{10^3 \text{ mL NaOH soln}}{0.988 \text{ mol NaOH}} = 77 \text{ mL NaOH soln}$$

Since the titration involves titrating a strong acid with a strong base, we should expect the pH to be 7.0 at the equivalence point. **Phenolphthalein** (pK_{In} = 8.7, pH range 8.3 - 10.0) is a suitable indicator, since there is a very rapid change in pH near the equivalence point of a strong acid-strong base titration (Fig 14.3).

14.62 ***See Section 14.4.***

For $HIn(aq) + H_2O(l) \rightleftarrows H_3O^+(aq) + In^-(aq)$,

$$K_{In} = \frac{[H_3O^+][In^-]}{[HIn]}, \quad -\log K_{In} = -\log[H_3O^+] - \log\left(\frac{[In^-]}{[HIn]}\right), \text{ and } pK_{In} = pH - \log\left(\frac{[In^-]}{[HIn]}\right)$$

(a) For pH to be one unit higher than pK_{In}, $\frac{[\mathbf{In^-}]}{[\mathbf{HIn}]} = \frac{\mathbf{10}}{\mathbf{1}}$.

$$pK_{In} = pH - \log\left(\frac{[In^-]}{[HIn]}\right) \quad pK_{In} = pH - \log\left(\frac{10}{1}\right) \quad pK_{In} = pH - 1$$

(b) For pH to be one unit lower than pK_{In}, $\frac{[\mathbf{In^-}]}{[\mathbf{HIn}]} = \frac{\mathbf{1}}{\mathbf{10}}$.

$$pK_{In} = pH - \log\left(\frac{[In^-]}{[HIn]}\right) \quad pK_{In} = pH - \log\left(\frac{1}{10}\right) \quad pK_{In} = pH + 1$$

14.64 ***See Section 14.4.***

For $HIn(aq) + H_2O(l) \rightleftarrows H_3O^+(aq) + In^-(aq)$,

$$K_{In} = \frac{[H_3O^+][In^-]}{[HIn]}, \quad -\log K_{In} = -\log[H_3O^+] - \log\left(\frac{[In^-]}{[HIn]}\right), \text{ and } pK_{In} = pH - \log\left(\frac{[In^-]}{[HIn]}\right)$$

So, $pH = pK_{In} + \log\left(\frac{[In^-]}{[HIn]}\right)$

When **1%** of methyl red is in the HIn form: $\mathbf{pH} = 5.00 + \log\left(\frac{0.99y}{0.01y}\right) = \mathbf{7.00}$

When **5%** of methyl red is in the HIn form: $\mathbf{pH} = 5.00 + \log\left(\frac{0.95y}{0.05y}\right) = \mathbf{6.28}$

When **95%** of methyl red is in the HIn form: $\mathbf{pH} = 5.00 + \log\left(\frac{0.05y}{0.95y}\right) = \mathbf{3.72}$

When **99%** of methyl red is in the HIn form: $\mathbf{pH} = 5.00 + \log\left(\frac{0.01y}{0.99y}\right) = \mathbf{3.00}$

Chapter 15: Thermodynamics

15.6 ***See Section 15.1.***

(a) $CaCO_3(s) + H_2SO_4(aq) \rightarrow CaSO_4(aq) + H_2O(l) + CO_2(g)$: There is an increase in mole of gas during this reaction causing the system to expand against an opposing pressure exerted by the surroundings. The sign of **w is negative**, since the system transfers energy to the surroundings.

(b) $HCl(g) + NH_3(g) \rightarrow NH_4Cl(s)$: There is a decrease in moles of gas during this reaction causing the system to contract under the influenece of pressure exerted by the surroundings. The sign of **w is positive**, since the surroundings transfers energy to the system as it compresses it.

15.8 ***See Section 15.1.***

Since the volumes of solids and liquids are small compared to the volumes of gases, the amount of PV work done during physical changes of state and chemical reactions involving gases can be calculated from changes in the numbers of moles of gases.

The volume of 1 mole of an ideal gas at 298 K can be calculated using $V = \frac{nRT}{P}$ and is 24.5 L.

(a) $CO_2(g) + NaOH(s) \rightarrow NaHCO_3(s)$ involves a decrease of one mole of gas.

Hence, $w = -P\Delta V$ $\quad w = -(1\text{ atm})(-24.5\text{ L}) = \mathbf{24.5\ L\cdot atm}$

(b) $2O_3(g) \rightarrow 3O_2(g)$ involves an increase of one mole of gas.

Hence, $w = -P\Delta V$ $\quad w = -(1\text{ atm})(+24.5\text{ L}) = \mathbf{-24.5\ L\cdot atm}$

Note: The reaction in (a) involves a decrease in the number of moles of gas, and the sign of w is positive. The reaction in (b) involves an increase in the number of moles of gas, and the sign of w is negative.

15.10 ***See Section 15.1 and the Solution for 15.8.***

(a) $Ni(CO)_4(g) \rightarrow Ni(s) + 4CO(g)$ involves an increase of three mole of gas.

Hence, $w = -P\Delta V$ $\quad w = -(1\text{ atm})\left[3 \times (+24.5\text{ L})\right] = \mathbf{-73.5\ L\cdot atm}$

(b) $2NO(g) \rightarrow N_2(g) + O_2(g)$ does not involve a change in moles of gases.
Hence, no pressure-volume work is done during this reaction.

15.12 ***See Section 15.1 and Example 15.1.***

Initial conditions:	$V_1 = 220$ L	$P_1 = 150$ atm
Final conditions:	$V_2 = ?$, unknown	$P_2 = P_{ext} = 1.0$ atm

Solving $P_1V_1 = P_2V_2$ for V_2 gives $V_2 = V_1 \times \frac{P_1}{P_2}$ $\qquad V_2 = 220\text{ L} \times \frac{150\text{ atm}}{1.0\text{ atm}} = 3.3 \times 10^4\text{ L}$

So, $w = -P\Delta V = -(1.0 \text{ atm})(3.3 \times 10^4 \text{ L} - 220 \text{ L}) = -3.3 \times 10^4$ L·atm.

Converting L·atm to kJ gives $w = -3.3 \times 10^4 \text{ L·atm} \times 101.3 \frac{\text{J}}{\text{L·atm}} \times \frac{1 \text{ kJ}}{10^3 \text{ J}} = \mathbf{-3.3 \times 10^3 \text{ kJ}}$

15.14 *See Sections 15.1 and 2 and Example 15.1.*

According to the First Law of Thermodynamics, $\Delta E = q + w$. In this case, $q = +4.35$ kJ and $w = -P\Delta V = -(0.94 \text{ atm})(4.5 \text{ L}) = -4.2$ L·atm.

Converting L·atm to kJ gives $w = -4.2 \text{ L·atm} \times 101.3 \frac{\text{J}}{\text{L·atm}} \times \frac{1 \text{ kJ}}{10^3 \text{ J}} = -0.43 \text{ kJ}$

and $\Delta E = 4.35 \text{ kJ} + (-0.43 \text{ kJ}) = \mathbf{3.92 \text{ kJ}}$.

15.16 *See Section 15.1 and 2 and Example 15.1.*

Initial conditions:	$V_1 = 1.00$ L	$P_1 = 9.00$ atm
Final conditions:	V_2 = ?, unknown	$P_2 = P_{ext} = 1.00$ atm

Solving $P_1V_1 = P_2V_2$ for V_2 gives $V_2 = V_1 \times \frac{P_1}{P_2}$ $\quad V_2 = 1.00 \text{ L} \times \frac{9.00 \text{ atm}}{1.00 \text{ atm}} = 9.00 \text{ L}$

So, $w = -P\Delta V = -(1.0 \text{ atm})(9.00 \text{ L} - 1.00 \text{ L}) = -8.00$ L·atm.

Converting L·atm to kJ gives $w = -8.00 \text{ L·atm} \times 101.3 \frac{\text{J}}{\text{L·atm}} \times \frac{1 \text{ kJ}}{10^3 \text{ J}} = \mathbf{-0.810 \text{ kJ}}$

Since $\Delta E = 0$ for an isothermal expansion and $\Delta E = q + w$, $\mathbf{q = +0.810 \text{ kJ}}$.

15.18 *See Section 15.1 and 2, Example 15.1 and the Solutions for 15.16.*

Following the plan outlined in the solution for 15.16 gives:

1) V_2 at the end of expansion against 3.00 atm = 3.00 L,
$w_1 = -P\Delta V = -(3.00 \text{ atm})(3.00 \text{ L} - 1.00 \text{ L}) = -6.00$ L·atm or -0.608 kJ and $q_1 = +0.608$ kJ.

2) V_2' at the end of expansion from (3.00 L, 3.00 atm) to (? L, 2.00 atm) = 4.50 L,
$w_2 = -(2.00 \text{ atm})(4.50 \text{ L} - 3.00 \text{ L}) = -3.00$ L·atm or -0.304 kJ and $q_2 = +0.304$ kJ.

3) V_2'' at the end of expansion from (4.50 L, 2.00 atm) to (? L, 1.00 atm) = 9.00 L,
$w_3 = -(1.00 \text{ atm})(9.00 \text{ L} - 4.50 \text{ L}) = -4.5$ L·atm or -0.456 kJ and $q_3 = +0.456$ kJ.

Hence, $\mathbf{w = w_1 + w_2 + w_3} = -0.608 \text{ kJ} - 0.304 \text{ kJ} - 0.456 \text{ kJ} = \mathbf{-1.368 \text{ kJ}}$,
and $\mathbf{q = q_1 + q_2 + q_3} = 0.608 \text{ kJ} + 0.304 \text{ kJ} + 0.456 \text{ kJ} = \mathbf{1.368 \text{ kJ}}$.

The inital and final states for the gas are the same in Excercises 15.16 and 18. However, expansion occurs in a one-step process in 15.16 and in a three-step process in 15.18:

15.16 (1.00 L, 9.00 atm) → (9.00 L, 1.00 atm)
15.18 (1.00 L, 9.00 atm) → (3.00 L, 3.00 atm) → (4.50 L, 2.00 atm) → (9.00 L, 1.00 atm)

Since the inital and final states are the same in both excercises and both involve isothermal expansions, $\Delta E = 0$ for both. However, $q = 0.810$ kJ and $w = -0.810$ kJ for the one-step expansion of 15.16 whereas $q = 1.368$ kJ and $w = -1.368$ kJ for the three-step expansion of 15.18. Similar calculations would show q for 15.16 $\neq$ q for 15.17 $\neq$ q for 15.18 and w for 15.16 $\neq$ w for 15.17 $\neq$ w for 15.18. **These calculations illustrate that heat and work (q and w) are path dependent and therefore not state functions whereas ΔE is independent of the pathway and therefore internal energy (E) is a state function.**

15.22 *See Section 15.2.*

Thermochemical equations corresponding to standard enthalpies (heats) of formation involve forming one mole of compound in its standard state from its constituent elements in their standard states. Hence,

(a) $\frac{1}{2}H_2(g)+\frac{1}{2}F_2(g)\rightarrow HF(g)$

(b) $H_2(g)+\frac{1}{8}S_8(s,\text{ rhombic})+2O_2(g)\rightarrow H_2SO_4(l)$ or $H_2(g)+S(s,\text{ rhombic})+2O_2(g)\rightarrow H_2SO_4(l)$

(c) $Al(s)+\frac{3}{2}O_2(g)+\frac{3}{2}H_2(g)\rightarrow Al(OH)_3(s)$

(d) $Na(s)+K(s)+\frac{1}{8}S_8(s,\text{ rhombic})+2O_2(g)\rightarrow NaKSO_4(s)$ or

$Na(s)+K(s)+S(s,\text{ rhombic})+2O_2(g)\rightarrow NaKSO_4(s)$

15.24 *See Section 15.2 and Appendix G.*

(a) For $C_6H_{12}O_6(s)\rightarrow 2CH_3CH_2OH(l)+2CO_2(g)$,

$$\Delta H^\circ_{rxn}=\left[2\Delta H^\circ_f CH_3CH_2OH(l)+2\Delta H^\circ_f CO_2(g)\right]-\left[\Delta H^\circ_f C_6H_{12}O(s)\right]$$

$$\Delta H^\circ_{rxn}=\left[\left(2\text{ mol }\times -277.69\text{ kJ}\cdot\text{mol}^{-1}\right)+\left(2\text{ mol }\times -393.51\text{ kJ}\cdot\text{mol}^{-1}\right)\right]-\left[1\text{ mol}\times -1268\text{ kJ}\cdot\text{mol}^{-1}\right]=\mathbf{-74\ kJ}$$

ΔH°_{rxn} is negative, and the reaction is exothermic.

(b) For $n-C_4H_{10}(g)+\frac{13}{2}O_2(g)\rightarrow 4CO_2(g)+5H_2O(l)$

$$\Delta H^\circ_{rxn}=\left[4\Delta H^\circ_f CO_2(g)+5\Delta H^\circ_f H_2O(l)\right]-\left[\Delta H^\circ_f n-C_4H_{10}(g)+\frac{13}{2}\Delta H^\circ_f O_2(g)\right]$$

$$\Delta H^\circ_{rxn}=\left[\left(4\text{ mol}\times -393.51\text{ kJ}\cdot\text{mol}^{-1}\right)+\left(5\text{ mol}\times -285.83\text{ kJ}\cdot\text{mol}^{-1}\right)\right]$$

$$-\left[\left(1\text{ mol}\times -124.73\text{ kJ}\cdot\text{mol}^{-1}\right)+\left(\frac{13}{2}\text{ mol}\times 0.\text{ kJ}\cdot\text{mol}^{-1}\right)\right]=\mathbf{-2878.46\ kJ}$$

ΔH°_{rxn} is negative, and the reaction is exothermic.

15.26 *See Section 15.2 and Appendic G.*

(a) For $NaOH(s) + CO_2(g)\rightarrow NaHCO_3(s)$, $\Delta H^\circ_{rxn}=\left[\Delta H^\circ_f NaHCO_3(s)\right]-\left[\Delta H^\circ_f NaOH(s)+\Delta H^\circ_f CO_2(g)\right]$

$$\Delta H^\circ_{rxn}=\left[\left(1\text{ mol }\times -950.81\text{ kJ}\cdot\text{mol}^{-1}\right)\right]-\left[\left(1\text{ mol}\times -425.61\text{ kJ}\cdot\text{mol}^{-1}\right)+\left(1\text{ mol}\times -393.51\text{ kJ}\cdot\text{mol}^{-1}\right)\right]=\mathbf{-131.69\ kJ}$$

ΔH°_{rxn} is negative, and the reaction is exothermic.

(b) For $2SO_2(g) + O_2(g) \rightarrow 2SO_3(g)$, $\Delta H^\circ_{rxn} = \left[2\Delta H^\circ_f SO_3(g)\right] - \left[2\Delta H^\circ_f SO_2(g) + \Delta H^\circ_f O_2(g)\right]$

$$\Delta H^\circ_{rxn} = \left[\left(2 \text{ mol} \times -395.72 \text{ kJ}\cdot\text{mol}^{-1}\right)\right] - \left[\left(2 \text{ mol} \times -296.83 \text{ kJ}\cdot\text{mol}^{-1}\right) + \left(1 \text{ mol} \times 0. \text{ kJ}\cdot\text{mol}^{-1}\right)\right] = \mathbf{-197.78 \text{ kJ}}$$

ΔH°_{rxn} is negative, and the reaction is exothermic.

15.28 *See Section 15.2 and Appendix G.*

(a) For C(graphite) $\rightarrow$ C(diamond), $\Delta H^\circ_{rxn} = \left[\Delta H^\circ_f C(\text{diamond})\right] - \left[\Delta H^\circ_f C(\text{graphite})\right]$

$$\Delta H^\circ_{rxn} = \left[\left(1 \text{ mol} \times 1.895 \text{ kJ}\cdot\text{mol}^{-1}\right)\right] - \left[1 \text{ mol} \times 0. \text{ kJ}\cdot\text{mol}^{-1}\right] = \mathbf{1.895 \text{ kJ}}$$

ΔH°_{rxn} is positive, and the reaction is endothermic.

(b) For $H_2O(l) \rightarrow H_2O(g)$, $\Delta H^\circ_{rxn} = \left[\Delta H^\circ_f H_2O(g)\right] - \left[\Delta H^\circ_f H_2O(l)\right]$

$$\Delta H^\circ_{rxn} = \left[\left(1 \text{ mol} \times -241.82 \text{ kJ}\cdot\text{mol}^{-1}\right)\right] - \left[\left(1 \text{ mol} \times -285.83 \text{ kJ}\cdot\text{mol}^{-1}\right)\right] = \mathbf{44.01 \text{ kJ}}$$

ΔH°_{rxn} is positive, and the reaction is endothermic.

15.30 *See Section 15.2.*

For $C_6H_{12}O_6(s) + 6O_2(g) \rightarrow 6CO_2(g) + 6H_2O(l)$,

$$\Delta H^\circ_{rxn} = \left[6\Delta H^\circ_f CO_2(g) + 6\Delta H^\circ_f H_2O(l)\right] - \left[\Delta H^\circ_f C_6H_{12}O_6(s) + 6\Delta H^\circ_f O_2(g)\right]$$

$$\Delta H^\circ_{rxn} = \left[\left(6 \text{ mol} \times -393.51 \text{ kJ}\cdot\text{mol}^{-1}\right) + \left(6 \text{ mol} \times -285.83 \text{ kJ}\cdot\text{mol}^{-1}\right)\right]$$

$$-\left[\left(1 \text{ mol} \times -1268 \text{ kJ}\cdot\text{mol}^{-1}\right) + \left(6 \text{ mol} \times 0. \text{ kJ}\cdot\text{mol}^{-1}\right)\right] = -2808 \text{ kJ}$$

For a 28 g sample of $C_6H_{12}O_6(s)$,

$$? \ \Delta H^\circ \text{ in kJ} = 28 \text{ g } C_6H_{12}O_6(s) \times \frac{1 \text{ mol } C_6H_{12}O_6(s)}{180.0 \text{ g } C_6H_{12}O_6(s)} \times \frac{-2808 \text{ kJ}}{1 \text{ mol } C_6H_{12}O_6(s)} = \mathbf{-4.4 \times 10^2 \text{ kJ}}$$

15.32 *See Section 15.3.*

(a) There is an increase in disorder as the test-tube shatters. Hence, the sign of the entropy change for the system is **positive**.

(b) There is an increase in disorder as a new deck of cards is shuffled. Hence, the sign of the entropy change for the system is **positive**.

(c) There is a decrease in disorder as steel is made from iron ore and carbon. Hence, the sign of the entropy change for the system is **negative**.

(d) There is an increase in disorder as a wooden fence rots. Hence, the sign of the entropy change for the system is **positive**.

15.34 ***See Section 15.3.***

(a) There is an increase in disorder as dry ice, CO_2 (s), melts. Hence, the sign of the entropy change for the system is **positive**.
Note: Dry ice sublimes under normal conditions. However, it does melt under high pressure conditions.

(b) There is a decrease in disorder when water freezes. Hence, the sign of the entropy change for the system is **negative**.

(c) There is an increase in disorder as gasoline evaporates. Hence, the sign of the entropy change for the system is **positive**.

15.36 ***See Section 15.3, Example 15.3 and Appendix G.***

(a) For $CO(g) + 2H_2(g) \rightarrow CH_3OH(l)$, there is a decrease in the number of moles of gas and formation of one mole of liquid and therefore an accompanying decrease in disorder. **The sign of the entropy change should be negative, and the magnitude of the entropy change should be rather large.**

$$\Delta S^\circ_{rxn} = \left[S^\circ CH_3OH(l)\right] - \left[S^\circ CO(g) + 2S^\circ H_2(g)\right]$$

$$\Delta S^\circ_{rxn} = \left[1\ \text{mol} \times 126.8\ \text{J}\cdot\text{mol}^{-1}\cdot\text{K}^{-1}\right] - \left[\left(1\ \text{mol} \times 197.56\ \text{J}\cdot\text{mol}^{-1}\cdot\text{K}^{-1}\right) + \left(2\ \text{mol} \times 130.57\ \text{J}\cdot\text{mol}^{-1}\cdot\text{K}^{-1}\right)\right]$$

$$= \mathbf{-331.9\ J\cdot K^{-1}}$$

(b) For $N_2(g) + 3H_2(g) \rightarrow 2NH_3(g)$, there is a decrease in the number of moles of gas and therefore an accompanying decrease in disorder. **The sign of the entropy change should be negative, and the magnitude of the entropy change should be moderately large.**

$$\Delta S^\circ_{rxn} = \left[2S^\circ NH_3(g)\right] - \left[2S^\circ N_2(g) + 3S^\circ H_2(g)\right]$$

$$\Delta S^\circ_{rxn} = \left[2\ \text{mol} \times 192.34\ \text{J}\cdot\text{mol}^{-1}\cdot\text{K}^{-1}\right] - \left[\left(1\ \text{mol} \times 191.50\ \text{J}\cdot\text{mol}^{-1}\cdot\text{K}^{-1}\right) + \left(3\ \text{mol} \times 130.57\ \text{J}\cdot\text{mol}^{-1}\cdot\text{K}^{-1}\right)\right]$$

$$= \mathbf{-198.53\ J\cdot K^{-1}}$$

(c) For $CH_4(g) + 2O_2(g) \rightarrow CO_2(g) + 2H_2O(l)$, there is a decrease in the number of moles of gas and formation of two moles of liquid and therefore an accompanying decrease in disorder. **The sign of the entropy change should be negative, and the magnitude of the entropy change should be rather large.**

$$\Delta S^\circ_{rxn} = \left[S^\circ CO_2(g) + 2S^\circ H_2O(l)\right] - \left[S^\circ CH_4(g) + 2S^\circ O_2(g)\right]$$

$$\Delta S^\circ_{rxn} = \left[\left(1\ \text{mol} \times 213.63\ \text{J}\cdot\text{mol}^{-1}\cdot\text{K}^{-1}\right) + \left(2\ \text{mol} \times 69.91\ \text{J}\cdot\text{mol}^{-1}\cdot\text{K}^{-1}\right)\right]$$

$$-\left[\left(1\ \text{mol} \times 186.15\ \text{J}\cdot\text{mol}^{-1}\cdot\text{K}^{-1}\right) + \left(2\ \text{mol} \times 205.03\ \text{J}\cdot\text{mol}^{-1}\cdot\text{K}^{-1}\right)\right] = \mathbf{-242.76\ J\cdot K^{-1}}$$

(d) For $CO(g) + H_2O(g) \rightarrow CO_2(g) + H_2(g)$, there is no change in the number of moles of gas or even in the complexity of the reactants and products. Hence, **the entropy change should be relatively close to zero.**

$$\Delta S^\circ_{rxn} = \left[S^\circ CO_2(g) + S^\circ H_2(g)\right] - \left[S^\circ CO(g) + S^\circ H_2O(g)\right]$$

$$\Delta S^\circ_{rxn} = \left[\left(1\text{ mol} \times 213.63\text{ J}\cdot\text{mol}^{-1}\cdot\text{K}^{-1}\right) + \left(1\text{ mol} \times 130.57\text{ J}\cdot\text{mol}^{-1}\cdot\text{K}^{-1}\right)\right]$$

$$-\left[\left(1\text{ mol} \times 197.56\text{ J}\cdot\text{mol}^{-1}\cdot\text{K}^{-1}\right) + \left(1\text{ mol} \times 188.72\text{ J}\cdot\text{mol}^{-1}\cdot\text{K}^{-1}\right)\right] = \mathbf{-42.08\ J\cdot K^{-1}}$$

15.38 *See Section 15.4.*

(a) For $Fe_2O_3(s) + 2Al(s) \rightarrow Al_2O_3(s) + 2Fe(s)$,

$$\Delta G^\circ_{rxn} = \left[\Delta G^\circ_f Al_2O_3(s) + 2\Delta G^\circ_f Fe(s)\right] - \left[\Delta G^\circ_f Fe_2O_3(s) + 2\Delta G^\circ_f Al(s)\right]$$

$$\Delta G^\circ_{rxn} = \left[\left(1\text{ mol} \times -1582.3\text{ kJ}\cdot\text{mol}^{-1}\right) + \left(2\text{ mol} \times 0.\text{ kJ}\cdot\text{mol}^{-1}\right)\right]$$

$$-\left[\left(1\text{ mol} \times -742.2\text{ kJ}\cdot\text{mol}^{-1}\right) + \left(2\text{ mol} \times 0.\text{ kJ}\cdot\text{mol}^{-1}\right)\right] = \mathbf{-840.1\ kJ}$$

ΔG°_{rxn} is negative, and the reaction is spontaneous under standard conditions.

(b) For $CO(g) + 2H_2(g) \rightarrow CH_3OH(l)$,

$$\Delta G^\circ_{rxn} = \left[\Delta G^\circ_f CH_3OH(l)\right] - \left[\Delta G^\circ_f CO(g) + 2\Delta G^\circ_f H_2(g)\right]$$

$$\Delta G^\circ_{rxn} = \left[\left(1\text{ mol} \times -166.35\text{ kJ}\cdot\text{mol}^{-1}\right)\right] - \left[\left(1\text{ mol} \times -137.15\text{ kJ}\cdot\text{mol}^{-1}\right) + \left(2\text{ mol} \times 0.\text{ kJ}\cdot\text{mol}^{-1}\right)\right] = \mathbf{-29.2\ kJ}$$

ΔG°_{rxn} is negative, and the reaction is spontaneous under standard conditions.

15.40 *See Section 15.4, Example 15.4 and Appendix G.*

(a) For $Zn(s) + H_2SO_4(l) \rightarrow ZnSO_4(s) + H_2(g)$,

$$\Delta G^\circ_{rxn} = \left[\Delta G^\circ_f ZnSO_4(s) + \Delta G^\circ_f H_2(g)\right] - \left[\Delta G^\circ_f Zn(s) + \Delta G^\circ_f H_2SO_4(l)\right]$$

$$\Delta G^\circ_{rxn} = \left[\left(1\text{ mol} \times -871.5\text{ kJ}\cdot\text{mol}^{-1}\right) + \left(1\text{mol} \times 0.\text{ kJ}\cdot\text{mol}^{-1}\right)\right]$$

$$-\left[\left(1\text{ mol} \times 0.\text{ kJ}\cdot\text{mol}^{-1}\right) + \left(1\text{ mol} \times 690.10\text{ kJ}\cdot\text{mol}^{-1}\right)\right] = \mathbf{-181.4\ kJ}$$

ΔG°_{rxn} is negative, and the reaction is spontaneous under standard conditions.

(b) For $Cu(s) + H_2SO_4(l) \rightarrow CuSO_4(s) + H_2(g)$,

$$\Delta G^\circ_{rxn} = \left[\Delta G^\circ_f CuSO_4(s) + \Delta G^\circ_f H_2(g)\right] - \left[\Delta G^\circ_f Cu(s) + \Delta G^\circ_f H_2SO_4(l)\right]$$

$$\Delta G^\circ_{rxn} = \left[\left(1 \text{ mol} \times -661.9 \text{ kJ}\cdot\text{mol}^{-1}\right) + \left(1 \text{ mol} \times 0.\ \text{kJ}\cdot\text{mol}^{-1}\right)\right]$$

$$-\left[\left(1 \text{ mol} \times 0.\ \text{kJ}\cdot\text{mol}^{-1}\right) + \left(1 \text{ mol} \times -690.10 \text{ kJ}\cdot\text{mol}^{-1}\right)\right] = \mathbf{28.2\ kJ}$$

ΔG°_{rxn} is positive, and the reaction is nonspontaneous under standard conditions.

15.42 *See Section 15.4.*

For $H_2(g) + CuO(s) \rightarrow H_2O(l) + Cu(s)$,

$$\Delta S^\circ_{rxn} = \left[S^\circ H_2O(l) + S^\circ Cu(s)\right] - \left[S^\circ H_2(g) + S^\circ CuO(s)\right]$$

$$\Delta S^\circ_{rxn} = \left[\left(1 \text{ mol} \times 69.91 \text{ J}\cdot\text{mol}^{-1}\cdot\text{K}^{-1}\right) + \left(1 \text{ mol} \times 33.15 \text{ J}\cdot\text{mol}^{-1}\cdot\text{K}^{-1}\right)\right]$$

$$-\left[\left(1 \text{ mol} \times 130.57 \text{ J}\cdot\text{mol}^{-1}\cdot\text{K}^{-1}\right) + \left(1 \text{ mol} \times 42.63 \text{ J}\cdot\text{mol}^{-1}\cdot\text{K}^{-1}\right)\right] = \mathbf{-70.14\ J\cdot K^{-1}}$$

$$\Delta G^\circ_{rxn} = \left[\Delta G^\circ_f H_2O(l) + \Delta G^\circ_f Cu(s)\right] - \left[\Delta G^\circ_f H_2(g) + \Delta G^\circ_f CuO(s)\right]$$

$$\Delta G^\circ_{rxn} = \left[\left(1 \text{ mol} \times -237.18 \text{ kJ}\cdot\text{mol}^{-1}\right) + \left(1 \text{ mol} \times 0.\ \text{kJ}\cdot\text{mol}^{-1}\right)\right]$$

$$-\left[\left(1 \text{ mol} \times 0.\ \text{kJ}\cdot\text{mol}^{-1}\right) + \left(1 \text{ mol} \times -129.7 \text{ kJ}\cdot\text{mol}^{-1}\right)\right] = -107.48 \text{ kJ}$$

Solving $\Delta G^\circ_{rxn} = -T\Delta S^\circ_{univ}$ for ΔS°_{univ} and substituting gives

$$\Delta S^\circ_{univ} = \frac{-\Delta G^\circ_{rxn}}{T} \qquad \Delta S^\circ_{univ} = \frac{-(-107.48 \text{ kJ})}{298 \text{ K}} = \mathbf{0.361\ kJ\cdot K^{-1}} \text{ or } \mathbf{361\ J\cdot K^{-1}}$$

Solving $\Delta S_{univ} = \Delta S_{surr} + \Delta S_{sys}$ for ΔS_{surr} and substituting gives

$$\Delta S_{surr} = \Delta S_{univ} - \Delta S_{sys} \qquad \Delta S_{surr} = 361 \text{ J}\cdot\text{K}^{-1} - \left(-70.14 \text{ J}\cdot\text{K}^{-1}\right) = \mathbf{431\ J\cdot K^{-1}}$$

15.44 *See Section 15.1 and 2.*

Enthalpy is related to internal energy H = E + PV. Internal energy, E, represents the total energy of the system which involves both kinetic and potential energy. Potential energy is dependent on coulombic forces of attraction and repulsions within substances. Since we can only know the internal energy of ideal monatomic gases experiencing no forces of attraction and repulsion, it is generally impossible to know the absolute enthalpies of substances. Hence, we can only measure changes in enthalpies.

15.46 *See Section 15.4 and Example 15.5.*

(a) For $H_2(g) + \frac{1}{2}O_2(g) \rightarrow H_2O(l)$,

$$\Delta H^\circ_{rxn} = \left[\Delta H^\circ_f H_2O(l)\right] - \left[\Delta H^\circ_f H_2(g) + \tfrac{1}{2}\Delta H^\circ_f O_2(g)\right]$$

$$\Delta H^\circ_{rxn} = \left[1 \text{ mol} \times -285.83 \text{ kJ}\cdot\text{mol}^{-1}\right] - \left[\left(1 \text{ mol} \times 0.\ \text{kJ}\cdot\text{mol}^{-1}\right) + \left(\tfrac{1}{2} \text{ mol} \times 0.\ \text{kJ}\cdot\text{mol}^{-1}\right)\right] = -285.83 \text{ kJ}$$

$$\Delta S^\circ_{rxn} = \left[S^\circ H_2O(l)\right] - \left[S^\circ H_2(g) + \tfrac{1}{2} S^\circ O_2(g)\right]$$

$$\Delta S^\circ_{rxn} = \left[1 \text{ mol} \times 69.91 \text{ J}\cdot\text{mol}^{-1}\cdot\text{K}^{-1}\right] - \left[\left(1 \text{ mol} \times 130.57 \text{ J}\cdot\text{mol}^{-1}\cdot\text{K}^{-1}\right) + \left(\tfrac{1}{2} \text{ mol} \times 205.03 \text{ J}\cdot\text{mol}^{-1}\cdot\text{K}^{-1}\right)\right]$$

$$= -163.18 \text{ J}\cdot\text{K}^{-1}$$

At 25°C, $\Delta G^\circ_{298} = \Delta H^\circ - (298 \text{ K})\Delta S^\circ$

$\Delta G^\circ_{298} = -285.83 \times 10^3 \text{ J} - (298 \text{ K})(-163.18 \text{ J}\cdot\text{K}^{-1}) = \mathbf{-2.3720 \times 10^5 \text{ J or } -237.20 \text{ kJ}}$

At 90°C, $\Delta G^\circ_{363} = \Delta H^\circ - (363 \text{ K})\Delta S^\circ$

$\Delta G^\circ_{363} = -285.83 \times 10^3 \text{ J} - (363 \text{ K})(-163.18 \text{ J}\cdot\text{K}^{-1}) = \mathbf{-2.2660 \times 10^5 \text{ J or } -226.60 \text{ kJ}}$

The reaction has a favorable enthalpy change and an unfavorable entropy change. At low temperatures, the favorable ΔH° term dominates the unfavorable $T\Delta S^\circ$ term. As temperature increases, the $T\Delta S^\circ$ term increases in importance. Hence, the value of ΔG° becomes less negative as temperature increases, and the reaction proceeds to a lesser extent.

(b) For $CO(g) + 2H_2(g) \rightarrow CH_3OH(l)$,

$$\Delta H^\circ_{rxn} = \left[\Delta H^\circ_f CH_3OH(l)\right] - \left[\Delta H^\circ_f CO(g) + 2\Delta H^\circ_f H_2(g)\right]$$

$$\Delta H^\circ_{rxn} = \left[1 \text{ mol} \times -238.66 \text{ kJ}\cdot\text{mol}^{-1}\right] - \left[\left(1 \text{ mol} \times -110.52 \text{ kJ}\cdot\text{mol}^{-1}\right) + \left(2 \text{ mol} \times 0. \text{ kJ}\cdot\text{mol}^{-1}\right)\right]$$

$$= -128.14 \text{ kJ}$$

$$\Delta S^\circ_{rxn} = \left[S^\circ CH_3OH(l)\right] - \left[S^\circ CO(g) + 2S^\circ H_2(g)\right]$$

$$\Delta S^\circ_{rxn} = \left[1 \text{ mol} \times 126.8 \text{ J}\cdot\text{mol}^{-1}\cdot\text{K}^{-1}\right] - \left[\left(1 \text{ mol} \times 197.56 \text{ J}\cdot\text{mol}^{-1}\cdot\text{K}^{-1}\right) + \left(2 \text{ mol} \times 130.57 \text{ J}\cdot\text{mol}^{-1}\cdot\text{K}^{-1}\right)\right]$$

$$= -331.9 \text{ J}\cdot\text{K}^{-1}$$

At -20°C, $\Delta G^\circ_{253} = \Delta H^\circ - (253 \text{ K})\Delta S^\circ$

$\Delta G^\circ_{253} = -128.14 \times 10^3 \text{ J} - (253 \text{ K})(-331.9 \text{ J}\cdot\text{K}^{-1}) = \mathbf{-4.42 \times 10^4 \text{ J or } -44.2 \text{ kJ}}$

At 50°C, $\Delta G^\circ_{323} = \Delta H^\circ - (323 \text{ K})\Delta S^\circ$

$\Delta G^\circ_{323} = -128.14 \times 10^3 \text{ J} - (323 \text{ K})(-331.9 \text{ J}\cdot\text{K}^{-1}) = \mathbf{-2.09 \times 10^4 \text{ J or } -20.9 \text{ kJ}}$

The reaction has a favorable enthalpy change and an unfavorable entropy change. At low temperatures, the favorable ΔH° term dominates the unfavorable $T\Delta S^\circ$ term. As temperature increases, the $T\Delta S^\circ$ term increases in importance. Hence, the value of ΔG° becomes less negative as temperature increases, and the reaction proceeds to a lesser extent.

15.48 ***See Section 15.4 and Example 15.5.***

(a) For $CH_3COOH(l) + NaOH(s) \rightarrow Na^+(aq) + CH_3COO^-(aq) + H_2O(l)$,

$$\Delta H^\circ_{rxn} = \left[\Delta H^\circ_f Na^+(aq) + \Delta H^\circ_f CH_3COO^-(aq) + \Delta H^\circ_f H_2O(l)\right] - \left[\Delta H^\circ_f CH_3COOH(l) + \Delta H^\circ_f NaOH(l)\right]$$

$$\Delta H^\circ_{rxn} = \left[\left(1\ mol \times -240.12\ kJ \cdot mol^{-1}\right) + \left(1\ mol \times -486.01\ kJ \cdot mol^{-1}\right) + \left(1\ mol \times -285.83\ kJ \cdot mol^{-1}\right)\right]$$

$$-\left[\left(1\ mol \times -484.5\ kJ \cdot mol^{-1}\right) + \left(1\ mol \times -425.61\ kJ \cdot mol^{-1}\right)\right] = -101.8\ kJ$$

$$\Delta S^\circ_{rxn} = \left[S^\circ Na^+(aq) + S^\circ CH_3COO^-(aq) + S^\circ H_2O(l)\right] - \left[S^\circ CH_3COOH(l) + S^\circ NaOH(s)\right]$$

$$\Delta S^\circ_{rxn} = \left[\left(1\ mol \times 59.0\ J \cdot mol^{-1} \cdot K^{-1}\right) + \left(1\ mol \times 86.6\ J \cdot mol^{-1} \cdot K^{-1}\right) + \left(1\ mol \times 69.91\ J \cdot mol^{-1} \cdot K^{-1}\right)\right]$$

$$-\left[\left(1\ mol \times 159.8\ J \cdot mol^{-1} \cdot K^{-1}\right) + \left(1\ mol \times 64.46\ J \cdot mol^{-1} \cdot K^{-1}\right)\right] = -8.8\ J \cdot K^{-1}$$

At 25°C, $\Delta G^\circ_{298} = \Delta H^\circ - (298\ K)\Delta S^\circ$

$$\Delta G^\circ_{298} = -101.8 \text{ x } 10^3\ J - (298\ K)\left(-8.8\ J \cdot K^{-1}\right) = \mathbf{-9.92 \text{ x } 10^4\ J \text{ or } -99.2\ kJ}$$

The reaction has a highly favorable enthalpy change and a slightly unfavorable entropy change. At low temperatures, the highly favorable ΔH° term dominates the slightly unfavorable $T\Delta S^\circ$. Hence, the value of ΔG°_{298} is negative, and the reaction is spontaneous in the forward direction.

(b) For $AgNO_3(s) + Cl^-(aq) \rightarrow AgCl(s) + NO_3^-(aq)$,

$$\Delta H^\circ_{rxn} = \left[\Delta H^\circ_f AgCl(s) + \Delta H^\circ_f NO_3^-(aq)\right] - \left[\Delta H^\circ_f AgNO_3(s) + \Delta H^\circ_f Cl^-(aq)\right]$$

$$\Delta H^\circ_{rxn} = \left[\left(1\ mol \times -127.07\ kJ \cdot mol^{-1}\right) + \left(1\ mol \times -205.0\ kJ \cdot mol^{-1}\right)\right]$$

$$-\left[\left(1\ mol \times -124.39\ kJ \cdot mol^{-1}\right) + \left(1\ mol \times -167.16\ kJ \cdot mol^{-1}\right)\right] = -40.5\ kJ$$

$$\Delta S^\circ_{rxn} = \left[S^\circ AgCl(s) + S^\circ NO_3^-(aq)\right] - \left[S^\circ AgNO_3(s) + S^\circ Cl^-(aq)\right]$$

$$\Delta S^\circ_{rxn} = \left[\left(1\ mol \times 96.2\ J \cdot mol^{-1} \cdot K^{-1}\right) + \left(1\ mol \times 146.4\ J \cdot mol^{-1} \cdot K^{-1}\right)\right]$$

$$-\left[\left(1\ mol \times 140.92\ J \cdot mol^{-1} \cdot K^{-1}\right) + \left(1\ mol \times 56.5\ J \cdot mol^{-1} \cdot K^{-1}\right)\right] = 45.2\ J \cdot K^{-1}$$

At 25°C, $\Delta G^\circ_{298} = \Delta H^\circ - (298\ K)\Delta S^\circ$

$$\Delta G^\circ_{298} = -40.5 \text{ x } 10^3\ J - (298\ K)\left(45.2\ J \cdot K^{-1}\right) = \mathbf{-5.40 \text{ x } 10^4\ J \text{ or } -54.0\ kJ}$$

The reaction has a favorable enthalpy change and a favorable entropy change. Hence, the reaction is spontaneous in the forward direction at all temperatures.

15.50 ***See Section 15.4 and Example 15.6.***

For $CH_3CHO(l) + \frac{5}{2}O_2(g) \rightarrow 2CO_2(g) + 2H_2O(l)$,

$$\Delta H^\circ_{rxn} = \left[2\Delta H^\circ_f CO_2(g) + 2\Delta H^\circ_f H_2O(l)\right] - \left[\Delta H^\circ_f CH_3CHO(l) + \tfrac{5}{2}\Delta H^\circ_f O_2(g)\right]$$

$$\Delta H^\circ_{rxn} = \left[\left(2\text{ mol} \times -393.51\text{ kJ}\cdot\text{mol}^{-1}\right) + \left(2\text{ mol} \times -285.83\text{ kJ}\cdot\text{mol}^{-1}\right)\right]$$

$$-\left[\left(1\text{ mol} \times -192.30\text{ kJ}\cdot\text{mol}^{-1}\right) + \left(\tfrac{5}{2}\text{ mol} \times 0.\text{ kJ}\cdot\text{mol}^{-1}\right)\right] = -1166.38\text{ kJ}$$

$$\Delta S^\circ_{rxn} = \left[2S^\circ CO_2(g) + 2S^\circ H_2O(l)\right] - \left[S^\circ CH_3CHO(l) + \tfrac{5}{2}S^\circ O_2(g)\right]$$

$$\Delta S^\circ_{rxn} = \left[\left(2\text{ mol} \times 213.63\text{ J}\cdot\text{mol}^{-1}\cdot\text{K}^{-1}\right) + \left(2\text{ mol} \times 69.91\text{ J}\cdot\text{mol}^{-1}\cdot\text{K}^{-1}\right)\right]$$

$$-\left[\left(1\text{ mol} \times 160.2\text{ J}\cdot\text{mol}^{-1}\cdot\text{K}^{-1}\right) + \left(\tfrac{5}{2}\text{ mol} \times 205.03\text{ J}\cdot\text{mol}^{-1}\cdot\text{K}^{-1}\right)\right] = -105.7\text{ J}\cdot\text{K}^{-1}$$

The reaction has a highly favorable enthalpy change and an unfavorable entropy change. At low temperatures, the favorable ΔH° term will dominate the unfavorable $T\Delta S^\circ$ term, and the reaction will be spontaneous. As temperature increases, the importance of the unfavorable $T\Delta S^\circ$ term will increase, and eventually the reaction will become nonspontaneous.

15.52 ***See Section 15.4 and Example 15.6.***

For $N_2(g) + 3H_2(g) \rightarrow 2NH_3(g)$,

$$\Delta H^\circ_{rxn} = \left[2\Delta H^\circ_f NH_3(g)\right] - \left[\Delta H^\circ_f N_2(g) + 3\Delta H^\circ_f H_2(g)\right]$$

$$\Delta H^\circ_{rxn} = \left[\left(2\text{ mol} \times -46.11\text{ kJ}\cdot\text{mol}^{-1}\right)\right] - \left[\left(1\text{ mol} \times 0.\text{ kJ}\cdot\text{mol}^{-1}\right) + \left(3\text{ mol} \times 0.\text{ kJ}\cdot\text{mol}^{-1}\right)\right] = -92.22\text{ kJ}$$

$$\Delta S^\circ_{rxn} = \left[2S^\circ NH_3(g)\right] - \left[S^\circ N_2(g) + 3S^\circ H_2(g)\right]$$

$$\Delta S^\circ_{rxn} = \left[\left(2\text{ mol} \times 192.34\text{ J}\cdot\text{mol}^{-1}\cdot\text{K}^{-1}\right)\right] - \left[\left(1\text{ mol} \times 191.50\text{ J}\cdot\text{mol}^{-1}\cdot\text{K}^{-1}\right) + \left(3\text{ mol} \times 130.57\text{ J}\cdot\text{mol}^{-1}\cdot\text{K}^{-1}\right)\right]$$
$$= -198.53\text{ J}\cdot\text{K}^{-1}$$

The reaction has a favorable enthalpy change and an unfavorable entropy change. At low temperatures, the favorable ΔH° term will dominate the unfavorable $T\Delta S^\circ$ term, and the reaction will be spontaneous. As temperature increases, the importance of the unfavorable $T\Delta S^\circ$ term will increase, and eventually the reaction will become nonspontaneous.

15.54 ***See Section 15.4 and Example 15.6.***

(a) For $CO(g) + Cl_2(g) \rightarrow COCl_2(g)$,

$$\Delta H^\circ_{rxn} = \left[\Delta H^\circ_f COCl_2(g)\right] - \left[\Delta H^\circ_f CO(g) + \Delta H^\circ_f Cl_2(g)\right]$$

$$\Delta H^\circ_{rxn} = \left[\left(1 \text{ mol} \times -218.8 \text{ kJ}\cdot\text{mol}^{-1}\right)\right] - \left[\left(1 \text{ mol} \times -110.52 \text{ kJ}\cdot\text{mol}^{-1}\right) + \left(1 \text{ mol} \times 0. \text{ kJ}\cdot\text{mol}^{-1}\right)\right] = -108.28 \text{ kJ}$$

$$\Delta S^\circ_{rxn} = \left[S^\circ COCl_2(g)\right] - \left[S^\circ CO(g) + S^\circ Cl_2(g)\right]$$

$$\Delta S^\circ_{rxn} = \left[\left(1 \text{ mol} \times 283.53 \text{ J}\cdot\text{mol}^{-1}\cdot\text{K}^{-1}\right)\right] - \left[\left(1 \text{ mol} \times 197.56 \text{ J}\cdot\text{mol}^{-1}\cdot\text{K}^{-1}\right) + \left(1 \text{ mol} \times 222.96 \text{ J}\cdot\text{mol}^{-1}\cdot\text{K}^{-1}\right)\right]$$

$$= -136.99 \text{ J}\cdot\text{K}^{-1}$$

At 25°C, $\Delta G^\circ_{298} = \Delta H^\circ - (298 \text{ K})\Delta S^\circ$

$\Delta G^\circ_{298} = -108.28 \times 10^3 \text{ J} - (298 \text{ K})(-136.99 \text{ J}\cdot\text{K}^{-1}) =$ **-6.746×10^4 J** or **-67.46 kJ**

At the temperature at which there is no net driving force in either direction, $\Delta G^\circ = 0$.
At this temperature, $0 = \Delta H^\circ - T\Delta S^\circ$. Hence,

$$T = \frac{\Delta H^\circ}{\Delta S^\circ} \qquad T = \frac{-108.28 \times 10^3 \text{ J}}{-136.99 \text{ J}\cdot\text{K}^{-1}} = 790 \text{ K}$$

The reaction is spontaneous in the direction written at 298 K ($\Delta G^\circ_{298} = -$) and experiences no net driving force in either direction at 790 K ($\Delta G^\circ_{790} = 0$). Above 790 K the reaction becomes nonspontaneous in the direction written. Hence, the reaction is spontaneous **up to 790 K**.

(b) For $NO(g) + \frac{1}{2}O_2(g) \rightarrow NO_2(g)$,

$$\Delta H^\circ_{rxn} = \left[\Delta H^\circ_f NO_2(g)\right] - \left[\Delta H^\circ_f NO(g) + \tfrac{1}{2}\Delta H^\circ_f O_2(g)\right]$$

$$\Delta H^\circ_{rxn} = \left[\left(1 \text{ mol} \times 33.18 \text{ kJ}\cdot\text{mol}^{-1}\right)\right] - \left[\left(1 \text{ mol} \times 90.25 \text{ kJ}\cdot\text{mol}^{-1}\right) + \left(\tfrac{1}{2} \text{ mol} \times 0. \text{ kJ}\cdot\text{mol}^{-1}\right)\right] = -57.07 \text{ kJ}$$

$$\Delta S^\circ_{rxn} = \left[S^\circ NO_2(g)\right] - \left[S^\circ NO(g) + \tfrac{1}{2}S^\circ O_2(g)\right]$$

$$\Delta S^\circ_{rxn} = \left[\left(1 \text{ mol} \times 239.95 \text{ J}\cdot\text{mol}^{-1}\cdot\text{K}^{-1}\right)\right] - \left[\left(1 \text{ mol} \times 210.65 \text{ J}\cdot\text{mol}^{-1}\cdot\text{K}^{-1}\right) + \left(\tfrac{1}{2} \text{ mol} \times 205.03 \text{ J}\cdot\text{mol}^{-1}\cdot\text{K}^{-1}\right)\right]$$

$$= -73.22 \text{ J}\cdot\text{K}^{-1}$$

At 25°C, $\Delta G^\circ_{298} = \Delta H^\circ - (298 \text{ K})\Delta S^\circ$

$\Delta G^\circ_{298} = -57.07 \times 10^3 \text{ J} - (298 \text{ K})(-73.22 \text{ J}\cdot\text{K}^{-1}) =$ **-3.52×10^3 J** or **-3.52 kJ**

At the temperature at which there is no net driving force in either direction, $\Delta G^\circ = 0$.
At this temperature, $0 = \Delta H^\circ - T\Delta S^\circ$. Hence,

$$T = \frac{\Delta H^\circ}{\Delta S^\circ} \qquad T = \frac{-57.07 \times 10^3 \text{ J}}{-73.22 \text{ J}\cdot\text{K}^{-1}} = 779 \text{ K}$$

The reaction is spontaneous in the direction written at 298 K ($\Delta G^\circ_{298} = -$) and experiences no net driving force in either direction at 779 K ($\Delta G^\circ_{779} = 0$). Above 779 K the reaction becomes nonspontaneous in the direction written. Hence, the reaction is spontaneous **up to 779 K.**

15.56 ***See Section 15.4 and Example 15.6.***

(a) For $\Delta H^\circ = +53.4$ kJ and $\Delta S^\circ = +112.4$ J·K^{-1},

$\Delta G^\circ_{298} = \Delta H^\circ - (298\ K)\Delta S^\circ$ $\quad$ $\Delta G^\circ_{298} = 53.4 \times 10^3\ J - (298\ K)(112.4\ J \cdot K^{-1}) = 1.99 \times 10^4\ J$ or 19.9 kJ

At the temperature at which there is no net driving force in either direction, $\Delta G^\circ = 0$.
At this temperature, $0 = \Delta H^\circ - T\Delta S^\circ$. Hence,

$$T = \frac{\Delta H^\circ}{\Delta S^\circ} \qquad T = \frac{53.4 \times 10^3\ J}{112.4\ J \cdot K^{-1}} = 475\ K$$

The reaction is nonspontaneous in the direction written at 298 K ($\Delta G^\circ_{298} = +$) and experiences no net driving force in either direction at 475 K ($\Delta G^\circ_{475} = 0$). Above 475 K the reaction becomes spontaneous in the direction written. Hence, the reaction is spontaneous **above 475 K.**

(b) For $\Delta H^\circ = -29.4$ kJ and $\Delta S^\circ = -91.2$ J·K^{-1},

$\Delta G^\circ_{298} = \Delta H^\circ - (298\ K)\Delta S^\circ$ $\quad$ $\Delta G^\circ_{298} = -29.4 \times 10^3\ J - (298\ K)(-91.2\ J \cdot K^{-1}) = -2.2 \times 10^3\ J$ or -22 kJ

At the temperature at which there is no net driving force in either direction, $\Delta G^\circ = 0$.
At this temperature, $0 = \Delta H^\circ - T\Delta S^\circ$. Hence,

$$T = \frac{\Delta H^\circ}{\Delta S^\circ} \qquad T = \frac{-29.4 \times 10^3\ J}{-91.2\ J \cdot K^{-1}} = 322\ K$$

The reaction is spontaneous in the direction written at 298 K ($\Delta G^\circ_{298} = -$) and experiences no net driving force in either direction at 322 K ($\Delta G^\circ_{322} = 0$). Above 322 K the reaction becomes nonspontaneous in the direction written. Hence, the reaction is spontaneous **up to 322 K.**

15.58 ***See Section 15.4 and Example 15.6.***

(a) For $NO_2(g) + N_2O(g) \rightarrow 3NO(g)$,

$$\Delta H^\circ_{rxn} = \left[3\Delta H^\circ_f NO(g)\right] - \left[\Delta H^\circ_f NO_2(g) + \Delta H^\circ_f N_2O(g)\right]$$

$$\Delta H^\circ_{rxn} = \left[\left(3\ mol \times 90.25\ kJ \cdot mol^{-1}\right)\right] - \left[\left(1\ mol \times 33.18\ kJ \cdot mol^{-1}\right) + \left(1\ mol \times 82.05\ kJ \cdot mol^{-1}\right)\right] = 155.52\ kJ$$

$$\Delta S^\circ_{rxn} = \left[3S^\circ NO(g)\right] - \left[S^\circ NO_2(g) + S^\circ N_2O(g)\right]$$

$$\Delta S^\circ_{rxn} = \left[\left(3\ mol \times 210.65\ J \cdot mol^{-1} \cdot K^{-1}\right)\right] - \left[\left(1\ mol \times 239.95\ J \cdot mol^{-1} \cdot K^{-1}\right) + \left(1\ mol \times 219.74\ J \cdot mol^{-1} \cdot K^{-1}\right)\right]$$
$$= 172.26\ J \cdot K^{-1}$$

$\Delta G^\circ_{400} = \Delta H^\circ - (400\ \text{K})\Delta S^\circ$ $\quad \Delta G^\circ_{400} = 155.52 \times 10^3\ \text{J} - (400\ \text{K})(172.26\ \text{J}\cdot\text{K}^{-1}) = \mathbf{8.66 \times 10^4\ J}$ or **86.6 kJ**

$\Delta G^\circ_{600} = \Delta H^\circ - (600\ \text{K})\Delta S^\circ$ $\quad \Delta G^\circ_{600} = 155.52 \times 10^3\ \text{J} - (600\ \text{K})(172.26\ \text{J}\cdot\text{K}^{-1}) = \mathbf{5.22 \times 10^4\ J}$ or **52.2 kJ**

Note: The reaction has an unfavorable enthalpy change and a favorable entropy change; the latter being due to the increase in moles of gas. At low temperatures, the unfavorable ΔH° term dominates the favorable $T\Delta S^\circ$ term. As temperature increases, the importance of the $T\Delta S^\circ$ increases. Eventually, the reaction will become spontaneous.

(b) For $2NH_3(g) \rightarrow N_2H_4(l) + H_2(g)$,

$$\Delta H^\circ_{rxn} = [\Delta H^\circ_f N_2H_4(l) + \Delta H^\circ_f H_2(g)] - [2\Delta H^\circ_f NH_3(g)]$$

$$\Delta H^\circ_{rxn} = [(1\ \text{mol} \times 50.63\ \text{kJ}\cdot\text{mol}^{-1}) + (1\ \text{mol} \times 0.\ \text{kJ}\cdot\text{mol}^{-1})] - [(2\ \text{mol} \times -46.11\ \text{kJ}\cdot\text{mol}^{-1})] = 142.85\ \text{kJ}$$

$$\Delta S^\circ_{rxn} = [S^\circ N_2H_4(l) + S^\circ H_2(g)] - [2S^\circ NH_3(g)]$$

$$\Delta S^\circ_{rxn} = [(1\ \text{mol} \times 121.21\ \text{J}\cdot\text{mol}^{-1}\cdot\text{K}^{-1}) + (1\ \text{mol} \times 130.57\ \text{J}\cdot\text{mol}^{-1}\cdot\text{K}^{-1})] - [(2\ \text{mol} \times 192.34\ \text{J}\cdot\text{mol}^{-1}\cdot\text{K}^{-1})]$$
$$= -132.90\ \text{J}\cdot\text{K}^{-1}$$

$\Delta G^\circ_{400} = \Delta H^\circ - (400\ \text{K})\Delta S^\circ$ $\quad \Delta G^\circ_{400} = 142.85 \times 10^3\ \text{J} - (400\ \text{K})(-132.90\ \text{J}\cdot\text{K}^{-1}) = \mathbf{1.96 \times 10^5\ J}$ or **196 kJ**

$\Delta G^\circ_{600} = \Delta H^\circ - (600\ \text{K})\Delta S^\circ$ $\quad \Delta G^\circ_{600} = 142.85 \times 10^3\ \text{J} - (600\ \text{K})(-132.90\ \text{J}\cdot\text{K}^{-1}) = \mathbf{2.23 \times 10^5\ J}$ or **223 kJ**

Note: The reaction has an unfavorable enthalpy change and an unfavorable entropy change; the latter being due to formation of one mole of liquid and one mole of gas from two moles of gas. The reaction is therefore nonspontaneous at all temperatures.

15.60 ***See Section 15.4.***

At low temperatures the ΔH° term dominates the $T\Delta S^\circ$ term in $\Delta G^\circ = \Delta H^\circ - T\Delta S^\circ$. As temperature increases, the importance of the $T\Delta S^\circ$ term increases and eventually dominates ΔH°. Since the reaction proceeds spontaneously at low temperatures, it must have a favorable free energy change and favorable enthalpy change. The signs of ΔG° and ΔH° must therefore be negative. Similarly, since the reaction becomes nonspontaneous in the forward direction at elevated temperatures, it must have an unfavorable entropy change. The sign of ΔS° must therefore be negative.

Note: Only reactions having like signs for both ΔH° and ΔS° exhibit temperature dependent spontaneity. This is due to the fact one of these is unfavorable when both have like signs.

15.62 ***See Section 15.3, 4 and 5, Example 15.6 and the Solution for 15.14.***

For $CO(g) + 2H_2(g) \rightarrow CH_3OH(g)$, $\Delta H^\circ_{rxn} = -90.14$ kJ and $\Delta S^\circ_{rxn} = -367.92\ \text{J}\cdot\text{K}^{-1}$.

$\Delta G^\circ_{298} = \Delta H^\circ - (298\ \text{K})\Delta S^\circ$ $\quad \Delta G^\circ_{298} = -90.14 \times 10^3\ \text{J} - (298\ \text{K})(-367.92\ \text{J}\cdot\text{K}^{-1}) = 1.95 \times 10^4$ J or 19.5 kJ.

(a) False: The spontaneous direction of a reaction is determined by the sign of ΔG°_{rxn}, not ΔH°_{rxn}.

(b) False: The sign of ΔG°_{298} is positive. Hence, the reaction is nonspontaneous at 298 K.

(c) False: The reaction has a favorable enthalpy change and an unfavorable entropy change. As temperature increases, the importance of the unfavorable $T\Delta S^\circ$ term in $\Delta G^\circ = \Delta H^\circ - T\Delta S^\circ$ will increase and the reaction will remain nonspontaneous in the direction written.

(d) False: The reaction becomes nonspontaneous before 298 K is reached.

(e) False: Since ΔH°_{rxn} is negative, ln K_{eq} will decrease at higher temperatures.

15.64 ***See Section 15.5.***

The effect of temperature change on an equilibrium constant is related to the sign of ΔH°_{rxn}. Since the equilibrium constant decreases as temperature increases, we can conclude that ΔH°_{rxn} is **negative**.

15.66 ***See Section 15.4.***

(a) $H_2O(s) \rightarrow H_2O(l)$ (b) $H_2O(l) \rightarrow H_2O(g)$ (c) $CO_2(s) \rightarrow CO_2(g)$

15.68 ***See Section 15.4.***

For $CH_3OH(l) \rightarrow CH_3OH(g)$,

$$\Delta H^\circ_{rxn} = \left[\Delta H^\circ_f CH_3OH(g)\right] - \left[\Delta H^\circ_f CH_3OH(l)\right]$$

$$\Delta H^\circ_{rxn} = \left[\left(1\ \text{mol} \times -200.66\ \text{kJ}\cdot\text{mol}^{-1}\right)\right] - \left[\left(1\ \text{mol} \times -238.66\ \text{kJ}\cdot\text{mol}^{-1}\right)\right] = 38.00\ \text{kJ}$$

$$\Delta S^\circ_{rxn} = \left[S^\circ CH_3OH(g)\right] - \left[S^\circ CH_3OH(l)\right]$$

$$\Delta S^\circ_{rxn} = \left[\left(1\ \text{mol} \times 239.70\ \text{J}\cdot\text{mol}^{-1}\cdot\text{K}^{-1}\right)\right] - \left[\left(1\ \text{mol} \times 126.8\ \text{J}\cdot\text{mol}^{-1}\cdot\text{K}^{-1}\right)\right] = 112.9\ \text{J}\cdot\text{K}^{-1}$$

$$\Delta G^\circ_{353} = \Delta H^\circ - (353\ \text{K})\Delta S^\circ \qquad \Delta G^\circ_{353} = 38.00 \times 10^3\ \text{J} - (353\ \text{K})\left(112.9\ \text{J}\cdot\text{K}^{-1}\right) = \mathbf{-1.85 \times 10^3\ J}\ \text{or}\ \mathbf{-1.85\ kJ}$$

The negative value of ΔG°_{353} indicates the vaporization of methanol is spontaneous at 80°C and 1 atm.

Note: See Solution for 15.70 below.

15.70 ***See Section 15.4 and the Solution for 15.68.***

At the normal boiling point, the liquid is in equilibrium with the vapor at one atmosphere, so ΔG° is zero. Hence, $0 = \Delta H^\circ - T\Delta S^\circ$ and

$$T = \frac{\Delta H^\circ}{\Delta S^\circ} \qquad T = \frac{38.00 \times 10^3\ \text{J}}{112.9\ \text{J}\cdot\text{K}^{-1}} = 336.6\ \text{K or } 63.4^\circ\text{C}$$

According to the Handbook of Chemistry and Physics, the normal boiling point of methanol is 64.96°C. Hence, the calculated and experimental values are in excellent agreement.

15.72 *See Section 15.5.*

For $H_2O(l) \rightarrow H_2O(g)$, $K_{eq} = P_{H_2O}$. Since $\ln K_{eq} = \ln P_{H_2O} = \frac{-\Delta H^\circ}{R}\left(\frac{1}{T}\right) + \frac{\Delta S^\circ}{R}$, we can plot $\ln P_{H_2O}$ versus $\frac{1}{T}$ to determine the values of ΔH° and ΔS°.

The slope of the line will be equal to $\frac{-\Delta H^\circ}{R}$ and the intercept will be $\frac{\Delta S^\circ}{R}$.

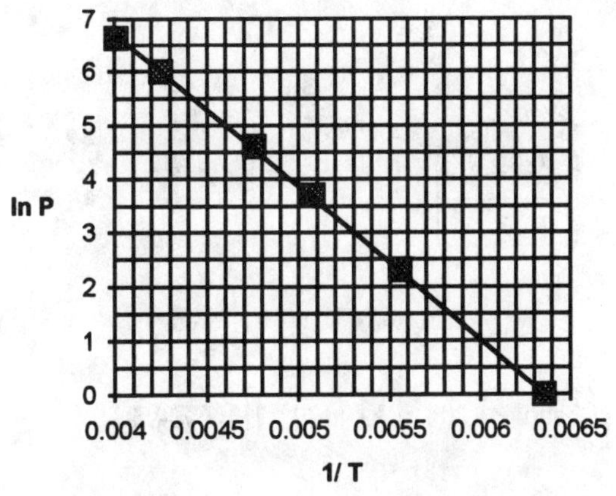

Temp °C	Temp K	$\frac{1}{T}$ K^{-1}	P_{H_2O} torr	$\ln P_{H_2O}$
-115.7	157.3	6.36×10^{-3}	1	0.
-93.3	179.7	5.56×10^{-3}	10	2.30
-76.2	196.8	5.08×10^{-3}	40	3.69
-62.7	210.3	4.76×10^{-3}	100	4.61
-37.8	235.2	4.25×10^{-3}	400	5.99
-23.7	249.3	4.01×10^{-3}	760	6.63

A linear regression analysis of the data yields a slope of -2.83×10^3 K and an intercept of 18.0 with a correlation coefficient of 0.9999 indicating excellent agreement between the experimental data and the equation.

$$\text{slope} = \frac{-\Delta H^\circ}{R} = -2.83 \times 10^3 \text{ K}$$

$$\Delta H^\circ = (2.83 \times 10^3 \text{ K})(8.314 \text{ J}\cdot\text{mol}^{-1}\cdot\text{K}^{-1}) = \mathbf{2.36 \times 10^4 \ J\cdot mol^{-1} \ or \ 23.6 \ kJ\cdot mol^{-1}}$$

$$\text{intercept} = \frac{\Delta S^\circ}{R} = 18.0 \qquad \Delta S^\circ = (18.0)(8.314 \text{ J}\cdot\text{mol}^{-1}\cdot\text{K}^{-1}) = \mathbf{150 \ J\cdot mol^{-1}\cdot K^{-1}}$$

Note: If your calculator does not have linear regression capability , you can use a computer graphics program to do the linear regression analysis. In the event neither is available, the slope can be approximated from the graph:

$$\text{slope} = \frac{0.75 - 5.3}{0.0061 \text{ K}^{-1} - 0.0045 \text{ K}^{-1}} = -2.84 \times 10^4 \text{ K}$$

The intercept cannot, however, be determined from a graph of reasonable size, since it occurs when $1/T = 0$. The value of the intercept must be determined by substituting $\ln P_{H_2O}$, T and ΔH° into the equation for several data points.

15.74 *See Section 15.5 and Example 15.7.*

(a) For PCl_3(g, 0.2 atm) + $Cl_2$9g, 0.5 atm) $\rightarrow$ PCl_5(g, 0.2 atm) at 438 K,

$$\Delta G = \Delta G^\circ + RT\ln Q \qquad \Delta G = \Delta G^\circ + (8.314 \text{ J}\cdot\text{mol}^{-1}\cdot\text{K}^{-1})(438 \text{ K})\ln\left(\frac{P_{PCl_5}}{P_{PCl_3}P_{Cl_2}}\right)$$

$$\Delta G^\circ_{rxn} = \left[\Delta G^\circ_f PCl_5(g)\right] - \left[\Delta G^\circ_f PCl_3(g) + \Delta G^\circ_f Cl_2(g)\right]$$

$$\Delta G^\circ_{rxn} = \left[\left(1\ \text{mol} \times -305.0\ \text{kJ}\cdot\text{mol}^{-1}\right)\right] - \left[\left(1\ \text{mol} \times -267.8\ \text{kJ}\cdot\text{mol}^{-1}\right) + \left(1\ \text{mol} \times 0.\ \text{kJ}\cdot\text{mol}^{-1}\right)\right] = -32.7\ \text{kJ}$$

Hence, $\Delta G = -37.2 \text{ x } 10^3\ \text{J} + \left(8.314\ \text{J}\cdot\text{mol}^{-1}\cdot\text{K}^{-1}\right)(438\ \text{K})\ln\left(\dfrac{0.2}{(0.2)(0.5)}\right) = \mathbf{-3.47 \text{ x } 10^4}$ **J or -34.7 kJ.**

The negative sign for ΔG indicates the reaction is **spontaneous** in the forward direction under the given conditions.

(b) For $COCl_2$(g, 0.1 atm) $\rightarrow$ CO(g, 1.0 atm) + Cl_2(g, 0.5 atm) at 350 K,

$$\Delta G = \Delta G^\circ + RT\ln Q \quad \Delta G = \Delta G^\circ + \left(8.314\ \text{J}\cdot\text{mol}^{-1}\cdot\text{K}^{-1}\right)(350\ \text{K})\ln\left(\frac{P_{CO}P_{Cl_2}}{P_{COCl_2}}\right)$$

$$\Delta G^\circ_{rxn} = \left[\Delta G^\circ_f CO(g) + \Delta G^\circ_f Cl_2(g)\right] - \left[\Delta G^\circ_f COCl_2(g)\right]$$

$$\Delta G^\circ_{rxn} = \left[\left(1\ \text{mol} \times -137.15\ \text{kJ}\cdot\text{mol}^{-1}\right) + \left(1\ \text{mol} \times 0.\ \text{kJ}\cdot\text{mol}^{-1}\right)\right] - \left[\left(1\ \text{mol} \times -204.6\ \text{kJ}\cdot\text{mol}^{-1}\right)\right] = 67.4\ \text{kJ}$$

Hence, $\Delta G = 67.4 \text{ x } 10^3\ \text{J} + \left(8.314\ \text{J}\cdot\text{mol}^{-1}\cdot\text{K}^{-1}\right)(350\ \text{K})\ln\left(\dfrac{(1.0)(0.5)}{0.1}\right) = \mathbf{7.21 \text{ x } 10^4}$ **J or 72.1 kJ.**

The positive sign for ΔG indicates the reaction is **nonspontaneous** in the forward direction under the given conditions. Under these conditions, the reverse reaction is spontaneous.

15.76 ***See Section 15.5 and Example 15.8.***

For $PCl_5(g) \rightleftarrows PCl_3(g) + Cl_2(g)$,

$$\Delta H^\circ_{rxn} = \left[\Delta H^\circ_f PCl_3(g) + \Delta H^\circ_f Cl_2(g)\right] - \left[\Delta H^\circ_f PCl_5(g)\right]$$

$$\Delta H^\circ_{rxn} = \left[\left(1\ \text{mol} \times -287.0\ \text{kJ}\cdot\text{mol}^{-1}\right) + \left(1\ \text{mol} \times 0.\ \text{kJ}\cdot\text{mol}^{-1}\right)\right] - \left[\left(1\ \text{mol} \times -374.9\ \text{kJ}\cdot\text{mol}^{-1}\right)\right] = 87.9\ \text{kJ}$$

$$\Delta S^\circ_{rxn} = \left[S^\circ PCl_3(g) + S^\circ Cl_2(g)\right] - \left[S^\circ PCl_5(g)\right]$$

$$\Delta S^\circ_{rxn} = \left[\left(1\ \text{mol} \times 311.67\ \text{J}\cdot\text{mol}^{-1}\cdot\text{K}^{-1}\right) + \left(1\ \text{mol} \times 222.96\ \text{J}\cdot\text{mol}^{-1}\cdot\text{K}^{-1}\right)\right] - \left[\left(1\ \text{mol} \times 364.47\ \text{J}\cdot\text{mol}^{-1}\cdot\text{K}^{-1}\right)\right]$$
$$= 170.16\ \text{J}\cdot\text{K}^{-1}$$

(a) At 25°C, $\Delta G^\circ_{298} = \Delta H^\circ - (298\ \text{K})\Delta S^\circ$

$$\Delta G^\circ_{298} = 87.9 \text{ x } 10^3\ \text{J} - (298\ \text{K})\left(170.16\ \text{J}\cdot\text{K}^{-1}\right) = 3.719 \text{ x } 10^4\ \text{J or } 37.19\ \text{kJ}$$

$$K_{eq} = e^{\frac{-\Delta G^\circ}{RT}} \text{ and } \frac{-\Delta G^\circ}{RT} = \frac{-3.719 \times 10^4 \text{ J}\cdot\text{mol}^{-1}}{(8.314 \text{ J}\cdot\text{mol}^{-1}\cdot\text{K}^{-1})(298 \text{ K})} = -15.0. \text{ Therefore, } K_{eq} = e^{-15.0} = 3.06 \times 10^{-7}.$$

(b) At 250°C, $\Delta G^\circ_{523} = \Delta H^\circ - (523 \text{ K})\Delta S^\circ$

$\Delta G^\circ_{523} = 87.9 \times 10^3 \text{ J} - (523 \text{ K})(170.16 \text{ J}\cdot\text{K}^{-1}) = -1.09 \times 10^3 \text{ J}$ or -1.09 kJ

$$K_{eq} = e^{\frac{-\Delta G^\circ}{RT}} \text{ and } \frac{-\Delta G^\circ}{RT} = \frac{-(-1.09 \times 10^3 \text{ J}\cdot\text{mol}^{-1})}{(8.314 \text{ J}\cdot\text{mol}^{-1}\cdot\text{K}^{-1})(298 \text{ K})} = 0.251. \text{ Therefore, } K_{eq} = e^{0.251} = 1.29.$$

Note: The reaction is endothermic, so increasing the temperature from 25°C to 298°C increases the value of K_{eq} from 3.06×10^{-7} to 1.29.

15.78 *See Section 15.5 and Example 15.8.*

At 10.0 K, $\Delta G^\circ_{10} = \Delta H^\circ - (10.0 \text{ K})\Delta S^\circ$

$\Delta G^\circ_{10} = 10.0 \times 10^3 \text{ J} - (10.0 \text{ K})(100 \text{ J}\cdot\text{K}^{-1}) = 9.0 \times 10^3 \text{ J}$ or 9.0 kJ

$$K_{eq} = e^{\frac{-\Delta G^\circ}{RT}} \text{ and } \frac{-\Delta G^\circ}{RT} = \frac{-(9.0 \times 10^3 \text{ J}\cdot\text{mol}^{-1})}{(8.314 \text{ J}\cdot\text{mol}^{-1}\cdot\text{K}^{-1})(10.0 \text{ K})} = -1.1 \times 10^2.$$

Therefore, $K_{eq} = e^{-1.1 \times 10^2} = \mathbf{1.7 \times 10^{-48}}$.

At 100 K, $\Delta G^\circ_{100} = \Delta H^\circ - (100 \text{ K})\Delta S^\circ$

$\Delta G^\circ_{100} = 10.0 \times 10^3 \text{ J} - (100 \text{ K})(100 \text{ J}\cdot\text{K}^{-1}) = 0.$ kJ

$$K_{eq} = e^{\frac{-\Delta G^\circ}{RT}} \text{ and } \frac{-\Delta G^\circ}{RT} = \frac{-(0. \text{ J}\cdot\text{mol}^{-1})}{(8.314 \text{ J}\cdot\text{mol}^{-1}\cdot\text{K}^{-1})(100 \text{ K})} = 0.. \text{ Therefore, } K_{eq} = e^{0.} = \mathbf{1}.$$

At 1,000 K, $\Delta G^\circ_{1,000} = \Delta H^\circ - (1,000 \text{ K})\Delta S^\circ$

$\Delta G^\circ_{1,000} = 10.0 \times 10^3 \text{ J} - (1,000 \text{ K})(100 \text{ J}\cdot\text{K}^{-1}) = -9.00 \times 10^4 \text{ J}$ or -90.0 kJ

$$K_{eq} = e^{\frac{-\Delta G^\circ}{RT}} \text{ and } \frac{-\Delta G^\circ}{RT} = \frac{-(-9.00 \times 10^4 \text{ J}\cdot\text{mol}^{-1})}{(8.314 \text{ J}\cdot\text{mol}^{-1}\cdot\text{K}^{-1})(1,000 \text{ K})} = 10.8. \text{ Therefore, } K_{eq} = e^{10.8} = \mathbf{4.9 \times 10^4}.$$

15.80 ***See Section 15.5 and Example 15.8.***

At 10.0 K, $\Delta G^\circ_{10} = \Delta H^\circ - (10.0\ K)\Delta S^\circ$

$\Delta G^\circ_{10} = -10.0 \times 10^3\ J - (10.0\ K)(100\ J \cdot K^{-1}) = \mathbf{-1.1 \times 10^4\ J\ or\ -11\ kJ}$

$$K_{eq} = e^{\frac{-\Delta G^\circ}{RT}} \text{ and } \frac{-\Delta G^\circ}{RT} = \frac{-(-1.1 \times 10^4\ J \cdot mol^{-1})}{(8.314\ J \cdot mol^{-1} \cdot K^{-1})(10.0\ K)} = 1.3 \times 10^2. \text{ Therefore, } K_{eq} = e^{1.3 \times 10^2} = \mathbf{2.9 \times 10^{57}}.$$

At 100 K, $\Delta G^\circ_{100} = \Delta H^\circ - (100\ K)\Delta S^\circ$

$\Delta G^\circ_{100} = -10.0 \times 10^3\ J - (100\ K)(100\ J \cdot K^{-1}) = \mathbf{-2.00 \times 10^4\ J\ or\ -20.0\ kJ}$

$$K_{eq} = e^{\frac{-\Delta G^\circ}{RT}} \text{ and } \frac{-\Delta G^\circ}{RT} = \frac{-(-2.00 \times 10^4\ J \cdot mol^{-1})}{(8.314\ J \cdot mol^{-1} \cdot K^{-1})(100\ K)} = 24.1. \text{ Therefore, } K_{eq} = e^{24.1} = \mathbf{2.93 \times 10^{10}}.$$

At 1,000 K, $\Delta G^\circ_{1,000} = \Delta H^\circ - (1,000\ K)\Delta S^\circ$

$\Delta G^\circ_{1,000} = -10.0 \times 10^3\ J - (1,000\ K)(100\ J \cdot K^{-1}) = \mathbf{-1.10 \times 10^5\ J\ or\ -110\ kJ}$

$$K_{eq} = e^{\frac{-\Delta G^\circ}{RT}} \text{ and } \frac{-\Delta G^\circ}{RT} = \frac{-(-1.10 \times 10^5\ J \cdot mol^{-1})}{(8.314\ J \cdot mol^{-1} \cdot K^{-1})(1,000\ K)} = 13.2. \text{ Therefore, } K_{eq} = e^{13.2} = \mathbf{5.40 \times 10^5}.$$

Note: The reaction is exothermic, so increasing the temperature from 10.0 K to 100 K to 1,000 K decreases the value of K_{eq} from 2.9×10^{56} to 28.0×10^{10} to 5.40×10^5.

15.82 ***See Section 15.5 and Example 15.7.***

For $CO(g) + Cl_2(g) \rightleftarrows COCl_2(g)$,

$$\Delta G^\circ_{rxn} = [\Delta G^\circ_f COCl_2(g)] - [\Delta G^\circ_f CO(g) + \Delta G^\circ_f Cl_2(g)]$$

$$\Delta G^\circ_{rxn} = [(1\ mol \times -204.6\ kJ \cdot mol^{-1})] - [(1\ mol \times -137.15\ kJ \cdot mol^{-1}) + (1\ mol \times 0.\ kJ \cdot mol^{-1})] = -67.5\ kJ$$

For $CO(g, 2.0\ atm) + Cl_2(g, 1.0\ atm) \rightleftarrows COCl_2(g, 0.10\ atm)$,

$$\Delta G = \Delta G^\circ + RT\ln Q \qquad \Delta G = \Delta G^\circ + (8.314\ J \cdot mol^{-1} \cdot K^{-1})(576\ K)\ln\left(\frac{P_{COCl_2}}{P_{CO}P_{Cl_2}}\right)$$

$$\Delta G = -67.5 \times 10^3\ J + (8.314\ J\cdot mol^{-1}\cdot K^{-1})(576\ K)\ln\left(\frac{0.10}{(2.0)(1.0)}\right) = \mathbf{-8.18 \times 10^4\ J}\text{ or }\mathbf{-81.8\ kJ}.$$

15.84 ***See Section 15.5.***

(a) For $N_2O(g) + H_2(g) \rightleftarrows N_2(g) + H_2O(l)$,

$$\Delta H^\circ_{rxn} = \left[\Delta H^\circ_f N_2(g) + \Delta H^\circ_f H_2O(l)\right] - \left[\Delta H^\circ_f N_2O(g) + \Delta H^\circ_f H_2(g)\right]$$

$$\Delta H^\circ_{rxn} = \left[(1\ mol \times 0.\ kJ\cdot mol^{-1}) + (1\ mol \times -285.83\ kJ\cdot mol^{-1})\right]$$

$$-\left[(1\ mol \times 82.05\ kJ\cdot mol^{-1}) + (1\ mol \times 0.\ kJ\cdot mol^{-1})\right] = -367.88\ kJ$$

Since the reaction is exothermic, we should expect the value of the equilibrium constant to **decrease** with increasing temperature.

(b) For $CO(g) + Cl_2(g) \rightleftarrows COCl_2(g)$,

$$\Delta H^\circ_{rxn} = \left[\Delta H^\circ_f COCl_2(g)\right] - \left[\Delta H^\circ_f CO(g) + \Delta H^\circ_f Cl_2(g)\right]$$

$$\Delta H^\circ_{rxn} = \left[(1\ mol \times -218.8\ kJ\cdot mol^{-1})\right] - \left[(1\ mol \times -110.52\ kJ\cdot mol^{-1}) + (1\ mol \times 0.\ kJ\cdot mol^{-1})\right] = -108.28\ kJ$$

Since the reaction is exothermic, we should expect the value of the equilibrium constant to **decrease** with increasing temperature.

(c) For $CO(g) + H_2O(g) \rightleftarrows CO_2(g) + H_2(g)$,

$$\Delta H^\circ_{rxn} = \left[\Delta H^\circ_f CO_2(g) + \Delta H^\circ_f H_2(g)\right] - \left[\Delta H^\circ_f CO(g) + \Delta H^\circ_f H_2O(g)\right]$$

$$\Delta H^\circ_{rxn} = \left[(1\ mol \times -393.15\ kJ\cdot mol^{-1}) + (1\ mol \times 0.\ kJ\cdot mol^{-1})\right]$$

$$-\left[(1\ mol \times -110.52\ kJ\cdot mol^{-1}) + (1\ mol \times -241.82\ kJ\cdot mol^{-1})\right] = -40.81\ kJ$$

Since the reaction is exothermic, we should expect the value of the equilibrium constant to **decrease** with increasing temperature.

(d) For $PCl_5(g) \rightleftarrows PCl_3(g) + Cl_2(g)$,

$$\Delta H^\circ_{rxn} = \left[\Delta H^\circ_f PCl_3(g) + \Delta H^\circ_f Cl_2(g)\right] - \left[\Delta H^\circ_f PCl_5(g)\right]$$

$$\Delta H^\circ_{rxn} = \left[(1\ mol \times -287.0\ kJ\cdot mol^{-1}) + (1\ mol \times 0.\ kJ\cdot mol^{-1})\right] - \left[(1\ mol \times -374.9\ kJ\cdot mol^{-1})\right] = 87.9\ kJ$$

Since the reaction is endothermic, we should expect the value of the equilibrium constant to **increase** with increasing temperature.

(e) For $2SO_2(g) + O_2(g) \rightleftarrows 2SO_3(g)$,

$$\Delta H^\circ_{rxn} = \left[2\Delta H^\circ_f SO_3(g)\right] - \left[2\Delta H^\circ_f SO_2(g) + \Delta H^\circ_f O_2(g)\right]$$

$\Delta H^\circ_{rxn} = \left[\left(2 \text{ mol} \times -395.72 \text{ kJ}\cdot\text{mol}^{-1}\right)\right] - \left[\left(2 \text{ mol} \times -296.83 \text{ kJ}\cdot\text{mol}^{-1}\right) + \left(1 \text{ mol} \times 0.0 \text{ kJ}\cdot\text{mol}^{-1}\right)\right] = -197.78 \text{ kJ}$

Since the reaction is exothermic, we should expect the value of the equilibrium constant to **decrease** with increasing temperature.

15.86 ***See Section 15.5 and Example 15.9.***

(a) For $CS_2(l) \rightarrow CS_2(g)$, $K_{eq} = P_{CS_2(g)}$ and

$$\Delta H^\circ_{rxn} = \left[\Delta H^\circ_f CS_2(g)\right] - \left[\Delta H^\circ_f CS_2(l)\right]$$

$$\Delta H^\circ_{rxn} = \left[\left(1 \text{ mol} \times 117.36 \text{ kJ}\cdot\text{mol}^{-1}\right)\right] - \left[\left(1 \text{ mol} \times 89.70 \text{ kJ}\cdot\text{mol}^{-1}\right)\right] = 27.66 \text{ kJ}$$

$$\Delta S^\circ_{rxn} = \left[S^\circ CS_2(g)\right] - \left[S^\circ CS_2(l)\right]$$

$$\Delta S^\circ_{rxn} = \left[\left(1 \text{ mol} \times 237.73 \text{ J}\cdot\text{mol}^{-1}\cdot\text{K}^{-1}\right)\right] - \left[\left(1 \text{ mol} \times 151.34 \text{ J}\cdot\text{mol}^{-1}\cdot\text{K}^{-1}\right)\right] = 86.39 \text{ J}\cdot\text{K}^{-1}$$

At $5^\circ C$, $\Delta G^\circ_{278} = \Delta H^\circ - (278 \text{ K})\Delta S^\circ$

$$\Delta G^\circ_{278} = 27.66 \text{ x } 10^3 \text{ J} - (278 \text{ K})\left(86.39 \text{ J}\cdot\text{K}^{-1}\right) = 3.64 \text{ x } 10^3 \text{ J}$$

$$K_{eq} = e^{\frac{-\Delta G^\circ}{RT}} \text{ and } \frac{-\Delta G^\circ}{RT} = \frac{-3.64 \text{ x } 10^3 \text{ J}\cdot\text{mol}^{-1}}{\left(8.314 \text{ J}\cdot\text{mol}^{-1}\cdot\text{K}^{-1}\right)(278 \text{ K})} = -1.57.$$

Therefore, $K_{eq} = P_{CS_2(g)} = e^{-1.57} = \mathbf{0.208 \text{ atm}}$.

(b) For $CCl_4(l) \rightarrow CCl_4(g)$, $K_{eq} = P_{CCl_4(g)}$ and

$$\Delta H^\circ_{rxn} = \left[\Delta H^\circ_f CCl_4(g)\right] - \left[\Delta H^\circ_f CCl_4(l)\right]$$

$$\Delta H^\circ_{rxn} = \left[\left(1 \text{ mol} \times -102.9 \text{ kJ}\cdot\text{mol}^{-1}\right)\right] - \left[\left(1 \text{ mol} \times -135.44 \text{ kJ}\cdot\text{mol}^{-1}\right)\right] = 32.5 \text{ kJ}$$

$$\Delta S^\circ_{rxn} = \left[S^\circ CCl_4(g)\right] - \left[S^\circ CCl_4(l)\right]$$

$$\Delta S^\circ_{rxn} = \left[\left(1 \text{ mol} \times 309.74 \text{ J}\cdot\text{mol}^{-1}\cdot\text{K}^{-1}\right)\right] - \left[\left(1 \text{ mol} \times 216.40 \text{ J}\cdot\text{mol}^{-1}\cdot\text{K}^{-1}\right)\right] = 93.34 \text{ J}\cdot\text{K}^{-1}$$

At $29^\circ C$, $\Delta G^\circ_{302} = \Delta H^\circ - (302 \text{ K})\Delta S^\circ$

$$\Delta G^\circ_{302} = 32.5 \text{ x } 10^3 \text{ J} - (302 \text{ K})\left(93.34 \text{ J}\cdot\text{K}^{-1}\right) = 4.31 \text{ x } 10^3 \text{ J}$$

$$K_{eq} = e^{\frac{-\Delta G^\circ}{RT}} \text{ and } \frac{-\Delta G^\circ}{RT} = \frac{-4.31 \text{ x } 10^3 \text{ J}\cdot\text{mol}^{-1}}{\left(8.314 \text{ J}\cdot\text{mol}^{-1}\cdot\text{K}^{-1}\right)(302 \text{ K})} = -1.72.$$

Therefore, $K_{eq} = P_{CCl_4(g)} = e^{-1.72} = \mathbf{0.179 \text{ atm}}$.

(c) For $C_6H_6(l) \rightarrow C_6H_6(g)$, $K_{eq} = P_{C_6H_6(g)}$ and

$$\Delta H^\circ_{rxn} = \left[\Delta H^\circ_f C_6H_6(g)\right] - \left[\Delta H^\circ_f C_6H_6(l)\right]$$

$$\Delta H^\circ_{rxn} = \left[\left(1 \text{ mol} \times 82.93 \text{ kJ}\cdot\text{mol}^{-1}\right)\right] - \left[\left(1 \text{ mol} \times 49.03 \text{ kJ}\cdot\text{mol}^{-1}\right)\right] = 33.90 \text{ kJ}$$

$$\Delta S^\circ_{rxn} = \left[S^\circ C_6H_6(g)\right] - \left[S^\circ C_6H_6(l)\right]$$

$$\Delta S^\circ_{rxn} = \left[\left(1 \text{ mol} \times 269.2 \text{ J}\cdot\text{mol}^{-1}\cdot\text{K}^{-1}\right)\right] - \left[\left(1 \text{ mol} \times 172.8 \text{ J}\cdot\text{mol}^{-1}\cdot\text{K}^{-1}\right)\right] = 96.4 \text{ J}\cdot\text{K}^{-1}$$

At 45°C, $\Delta G^\circ_{318} = \Delta H^\circ - (318 \text{ K})\Delta S^\circ$

$$\Delta G^\circ_{318} = 33.9 \times 10^3 \text{ J} - (318 \text{ K})\left(96.4 \text{ J}\cdot\text{K}^{-1}\right) = 3.24 \times 10^3 \text{ J}$$

$$K_{eq} = e^{\frac{-\Delta G^\circ}{RT}} \text{ and } \frac{-\Delta G^\circ}{RT} = \frac{-3.24 \times 10^3 \text{ J}\cdot\text{mol}^{-1}}{\left(8.314 \text{ J}\cdot\text{mol}^{-1}\cdot\text{K}^{-1}\right)(318 \text{ K})} = -1.23.$$

Therefore, $K_{eq} = P_{C_6H_6(g)} = e^{-1.23} = \mathbf{0.292 \text{ atm}}$.

15.98 ***See Sections 3.2 and 15.5, Example 3.12 and Insights into Chemistry Box in Section 15.5.***

Assume the sample has a mass of 100.0 g and therefore contains 82.7 g C and 17.3 g H.

$$? \text{ mol C} = 82.7 \text{ g C} \times \frac{1 \text{ mol C}}{12.011 \text{ g C}} = 6.89 \text{ mol C} \qquad \text{relative mol C} = \frac{6.89 \text{ mol C}}{6.89} = 1.00 \text{ mol C}$$

$$? \text{ mol H} = 17.3 \text{ g H} \times \frac{1 \text{ mol H}}{1.008 \text{ g H}} = 17.2 \text{ mol H} \qquad \text{relative mol H} = \frac{17.2 \text{ mol H}}{6.89} = 2.50 \text{ mol H}$$

Multiplying each relative mol by the same smallest integer that gives whole numbers of relative moles of atoms for each yields:

relative mol C = 1.00 mol C x 2 = 2.00 mol C
relative mol H = 2.50 mol H x 2 = 5.00 mol H

The empirical formula is C_2H_5, and the empirical formula weight is: $\left(2[C] \times 12.0\right) + \left(5[H] \times 1.0\right) = 29.0$. Truoton's rule can be used to estimate the molar enthalpy of vaporization per gram, and this value can be used with the given enthalpy of vaporization per gram value to estimate the molar mass and molecular weight of the carbon-hydrogen compound. Applying Trouton's rule for hydrocarbons, gives

$\Delta H^\circ_{vap} = T_b \times 80$ $\qquad$ $\Delta H^\circ_{vap} = 272.5 \times 80 = 2.2 \times 10^4 \text{ J}\cdot\text{mol}^{-1}$ or 22 kJ·mol^{-1},

and calculating molar mass yields $\mathcal{M} = 2.2 \times 10^4 \frac{\text{J}}{\text{mol}} \times \frac{1 \text{ g}}{384 \text{ J}} = 57 \text{ g}\cdot\text{mol}^{-1}$.

Hence, $n = \dfrac{\text{molecular weight}}{\text{empirical formula weight}}$ $\quad n = \dfrac{57}{29.0} \cong 2.$

The molecular formula is $\mathbf{C_4H_{10}}$. The compound is butane.

15.100 ***See Section 15.5.***

Applying the equation $\ln K_{eq} = \dfrac{\Delta S^\circ}{R} - \dfrac{\Delta H^\circ}{RT}$ at temperatures T_2 and T_1 yields

$$\ln K_{eq_2} = \frac{\Delta S^\circ}{R} - \frac{\Delta H^\circ}{RT_2} \quad \text{and} \quad \ln K_{eq_1} = \frac{\Delta S^\circ}{R} - \frac{\Delta H^\circ}{RT_1}.$$

Subtracting the second equation from the first gives $\ln K_{eq_2} - \ln K_{eq_1} = \left(\dfrac{\Delta S^\circ}{R} - \dfrac{\Delta H^\circ}{RT_2}\right) - \left(\dfrac{\Delta S^\circ}{R} - \dfrac{\Delta H^\circ}{RT_1}\right)$

and rearranging yields $\ln\left(\dfrac{K_{eq_2}}{K_{eq_1}}\right) = \dfrac{\Delta H^\circ}{R}\left(\dfrac{1}{T_1} - \dfrac{1}{T_2}\right)$ and $\ln\left(\dfrac{K_{eq_2}}{K_{eq_1}}\right) = \dfrac{\Delta H^\circ}{R}\left(\dfrac{T_2 - T_1}{T_1T_2}\right)$

In this case, $K_{eq_1} = 1.3$ at $T_1 = 779$ K and $K_{eq_2} = 0.78$ at $T_2 = 803$ K.

Solving for ΔH° and substituting gives

$$\Delta H^\circ = \left(\frac{RT_1T_2}{T_2 - T_1}\right)\ln\left(\frac{K_{eq_2}}{K_{eq_1}}\right) \qquad \Delta H^\circ = \frac{(8.314\ \text{J}\cdot\text{mol}^{-1}\cdot\text{K}^{-1})(779\ \text{K})(803\ \text{K})}{(803\ \text{K} - 779\ \text{K})}\ln\left(\frac{0.78}{1.3}\right) = \mathbf{-1.1 \times 10^5\ J}$$

Solving $\ln K_{eq} = \dfrac{\Delta S^\circ}{R} - \dfrac{\Delta H^\circ}{RT}$ for ΔS° yields $\Delta S^\circ = \dfrac{\Delta H^\circ}{T} + R\ln K_{eq}$.

Substituting data for 779 K gives $\Delta S^\circ = \dfrac{-1.1 \times 10^5\ \text{J}}{779\ \text{K}} + (8.314\ \text{J}\cdot\text{mol}^{-1}\cdot\text{K}^{-1})(\ln\ 1.3) = \mathbf{-1.4 \times 10^2\ J\cdot K^{-1}}$

At 298 K, $\Delta G^\circ_{298} = \Delta H^\circ - (298\ \text{K})\Delta S^\circ$

$$\Delta G^\circ_{298} = -1.1 \times 10^5\ \text{J} - (298\ \text{K})(-1.4 \times 10^2\ \text{J}\cdot\text{K}^{-1}) = -7 \times 10^4\ \text{J}$$

At the temperature at which there is no net driving force in either direction under standard conditions $\Delta G^\circ = 0$. At this temperature, $0 = \Delta H^\circ - T\Delta S^\circ$. Hence,

$$T = \frac{\Delta H^\circ}{\Delta S^\circ} \qquad T = \frac{-1.1 \times 10^5\ \text{J}}{-1.4 \times 10^2\ \text{J}\cdot\text{K}^{-1}} = 786\ \text{K}$$

The reaction is spontaneous in the direction written at 298 K ($\Delta G^\circ_{298} = -$) and experiences no net driving force in either direction at 786 K ($\Delta G^\circ_{786} = 0$). Above 786 K, the reaction becomes nonspontaneous in the direction written.

Hence, the reaction is **spontaneous up to 786 K.**

Note: This temperature is only slightly different than the temperature of 790 K that was obtained from calculations involving the more precise data in 15.54.

Note: The decrease in K_{eq} from 1.3 to 0.78 with the increase in temperature from 506°C to 530°C suggests the reaction is exothermic. This is confirmed by $\Delta H^\circ = -1.1 \times 10^5$ J.

15.102 ***See Section 15.4 and Example 15.6.***

For $2H_2O(g) \rightarrow 2H_2(g) + O_2(g)$,

$$\Delta H^\circ_{rxn} = \left[2\Delta H^\circ_f H_2(g) + \Delta H^\circ_f O_2(g)\right] - \left[2\Delta H^\circ_f H_2O(g)\right]$$

$$\Delta H^\circ_{rxn} = \left[\left(2 \text{ mol} \times 0. \text{ kJ} \cdot \text{mol}^{-1}\right) + \left(1 \text{ mol} \times 0. \text{ kJ} \cdot \text{mol}^{-1}\right)\right] - \left[\left(2 \text{ mol} \times -241.82 \text{ kJ} \cdot \text{mol}^{-1}\right)\right] = 483.64 \text{ kJ}$$

$$\Delta S^\circ_{rxn} = \left[2S^\circ H_2(g) + S^\circ O_2(g)\right] - \left[2S^\circ H_2O(g)\right]$$

$$\Delta S^\circ_{rxn} = \left[\left(2 \text{ mol} \times 130.57 \text{ J} \cdot \text{mol}^{-1} \cdot \text{K}^{-1}\right) + \left(1 \text{ mol} \times 205.03 \text{ J} \cdot \text{mol}^{-1} \cdot \text{K}^{-1}\right)\right] - \left[\left(2 \text{ mol} \times 188.72 \text{ J} \cdot \text{mol}^{-1} \cdot \text{K}^{-1}\right)\right]$$
$$= 88.13 \text{ J} \cdot \text{K}^{-1}$$

$$\Delta G^\circ_{298} = \Delta H^\circ - (298 \text{ K})\Delta S^\circ \qquad \Delta G^\circ_{298} = 483.64 \times 10^3 \text{ J} - (298 \text{ K})\left(88.13 \text{ J} \cdot \text{K}^{-1}\right) = 4.57 \times 10^5 \text{ J}$$

At the temperature at which there is no net driving force in either direction under standard conditions $\Delta G^\circ = 0$. At this temperature, $0 = \Delta H^\circ - T\Delta S^\circ$. Hence,

$$T = \frac{\Delta H^\circ}{\Delta S^\circ} \qquad T = \frac{483.64 \times 10^3 \text{ J}}{88.13 \text{ J} \cdot \text{K}^{-1}} = 5.84 \times 10^3 \text{ K}$$

The reaction is nonspontaneous in the direction written at 298 K (ΔG°_{298} = +) and experiences no net driving force in either direction at 5.48×10^3 K ($\Delta G^\circ_{5.48 \times 10^3} = 0$). Above 5.48×10^3 K the reaction becomes spontaneous in the direction written. Hence, the reaction is spontaneous **above 5.48×10^3 K.**

15.104 ***See Section 15.5 and Example 15.8.***

$$K_{eq} = e^{\frac{-\Delta G^\circ}{RT}} \text{ and } \frac{-\Delta G^\circ}{RT} = \frac{-840.1 \times 10^3 \text{ J} \cdot \text{mol}^{-1}}{\left(8.314 \text{ J} \cdot \text{mol}^{-1} \cdot \text{K}^{-1}\right)(298 \text{ K})} = 3.39 \times 10^2. \text{ Therefore, } K_{eq} = e^{3.39 \times 10^2}.$$

The value of K_{eq} is greater than 10^{99} can cannot be calculated using $K_{eq} = e^{3.39 \times 10^2}$ on most calculators.

However, $K_{eq} = 10^{\frac{-\Delta G^\circ}{2.303RT}}$ can be used to obtain the value of K_{eq}: $K_{eq} = 10^{147.20} = 10^{.20} \times 10^{147} = 1.6 \times 10^{147}$.

The equilibrium constant tells us $Al^{3+} - O^{2-}$ interactions are stonger than $Fe^{3+} - O^{2-}$ interactions. This is to be expected, since Al^{3+} is smaller than Fe^{3+}.

Chapter 16: Kinetics

16.2 ***See Section 16.1 and Example 16.1.***

(a) For chemical reactions, rate is usually expressed in terms of change in concentration per unit time. For the reaction HOOC-COOH(g) → HCOOH(g) + CO_2(g), the rate of reaction can be expressed as

$$\text{rate} = \frac{-\Delta[\text{HOOC-COOH}]}{\Delta t} = \frac{\Delta[\text{HCOOH}]}{\Delta t} = \frac{\Delta[\text{CO}_2]}{\Delta t}$$

(b) The average rate of reaction between 10 and 30 seconds is given by

$$\text{ave rate} = \frac{-\left([\text{HOOC-COOH}]_{30} - [\text{HOOC-COOH}]_{10}\right)}{(30\text{ s} - 10\text{ s})}$$

$$\text{ave rate} = \frac{-\left(0.016\text{ mol}\cdot\text{L}^{-1} - 0.034\text{ mol}\cdot\text{L}^{-1}\right)}{(30\text{ s} - 10\text{ s})} = \frac{-\left(-0.018\text{ mol}\cdot\text{L}^{-1}\right)}{20\text{ s}} = \mathbf{9.0 \times 10^{-4}\ mol\cdot L^{-1}\cdot s^{-1}}$$

(c) The instantaneous rate of reaction at 20 seconds is equal to the negative of the slope of the tangent to the [HOOC-COOH] versus time curve at 20 seconds. This is given by

$$\text{rate} = -\text{slope} = \frac{-\left(0.000\text{ mol}\cdot\text{L}^{-1} - 0.040\text{ mol}\cdot\text{L}^{-1}\right)}{(50\text{ s} - 0.\text{ s})} = \frac{-\left(-0.040\text{ mol}\cdot\text{L}^{-1}\right)}{50\text{ s}} = \mathbf{8.0 \times 10^{-4}\ mol\cdot L^{-1}\cdot s^{-1}}$$

(d) The initial rate of reaction is equal to the negative of the slope of the tangent to the [HOOC-COOH] versus time curve at 0 seconds. This is given by

$$\text{rate} = -\text{slope} = \frac{-\left(0.000\text{ mol}\cdot\text{L}^{-1} - 0.050\text{ mol}\cdot\text{L}^{-1}\right)}{(20\text{ s} - 0.\text{ s})} = \mathbf{2.5 \times 10^{-3}\ mol\cdot L^{-1}\cdot s^{-1}}$$

(e) The instantaneous rate of reaction at 40 seconds is equal to the negative of the slope of the tangent to the [HOOC-COOH] versus time curve at 40 seconds. This is given by

$$\text{rate} = -\text{slope} = \frac{-\left(0.000\text{ mol}\cdot\text{L}^{-1} - 0.027\text{ mol}\cdot\text{L}^{-1}\right)}{(67.5\text{ s} - 0.\text{ s})} = \mathbf{4.0 \times 10^{-4}\ mol\cdot L^{-1}\cdot s^{-1}}$$

16.4 ***See Section 16.1 and Example 16.1.***

(a) For chemical reactions, rate is usually expressed in terms of change in concentration per unit time.

For the reaction $\text{NO}(g) + \frac{1}{2}\text{Cl}_2(g) \rightarrow \text{NOCl}(g)$, rate of reaction can be expressed as $\text{rate} = \frac{\Delta[\text{NOCl}(g)]}{\Delta t}$.

(b) The average rate of reaction between 40 and 120 seconds is given by

$$\text{ave rate} = \frac{\left([\text{NOCl}]_{120} - [\text{NOCl}]_{40}\right)}{(120\text{ s} - 40\text{ s})}$$

$$\text{ave rate} = \frac{\left(0.575\text{ mol}\cdot\text{L}^{-1} - 0.450\text{ mol}\cdot\text{L}^{-1}\right)}{(120\text{ s} - 40\text{ s})} = \frac{\left(0.125\text{ mol}\cdot\text{L}^{-1}\right)}{80\text{ s}} = \mathbf{1.6 \times 10^{-3}\ mol\cdot L^{-1}\cdot s^{-1}}$$

(c) The instantaneous rate of reaction at 80 seconds is equal to the slope of the tangent to the [NOCl] versus time curve at 80 seconds. This is given by

$$\text{rate} = \text{slope} = \frac{\left(0.700\text{ mol}\cdot\text{L}^{-1} - 0.433\text{ mol}\cdot\text{L}^{-1}\right)}{(200\text{ s} - 0.\text{ s})} = \frac{\left(0.267\text{ mol}\cdot\text{L}^{-1}\right)}{200\text{ s}} = \mathbf{1.3 \times 10^{-3}\ mol\cdot L^{-1}\cdot s^{-1}}$$

(d) The instantaneous rate of reaction at 60 seconds in terms of NOCl is equal to the slope of the tangent to the [NOCl] versus time curve at 60 seconds. This is given by

$$\text{rate} = \text{slope} = \frac{\left(0.70\text{ mol}\cdot\text{L}^{-1} - 0.38\text{ mol}\cdot\text{L}^{-1}\right)}{(160\text{ s} - 0.\text{ s})} = \frac{\left(0.32\text{ mol}\cdot\text{L}^{-1}\right)}{160\text{ s}} = \mathbf{2.0 \times 10^{-3}\ mol\cdot L^{-1}\cdot s^{-1}}$$

$$\text{Hence, } \frac{-\Delta[\text{Cl}_2]}{\Delta t} = \frac{\Delta[\text{NOCl}]}{\Delta t} \times \frac{\frac{1}{2}\text{ mol Cl}_2}{1\text{ mol NOCl}} = 2.0 \times 10^{-3}\text{ mol}\cdot\text{L}^{-1}\cdot\text{s}^{-1} \times \frac{\frac{1}{2}\text{ mol Cl}_2}{1\text{ mol NOCl}} = \mathbf{1.0 \times 10^{-3}\ mol\cdot L^{-1}\cdot s^{-1}}$$

16.8 ***See Section 16.1 and Example 16.2.***

(a) The stoichiometry of the reaction $N_2O_4(g) \rightarrow 2NO_2(g)$ tells us $[NO_2]$ increases twice as fast as $[N_2O_4]$ decreases. Hence,

Time, μs	$[N_2O_4], M$	$[NO_2], M$
0.000	0.050	0.000
20.00	0.033	**0.034**
40.00	**0.025**	0.050
60.00	0.020	**0.060**

(b) The instantaneous rate of reaction at 30 μs can be determined from the negative of the slope of the tangent to the $[N_2O_4]$ versus time curve at 30 μs.

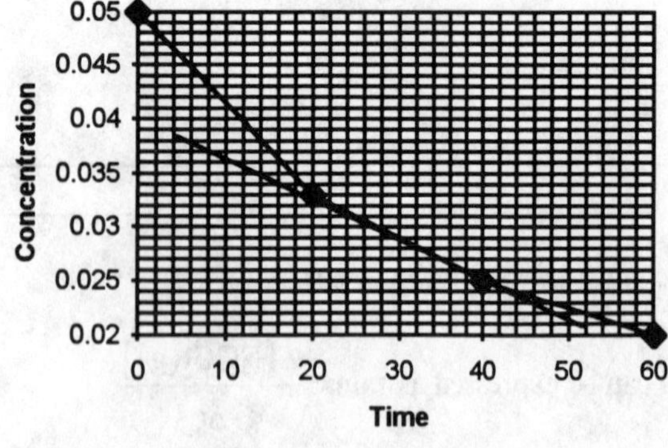

This gives:

$$\text{rate} = \frac{-(0.020\ M - 0.040\ M)}{(55\text{ s} - 0.\text{ s})} = 3.6 \times 10^{-4}\text{ mol}\cdot\text{L}^{-1}\cdot\text{s}^{-1}$$

16.10 *See Section 16.1 and Example 16.2.*

(a) The stoichiometry of the reaction $2NO(g) + Cl_2(g) \rightarrow 2NOCl(g)$ tells us $[Cl_2]$ decreases one-half as fast as $[NOCl]$ increases. Mathematically,

$$\frac{-\Delta[Cl_2]}{\Delta t} = \frac{\Delta[NOCl]}{\Delta t} \times \frac{1 \text{ mol } Cl_2}{2 \text{ mol NOCl}} \qquad \frac{-\Delta[Cl_2]}{\Delta t} = 0.030 \text{ mol} \cdot L^{-1} \cdot s^{-1} \times \frac{1 \text{ mol } Cl_2}{2 \text{ mol NOCl}} = \mathbf{0.015\ mol \cdot L^{-1} \cdot s^{-1}}$$

(b) rate of reaction $= \left(\frac{1}{2}\right)\frac{-\Delta[NO]}{\Delta t} = \frac{-\Delta[Cl_2]}{\Delta t} = \left(\frac{1}{2}\right)\frac{\Delta[NOCl]}{\Delta t} = \mathbf{0.015\ mol \cdot L^{-1} \cdot s^{-1}}$

16.12 *See Section 16.1 and Example 16.2.*

(a) The stoichiometry of the reaction $2CrO_2^- + 3H_2O_2 + 2OH^- \rightarrow 2CrO_4^{2-} + 4H_2O$ can be used to determine the rates at which the other species change concentration.

$$\frac{-\Delta[CrO_2^-]}{\Delta t} = 0.0050 \text{ mol} \cdot L^{-1} \cdot s^{-1} \times \frac{2 \text{ mol } CrO_2^-}{2 \text{ mol } CrO_4^{2-}} = \mathbf{0.0050\ mol \cdot L^{-1} \cdot s^{-1}}$$

$$\frac{-\Delta[H_2O_2]}{\Delta t} = 0.0050 \text{ mol} \cdot L^{-1} \cdot s^{-1} \times \frac{3 \text{ mol } H_2O_2}{2 \text{ mol } CrO_4^{2-}} = \mathbf{0.0075\ mol \cdot L^{-1} \cdot s^{-1}}$$

$$\frac{-\Delta[OH^-]}{\Delta t} = 0.0050 \text{ mol} \cdot L^{-1} \cdot s^{-1} \times \frac{2 \text{ mol } OH^-}{2 \text{ mol } CrO_4^{2-}} = \mathbf{0.0050\ mol \cdot L^{-1} \cdot s^{-1}}$$

$$\frac{\Delta[H_2O]}{\Delta t} = 0.0050 \text{ mol} \cdot L^{-1} \cdot s^{-1} \times \frac{4 \text{ mol } H_2O}{2 \text{ mol } CrO_4^{2-}} = \mathbf{0.0100\ mol \cdot L^{-1} \cdot s^{-1}}$$

(b) rate of reaction $= \frac{\Delta[CrO_4^{2-}]}{2\Delta t} \qquad \frac{\Delta[CrO_4^{2-}]}{2\Delta t} = \frac{0.0050 \text{ mol} \cdot L^{-1} \cdot s^{-1}}{2} = \mathbf{0.0025\ mol \cdot L^{-1} \cdot s^{-1}}$

16.14 *See Section 16.2 and Example 16.3.*

Experiments 1, 4 and 5 have the same concentration of H_2 and can be used to determine the order with respect to NO.

Expt.	Initial $[NO]$, M	Initial Rate, $mol \cdot L^{-1} \cdot s^{-1}$	Relative Concentration*	Relative Rate**
1	0.10	3.8	1.0	1.0
4	0.20	15.3	2.0	4.0
5	0.30	34.2	3.0	9.0

* Obtained by dividing each concentration by the smallest concentration.

** Obtained by dividing each initial rate by the smallest initial rate.

As the relative concentration of NO increases from 1.0 to 2.0 to 3.0, the relative rate of reaction increases from 1.0 to 4.0 to 9.0. This indicates that the rate of reaction is proportional to the concentration of NO raised to the second power. The reaction is therefore second order with respect to NO.

Experiments 1, 2 and 3 have the same concentration of NO and can be used to determine the order with respect to H_2.

Expt.	Initial $[H_2]$, M	Initial Rate, $mol{\cdot}L^{-1}{\cdot}s^{-1}$	Relative Concentration	Relative Rate
1	0.10	3.8	1.0	1.0
2	0.20	7.7	2.0	2.0
3	0.30	11.4	3.0	3.0

As the relative concentration of H_2 increases from 1 to 2 to 3, the relative rate of reaction increases from 1.0 to 2.0 to 3.0. This indicates that the rate of reaction is proportional to the concentration of H_2 raised to the first power. The reaction is therefore first order with respect to H_2.

The rate law is rate= $k[NO]^2[H_2]$, and using data for Expt. 1 in the original table yields

$$k = \frac{rate}{[NO]^2[H_2]} \qquad k = \frac{3.8\ mol \cdot L^{-1} \cdot s^{-1}}{\left(0.10\ mol \cdot L^{-1}\right)^2\left(0.10\ mol \cdot L^{-1}\right)} = 3.8 \times 10^3\ L^2 \cdot mol^{-2} \cdot s^{-1}$$

This gives **rate**= $\mathbf{3.8 \times 10^3\ L^2 \cdot mol^{-2} \cdot s^{-1}[NO]^2[H_2]}$ as the rate law.

16.16 ***See Section 16.2 and Example 16.3.***

Experiments 1 and 2 have the same concentration of NO and can be used to determine the order with respect to H_2.

Expt.	Initial $[H_2]$, M	Initial Rate, $mol{\cdot}L^{-1}{\cdot}s^{-1}$	Relative Concentration*	Relative Rate**
1	0.10	0.051	1.0	1.0
2	0.20	0.100	2.0	2.0

* Obtained by dividing each concentration by the smallest concentration.

** Obtained by dividing each initial rate by the smallest initial rate.

As the relative concentration of H_2 increases from 1.0 to 2.0 , the relative rate of reaction increases from 1.0 to 2.0. This indicates that the rate of reaction is proportional to the concentration of H_2 raised to the first power. The reaction is therefore first order with respect to H_2.

There are no experiments which have the same concentration of H_2 and can be easily used to determine the order with respect to NO. Instead, Experiments 2 and 3 having changing $[NO]$ and $[H_2]$ are used.

Expt.	Initial $[NO]$, M	Initial $[H_2]$, M	Initial Rate $mol{\cdot}L^{-1}{\cdot}s^{-1}$	Relative $[NO]$	Relative $[H_2]$	Relative Rate	Relative Rate / Relative $[H_2]$
2	0.12	0.20	0.100	1.0	1.0	1.0	1.0
3	0.25	0.30	0.313	2.1	1.5	3.13	2.1

The change in Relative Rate/ Relative $[H_2]$ gives the change in Relative Rate that is due to change in Relative $[NO]$, since the reaction has been shown to be first order with respect to H_2. As the relative concentration of NO changes from 1.0 to 2.1 there is an accompanying change in Relative Rate/ Relative $[H_2]$ from 1.0 to 2.1. This indicates the rate of reaction is proportional to the concentration of NO raised to the first power. The reaction is therefore first order with respect to NO.

The rate law is rate= $k[NO][H_2]$, and using data for Expt. 1 in the original table yields

$$k = \frac{\text{rate}}{[NO][H_2]} \qquad k = \frac{0.0051\ \text{mol}\cdot\text{L}^{-1}\cdot\text{s}^{-1}}{(0.12\ \text{mol}\cdot\text{L}^{-1})(0.10\ \text{mol}\cdot\text{L}^{-1})} = 4.2\ \text{L}\cdot\text{mol}^{-1}\cdot\text{s}^{-1}$$

This gives **rate= $4.2\ \text{L}\cdot\text{mol}^{-1}\cdot\text{s}^{-1}[NO][H_2]$** as the rate law.

16.18 ***See Section 16.2 and Example 16.3.***

Experiments 1 and 2 have the same concentration of O_3 and can be used to determine the order with respect to NO_2.

Expt.	Initial $[NO_2]$, *M*	Initial Rate, mol·L^{-1}·s^{-1}	Relative $[NO_2]$	Relative Rate
1	2.0×10^{-6}	2.1×10^{-7}	1.0	1.0
2	3.0×10^{-6}	3.1×10^{-7}	1.5	1.5

As the relative concentration of NO_2 increases from 1.0 to 1.5 , the relative rate of reaction increases from 1.0 to 1.5. This indicates that the rate of reaction is proportional to the concentration of NO_2 raised to the first power. The reaction is therefore first order with respect to NO_2.

Experiments 3 and 4 have the same concentration of NO_2 and can be used to determine the order with respect to O_3.

Expt.	Initial $[O_3]$, *M*	Initial Rate, mol·L^{-1}·s^{-1}	Relative $[O_3]$	Relative Rate
3	3.0×10^{-6}	6.2×10^{-7}	1.0	1.0
4	4.0×10^{-6}	8.3×10^{-7}	1.3	1.3

As the relative concentration of O_3 increases from 1.0 to 1.3, the relative rate of reaction increases from 1.0 to 1.3. This indicates that the rate of reaction is proportional to the concentration of O_3 raised to the first power. The reaction is therefore first order with respect to O_3.

The rate law is rate= $k[NO_2][O_3]$, and using data for Expt. 1 in the original table yields

$$k = \frac{\text{rate}}{[NO_2][O_3]} \qquad k = \frac{2.1 \times 10^{-7}\ \text{mol}\cdot\text{L}^{-1}\cdot\text{s}^{-1}}{(2.0 \times 10^{-6}\ \text{mol}\cdot\text{L}^{-1})(2.0 \times 10^{-6}\ \text{mol}\cdot\text{L}^{-1})} = 5.2 \times 10^{4}\ \text{L}\cdot\text{mol}^{-1}\cdot\text{s}^{-1}$$

This gives **rate= $5.2 \times 10^{4}\ \text{L}\cdot\text{mol}^{-1}\cdot\text{s}^{-1}[NO_2][O_3]$** as the rate law.

16.20 *See Section 16.2 and Example 16.3.*

Experiments 1 and 3 have the same concentration of Br_2 and can be used to determine the order with respect to NO.

Expt.	Initial $[NO]$, M	Initial Rate, $mol{\cdot}L^{-1}{\cdot}s^{-1}$	Relative $[NO]$	Relative Rate
1	1.1×10^{-5}	0.37	1.0	1.0
2	3.5×10^{-5}	1.2	3.2	3.2

As the relative concentration of NO increases from 1.0 to 3.2 , the relative rate of reaction increases from 1.0 to 3.2. This indicates that the rate of reaction is proportional to the concentration of NO raised to the first power. The reaction is therefore first order with respect to NO.

There are no experiments which have the same concentration of NO and can be easily used to determine the order with respect to Br_2. Instead, Experiments 1 and 4 having changing $[NO]$ and $[Br_2]$are used.

Expt.	Initial $[NO]$, M	Initial $[Br_2]$, M	Initial Rate $mol{\cdot}L^{-1}{\cdot}s^{-1}$	Relative $[NO]$	Relative $[Br_2]$	Relative Rate	$\frac{\text{Relative Rate}}{\text{Relative }[NO]}$
1	1.1×10^{-5}	1.2×10^{-5}	0.37	1.0	1.0	1.0	1.0
4	4.0×10^{-5}	3.0×10^{-5}	3.4	3.6	2.5	9.2	2.5

The change in Relative Rate/ Relative $[NO]$ gives the change in Relative Rate that is due to change in Relative $[Br_2]$, since the reaction has been shown to be first order with respect to NO. As the relative concentration of Br_2 changes from 1.0 to 2.5 there is an accompanying change in Relative Rate/ Relative $[NO]$ from 1.0 to 2.5. This indicates the rate of reaction is proportional to the concentration of Br_2 raised to the first power. The reaction is therefore first order with respect to Br_2.

The rate law is rate= $k[NO][Br_2]$, and using data for Expt. 1 in the original table yields

$$k = \frac{\text{rate}}{[NO][Br_2]} \qquad k = \frac{0.37 \text{ atm}\cdot\text{s}^{-1}}{(1.1 \times 10^{-5} \text{ atm})(1.2 \times 10^{-5} \text{ atm})} = 2.8 \times 10^9 \text{ atm}^{-1}\cdot\text{s}^{-1}$$

This gives **rate= 2.8×10^9 $atm^{-1}\cdot s^{-1}[NO][Br_2]$** as the rate law.

16.22 *See Section 16.3 and Example 16.5 and 6.*

The given data can be used to solve $\ln[A] = -kt + \ln[A]_o$ for k. The value of k can then be used to determine the amount of time that would be required for the concentration to decrease from 0.235 M to 0.100 M.
Solving $\ln[A] = -kt + \ln[A]_o$ for k and substituting gives

$$k = \frac{\ln[A]_o - \ln[A]}{t} \qquad k = \frac{\ln(0.451) - \ln(0.235)}{131 \text{ s}} = 4.98 \times 10^{-3} \text{ s}^{-1}$$

Solving $\ln[A] = -kt + \ln[A]_o$ for t and substituting gives

$$t = \frac{\ln[A]_o - \ln[A]}{k} \qquad t = \frac{\ln(0.235) - \ln(0.100)}{4.98 \times 10^{-3} \text{ s}^{-1}} = \mathbf{1.72 \times 10^2 \text{ s}}$$

16.24	***See Section 16.3 and Example 16.4.***

Time,s	[Reactant], M	ln[Reactant]	$\frac{1}{[\text{Reactant}]}$, M^{-1}
0	0.250	-1.386	4.00
1	0.216	-1.532	4.63
2	0.182	-1.704	5.49
3	0.148	-1.911	6.76
4	0.114	-2.172	8.77
5	0.080	-2.53	12

For a zero order reaction, a plot of [Reactant] vs. time gives a straight line.

For a first order reaction, a plot of ln[Reactant] vs. time gives a straight line.

For a second order reaction ,a plot of $\frac{1}{[\text{Reactant}]}$ vs. time gives a straight line.

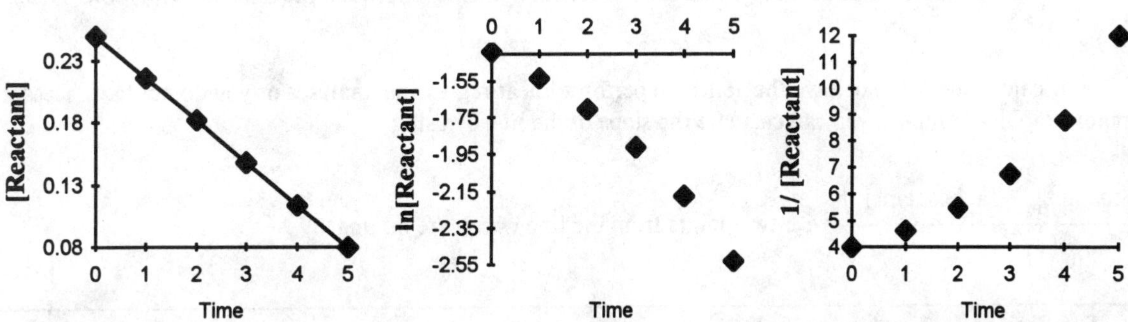

The data give a straight line for a **zero order** reaction. The rate consant for a zero order reaction is equal to the negative of the slope of the line. A linear regression analysis gives a slope of -0.034, and a correlation coefficient of 1.000 indicating excellent agreement between the data and the equation. Hence, **k = 0.034 mol·L^{-1}·s^{-1}**.

Note: If your calculator does not have the ability to perform linear regression analyses or you do not have acess to a computer graphics program, you can caculate the slope of the line by using

$$\frac{([\text{Reactant}]_2 - [\text{Reactant}]_1)}{(t_2 - t_1)}$$ with two points from the line which are far apart.

16.26	***See Section 16.3 and Example 16.4.***

Time,s	[Reactant], M	ln[Reactant]	$\frac{1}{[\text{Reactant}]}$, M^{-1}
0.001	0.220	-1.51	4.5
0.002	0.140	-1.97	7.1
0.003	0.080	-2.53	12
0.004	0.050	-3.00	20
0.005	0.030	-3.51	33

For a zero order reaction, a plot of [Reactant] vs time gives a straight line.

For a first order reaction, a plot of ln[Reactant] vs. time gives a straight line.

For a second order reaction ,a plot of $\dfrac{1}{[\text{Reactant}]}$ vs. time gives a straight line.

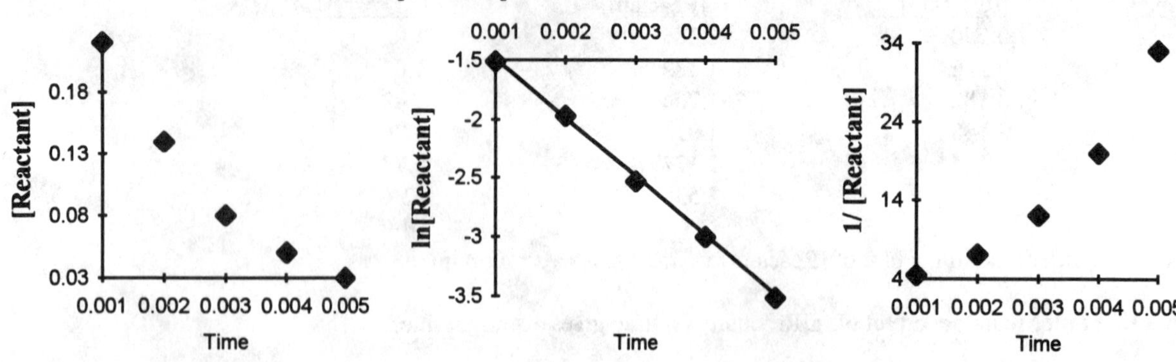

The data give a straight line plot for a **first order** reaction. The rate constant for a first order reaction is equal to the negative of the slope of the line when ln[conc] is plotted vs. time. A linear regression analysis gives a slope of -503, and a correlation coefficient of 0.9996 indicates excellent agreement between the data and equation. Hence, **k = 503 s^{-1}**.

Note: If your calculator does not have the ability to perform linear regression analyses or you do not have acess to a computer graphics program, you can caculate the slope of the line by using

$$\frac{\left(\ln[\text{Reactant}]_2 - \ln[\text{Reactant}]_1\right)}{\left(t_2 - t_1\right)}$$ with two points from the line which are far apart.

16.28 ***See Section 16.3 and Example 16.4.***

Time,s	[NOCl], *M*	ln[NOCl]	$\dfrac{1}{[\text{NOCl}]}$, M^{-1}
0	0.100	-2.303	10.0
30	0.064	-2.75	15.6
60	0.047	-3.06	21.3
100	0.035	-3.35	28.6
200	0.021	-3.86	47.6
300	0.015	-4.20	66.7
400	0.012	-4.42	83.3

For a zero order reaction, a plot of [NOCl] vs time gives a straight line.

For a first order reaction, a plot of ln[NOCl] vs. time gives a straight line.

For a second order reaction ,a plot of $\dfrac{1}{[\text{NOCl}]}$ vs. time gives a straight line.

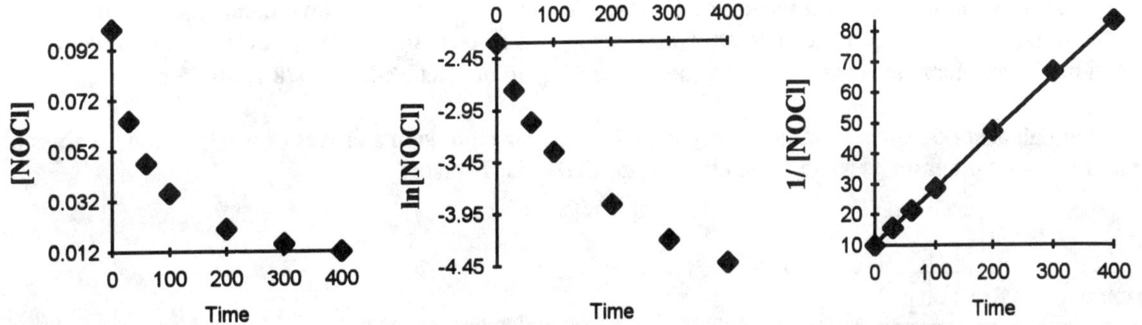

The data give a straight line plot for a **second order** reaction. The rate constant for a second order reaction is equal to the slope of the line. A linear regression analysis gives a slope of 0.185, and a correlation coefficient of 0.99998 indicates excellent agreement between the data and equation. Hence, $\mathbf{k = 0.185\ L{\cdot}mol^{-1}{\cdot}s^{-1}}$.

Note: If your calculator does not have the ability to perform linear regression analyses or you do not have acess to a computer graphics program, you can caculate the slope of the line by using

$$\frac{\left(\frac{1}{[\text{Reactant}]_2} - \frac{1}{[\text{Reactant}]_1}\right)}{(t_2 - t_1)}$$ with two points from the line which are far apart.

16.30 *See Section 16.3 and Example 16.4.*

Time,s	[Reactant], M	ln[Reactant]	$\frac{1}{[\text{Reactant}]}$, M^{-1}
0.00	0.7014	-0.3547	1.426
10.00	0.4534	-0.7910	2.206
20.00	0.3304	-1.107	3.027
40.00	0.2181	-1.523	4.585
60.00	0.1638	-1.809	6.105
80.00	0.1297	-2.043	7.710
100.00	0.1084	-2.222	9.225

For a zero order reaction, a plot of [Reactant] vs time gives a straight line.

For a first order reaction, a plot of ln[Reactant] vs. time gives a straight line.

For a second order reaction ,a plot of $\frac{1}{[\text{Reactant}]}$ vs. time gives a straight line.

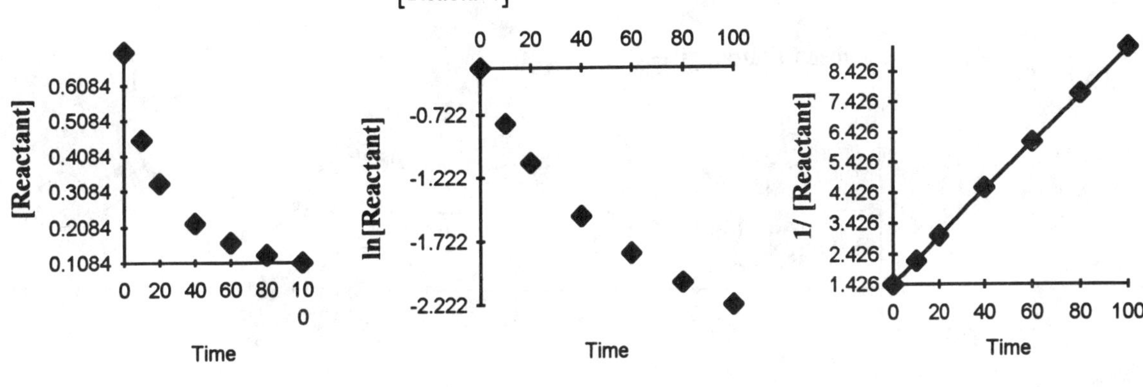

The data give a straight line plot for a **second order** reaction. The rate constant for a second order reaction is equal to the slope of the line. A linear regression analysis gives a slope of 0.078, and a correlation coefficient of 0.99998 indicates excellent agreement between the data and equation. Hence, **k = 0.078 L·mol^{-1}·s^{-1}**.

Note: If your calculator does not have the ability to perform linear regression analyses or you do not have acess to a computer graphics program, you can caculate the slope of the line by using

$$\frac{\left(\frac{1}{[\text{Reactant}]_2} - \frac{1}{[\text{Reactant}]_1}\right)}{(t_2 - t_1)}$$ with two points from the line which are far apart.

16.34 ***See Section 16.3.***

(a) Since the reaction is first order, $t_{1/2} = \frac{0.693}{k}$. Solving for k and substituting gives

$$k = \frac{0.693}{t_{1/2}} \qquad k = \frac{0.693}{24.5 \text{ min}} = \mathbf{2.83 \times 10^{-2}\ min^{-1}}$$

(b) Solving $\ln[A] = -kt + \ln[A]_o$ for t and substituting gives

$$t = \frac{\ln[A]_o - \ln[A]}{k} \qquad t = \frac{\ln(0.15) - \ln(0.015)}{2.83 \times 10^{-2} \text{ min}^{-1}} = \mathbf{81.4\ min} \text{ or } \mathbf{4.88 \times 10^3\ sec}$$

16.36 ***See Section 16.3.***

Since the reaction is first order, $t_{1/2} = \frac{0.693}{k}$. Solving $\ln[A] = -kt + \ln[A]_o$ for k and substituting gives

$$k = \frac{\ln[A]_o - \ln[A]}{t} \qquad k = \frac{\ln(0.012) - \ln(0.0082)}{66.2 \text{ s}} = 5.75 \times 10^{-3} \text{ s}^{-1}$$

Hence, $t_{1/2} = \frac{0.693}{5.75 \times 10^{-3} \text{ s}^{-1}} = \mathbf{120\ s}$

16.38 ***See Section 16.3.***

Solving $\frac{1}{[A]} = \frac{1}{[A]_o} + kt$ for t and substituting gives

$$t = \frac{\frac{1}{[A]} - \frac{1}{[A]_o}}{k} \qquad t = \frac{\left(\frac{1}{1.0 \text{ torr} \times \frac{1 \text{ atm}}{760 \text{ torr}}}\right) - \left(\frac{1}{21.0 \text{ torr} \times \frac{1 \text{ atm}}{760 \text{ torr}}}\right)}{30.6 \text{ atm}^{-1} \cdot \text{s}^{-1}} = \mathbf{24\ s}$$

Note: The conversion factors for converting reactant pressures to *M* and the units of k to *M* would be identical and cancel, since one of these conversion factors would be in the numerator of the expression for t and the other in the denominator of the same expression.

16.44 ***See Section 16.4 and Example 16.11.***

The Arrhenius equation can be written as $\ln k = \frac{-E_a}{R}\left(\frac{1}{T}\right) + \ln A$.

A plot of ln k vs. 1/ T will have a slope of $-E_a/R$ and an intercept of ln A. Determine the value of the slope of the line, and solve to E_a.

$k, L\cdot mol^{-1}\cdot s^{-1}$	ln k	T, K	$1/T, K^{-1}$
0.36×10^6	12.79	500	0.00200
3.7×10^6	15.12	550	0.00182
27×10^6	17.11	600	0.00167

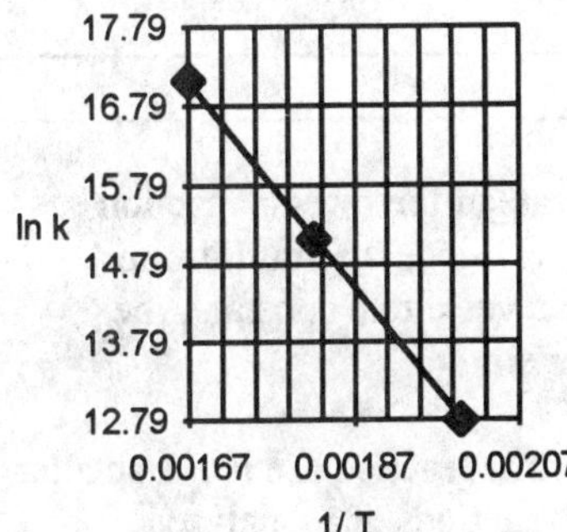

A linear regression analysis gives a slope of -1.31×10^4 and an intercept of 38.95 with a correlation coefficient of 0.99998 indicating excellent agreement between the data and equation. Hence,

$$\frac{-E_a}{R} = -1.31 \times 10^4\ K^{-1} \quad -E_a = \frac{-1.31 \times 10^4\ K^{-1}}{8.314\ J\cdot mol^{-1}\cdot K^{-1}} = \mathbf{1.09 \times 10^5\ J\cdot mol^{-1}}$$

$$A = \text{inv}\ln 38.95 = \mathbf{7.83 \times 10^{16}}$$

Note: If your calculator does not have the ability to perform linear regression analyses or you do not have acess to a computer graphics program, you can caculate the slope of the line by using

$\frac{(\ln k_2 - \ln k_1)}{(t_2 - t_1)}$ with two points from the line which are far apart.

The slope of the line can then be used to calculate the value of E_a, and this value can be used with the given data to solve the equation for lnA and therefore A. It isn't feasible to obtain lnA from the graph, since the y-intercept occurs when x = 0 and this would require a graph of unreasonable size.

16.46 ***See Section 16.4 and Example 16.12.***

Known Quantities: $k_1 = k_1$ $k_2 = 2k_1$ $T_1 = 298$ K $T_2 = 313$ K

Solving $\ln\left(\frac{k_1}{k_2}\right) = \frac{-E_a}{R}\left(\frac{1}{T_1} - \frac{1}{T_2}\right) = \frac{-E_a}{R}\left(\frac{T_2 - T_1}{T_1 T_2}\right)$ for E_a and substituting gives

$$E_a = \frac{-RT_1T_2}{(T_2 - T_1)}\ln\left(\frac{k_1}{k_2}\right) \qquad E_a = \frac{-(8.314\ J\cdot mol^{-1}\cdot K^{-1})(298\ K)(313\ K)}{(313\ K - 298\ K)}\ln\left(\frac{k_1}{2k_1}\right) = \mathbf{3.58 \times 10^4\ J\cdot mol^{-1}}$$

16.48 ***See Section 16.4.***

Known Quantities: $E_a = 274\ kJ{\cdot}mol^{-1}$ $T_1 = 500 + 273 = 773\ K$ $T_2 = 550 + 273 = 823\ K$

Substituting in $\ln\left(\frac{k_1}{k_2}\right) = \frac{-E_a}{R}\left(\frac{1}{T_1} - \frac{1}{T_2}\right) = \frac{-E_a}{R}\left(\frac{T_2 - T_1}{T_1T_2}\right)$ and solving for $\frac{k_1}{k_2}$ gives,

$$\ln\left(\frac{k_1}{k_2}\right) = \frac{-274 \times 10^3\ J \cdot mol^{-1}}{8.314\ J \cdot mol^{-1} \cdot K^{-1}}\left(\frac{823\ K - 773\ K}{(773\ K)(823\ K)}\right) = -2.59 \qquad \frac{k_1}{k_2} = e^{-2.59} = \text{invln}(-2.59) = 7.5 \times 10^{-2}$$

So, $\frac{k_2}{k_1} = \frac{1}{7.5 \times 10^{-2}} = 13$, meaning the reaction increases by a factor of 13 when the temperature increases from 500°C to 550°C.

16.50 ***See Sections 16.4 and 6.***

Increasing the temperature increases the rate of the forward reaction more than the rate of the reverse reaction, since the forward reaction of an endothermic process has the higher activation energy. Since the equilibrium constant for the same process is equal to the ratio of the forward rate constant to the reverse rate constant, the equilibrium constant for an endothermic reaction also increases with increasing temperature.

The activation energies for the forward and reverse reactions of a catalyzed endothermic reaction are much smaller than those for the uncatalyzed reaction. Hence, the rates of the forward and reverse reactions of a catalyzed endothermic reaction are much less sensitive to increases in temperature and must be affected equally because the value of K_{eq} is not changed by adding a catalyst.

16.52 ***See Section 16.4.***

Known Quantities: $E_a = 29 \times 10^3\ J{\cdot}mol^{-1}$ $T_1 = 13 + 273 = 286\ K$ $T_2 = 29 + 273 = 302\ K$

Substituting in $\ln\left(\frac{k_1}{k_2}\right) = \frac{-E_a}{R}\left(\frac{1}{T_1} - \frac{1}{T_2}\right) = \frac{-E_a}{R}\left(\frac{T_2 - T_1}{T_1T_2}\right)$ and solving for $\frac{k_1}{k_2}$ gives,

$$\ln\left(\frac{k_1}{k_2}\right) = \frac{-29 \times 10^3\ J \cdot mol^{-1}}{8.314\ J \cdot mol^{-1} \cdot K^{-1}}\left(\frac{302\ K - 286\ K}{(286\ K)(302\ K)}\right) = -0.65 \qquad \frac{k_1}{k_2} = e^{-.065} = \text{invln}(-0.65) = 0.52$$

So, $\frac{k_1}{k_2} = \frac{1}{0.52} = 1.92$, meaning the reaction goes 92% faster when the temperature increases from 13°C to 29°C.

16.54 ***See Section 16.4.***

Rates of reaction with high activation energies are influenced more by temperature changes than are rates of reaction with low activation energies. This is due to the fact that equal temperature changes cause greater increases in the fraction of molecules with sufficient kinetic energy to react (i.e. with $KE \geq E_a$) the higher the activation energy.

Note: This is discussed in the last paragraph of Section 16.4, and a comparison of effect is given there.

16.58 ***See Section 16.4.***

For the catalyzed path, $k_{cat} = Ae^{\frac{-E_a}{RT}}$ $\quad k_{cat} = Ae^{\frac{-28 \times 10^3\ J\cdot mol^{-1}}{(8.314\ J\cdot mol^{-1}\cdot K^{-1})(298\ K)}}$

For the uncatalyzed path, $k = Ae^{\frac{-E_a}{RT}}$ $\quad k = Ae^{\frac{-72 \times 10^3\ J\cdot mol^{-1}}{(8.314\ J\cdot mol^{-1}\cdot K^{-1})(298\ K)}}$

Hence, $\dfrac{k_{cat}}{k} = \dfrac{Ae^{\frac{-E_a}{RT}}}{Ae^{\frac{-E_a}{RT}}}$ $\quad \dfrac{k_{cat}}{k} = \dfrac{e^{\frac{-28 \times 10^3\ J\cdot mol^{-1}}{(8.314\ J\cdot mol^{-1}\cdot K^{-1})(298\ K)}}}{e^{\frac{-72 \times 10^3\ J\cdot mol^{-1}}{(8.314\ J\cdot mol^{-1}\cdot K^{-1})(298\ K)}}} = \dfrac{1.2 \times 10^{-5}}{2.4 \times 10^{-13}} = 5.0 \times 10^7$

For $H_2O_2(aq) \rightarrow H_2O(l) + \frac{1}{2}O_2(g)$,

$\Delta E \cong \Delta H^\circ = \Delta H_f^\circ H_2O(l) + \frac{1}{2}\Delta H_f^\circ O_2(g) - \Delta H_f^\circ H_2O_2(aq)$

$\Delta H^\circ = \left[\left(1\ mol \times -285.3\ kJ\cdot mol^{-1}\right) + \left(\frac{1}{2}\ mol \times 0.\ kJ\cdot mol^{-1}\right)\right] - \left[\left(1\ mol \times -191.17\ kJ\cdot mol^{-1}\right)\right] = -94.13\ kJ$

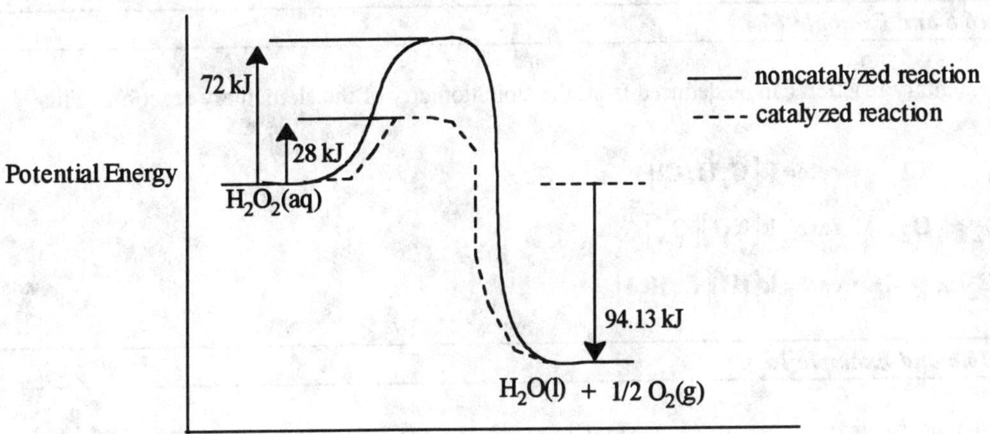

16.60 ***See Sections 16.4 and 6.***

The reaction $NO(g) + N_2O(g) \rightleftarrows N_2(g) + NO_2(g)$ is an exothermic reaction having a lower activation energy for the forward reaction than the reverse reaction. Hence, increasing the temperature increases the rate of the reverse reaction more than the forward reaction. This causes the ratio of the forward rate constant to the reverse constant, the equilibrium constant, of an exothermic reaction to decrease with increasing temperature.

16.62 See Section 16.4 and 6.

The reaction $NOCl(g) + NO_2(g) \rightleftarrows NO(g) + NO_2Cl(g)$ is an endothermic reaction having a higher activation energy for the forward reaction than the reverse reaction. Hence, increasing the temperature increases the rate of the forward reaction more than the reverse reaction. This causes the ratio of the forward rate constant to the reverse rate constant, the equilibrium constant, of an endothermic reaction to increase with increasing temperature.

16.66 See Section 16.6 and Example 16.13.

$$NO_2(g) + NO_2(g) \rightarrow NO_3(g) + NO(g)$$
$$NO_3(g) + CO(g) \rightarrow NO_2(g) + CO_2(g)$$

$$NO_2(g) + CO(g) \rightarrow NO(g) + CO_2(g)$$

16.68 See Section 16.6 and Example 16.4.

The rate law for an elementary reaction can be deduced from the stoichiometry of the elementary reaction. This gives:

(a) $HCl \rightarrow H + Cl$ **rate= $k[HCl]$**

(b) $H_2 + Cl \rightarrow HCl + H$ **rate= $k[H_2][Cl]$**

(c) $2NO_2 \rightarrow N_2O_4$ **rate= $k[NO_2]^2$**

16.70 See Section 16.6 and Example 16.4.

The rate law for an elementary reaction can be deduced from the stoichiometry of the elementary reaction. This gives:

(a) $C_2H_5Cl \rightarrow C_2H_4 + HCl$ **rate= $k[C_2H_5Cl]$**

(b) $NO + O_3 \rightarrow NO_2 + O_2$ **rate= $k[NO][O_3]$**

(c) $HI + C_2H_5I \rightarrow C_2H_6 + I_2$ **rate= $k[HI][C_2H_5I]$**

16.72 See Section 16.6 and Example 16.4.

Since the first step in the mechanism

$$NO_2Cl \rightarrow NO_2 + Cl$$
$$NO_2Cl + Cl \rightarrow NO_2 + Cl_2$$

$$2NO_2Cl \rightarrow 2NO_2 + Cl_2$$

is a unimolecular elementary reaction and the reaction is a first-order reaction , **the first step is the rate-limiting step.**

16.74 See Section 16.6 and Examples 16.14 and 15.

The rate law for Mechanism I $NO_2\ (g) + CO(g) \rightarrow NO(g) + CO_2(g)$

would be: **rate= $k[NO_2][CO]$**.

The rate law for Mechanism II $NO_2(g) + NO_2(g) \rightarrow NO_3(g) + NO(g)$ slow
$NO_3(g) + CO(g) \rightarrow NO_2(g) + CO_2(g)$ fast

would be: **rate=** $\mathbf{k[NO_2][NO_2] = k[NO_2]^2}$.

16.76 *See Section 16.6.*

$2NO(g) \rightleftarrows N_2O_2$ fast equilibrium, $K_{eq} = \dfrac{[N_2O_2]}{[NO]^2}$

$N_2O_2(g) + Cl_2(g) \rightarrow 2NOCl$ slow step

The rate of reaction is limited by the rate of the slowest step, step 2. Hence, rate= $k[N_2O_2][Cl_2]$. However, **N_2O_2 is an intermediate**, and it is customary to express the concentrations of intermediates in terms of concentrations of reactants and when appropriate products. The equilibrium expression indicates $[N_2O_2] = K_{eq}[NO]^2$. Substituting for $[N_2O_2]$ gives **rate=** $\mathbf{kK_{eq}[NO]^2[Cl_2]}$. Hence, this mechanism is consistent with the experimental rate law.

16.78 *See Section 16.6 and Examples 16.14 and 15.*

The rate law for Mechanism I $CH_3Br \xrightarrow{slow} CH_3^+ + Br^-$

$CH_3^+ + OH^- \xrightarrow{fast} CH_3OH$

would be: **rate=** $\mathbf{k[CH_3Br]}$.

The rate law for Mechanism II $CH_3Br + OH^- \rightarrow Br^- + CH_3OH$

would be: **rate=** $\mathbf{k[CH_3Br][OH^-]}$.

16.80 *See Section 16.3 and Example 16.4.*

Time,s	$[H_2O_2]$	$\ln[H_2O_2]$	$\dfrac{1}{[H_2O_2]}$
0	0.0334	-3.399	29.9
10	0.0300	-3.507	33.3
20	0.0283	-3.565	35.3
30	0.0249	-3.692	40.2
40	0.0198	-3.922	50.5
50	0.0164	-4.110	61.0
60	0.0130	-4.343	76.9

For a zero order reaction, a plot of $[H_2O_2]$ vs time gives a straight line.

For a first order reaction, a plot of $\ln[H_2O_2]$ vs. time gives a straight line.

For a second order reaction ,a plot of $\dfrac{1}{[H_2O_2]}$ vs. time gives a straight line.

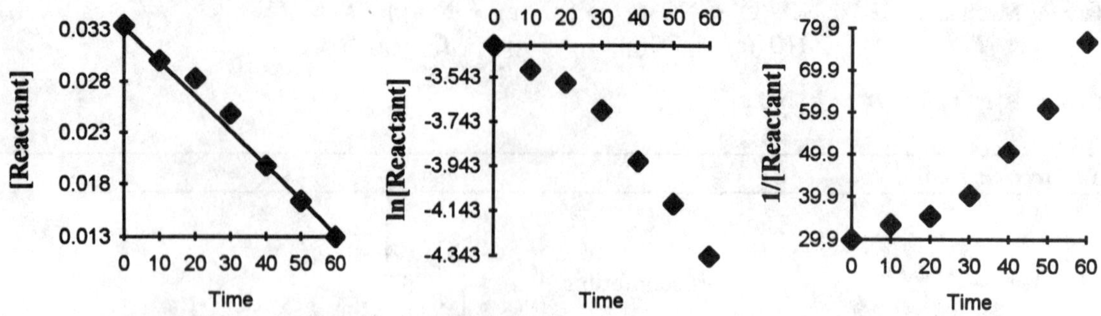

The data give a straight line for a **zero order** reaction. The rate constant for a zero order reaction is equal to the negative of the slope of the line. A linear regression analysis gives a slope of -0.00035, and a correlation coefficient of .9941 indicating reasonably good agreement between the data and the equation. Hence, **$k = 3.5 \times 10^{-4}$ mol·L^{-1}·s^{-1}**.

Note: If your calculator does not have the ability to perform linear regression analyses or you do not have acess to a computer graphics program, you can caculate the slope of the line by using

$$\frac{([\text{Reactant}]_2 - [\text{Reactant}]_1)}{(t_2 - t_1)}$$ with two points from the line which are far apart.

16.82 ***See Section 16.6.***

(a) $2NO(g) \rightleftarrows N_2O_2(g)$ fast equilibrium, $K_{eq} = \frac{[N_2O_2]}{[NO]^2}$

$N_2O_2(g) + O_2(g) \rightarrow 2NO_2$ slow

The rate of reaction is limited by the rate of the slowest step, step2. Hence, rate= $k[N_2O_2][O_2]$. However, N_2O_2 is an intermediate, and it is customary to express the concentrations of intermediates in terms of reactants and when appropriate products. The equilibrium expression indicates $[N_2O_2] = K_{eq}[NO]^2$. Substituting for $[N_2O_2]$ gives **rate= $kK_{eq}[NO]^2[O_2]$**.

(b) The first step in the mechanism involves the formation of a nitrogen-nitrogen bond and is likely to be exothermic. According to the Principle of LeChatelier, an increase in temperature would cause a decrease in the concentration of N_2O_2 and therefore a decrease in the concentration of N_2O_2 molecules available for reaction in step 2. The decrease in rate caused by the decrease in concentration of N_2O_2 must outweigh the increase in rate that would be expected for step 2 based on an increase in temperature.

16.84 ***See Section 16.3 and Example 16.4.***

$HCOOH(g) \rightarrow CO_2(g) + H_2(g)$

Since the stoichiometric coefficients are all one, we can represent the decrease in the pressure of HCOOH(g) during each time interval by -y and the increase in the pressure of $CO_2(g)$ and $H_2(g)$ each by +y during the same time interval. This gives $P_T = P_{HCOOH} + P_{CO_2} + P_{H_2}$ and $P_T = (220 - y) + y + y = 220 + y$ at each point during the reaction. Solving for y and using (220-y) for P_{HCOOH} gives

Time,s	P_{HCOOH}	$\ln(P_{HCOOH})$
0	220	5.394
50	96	4.56
100	61	4.11
150	32	3.47
200	17	2.83
250	9	2.2
300	5	1.6

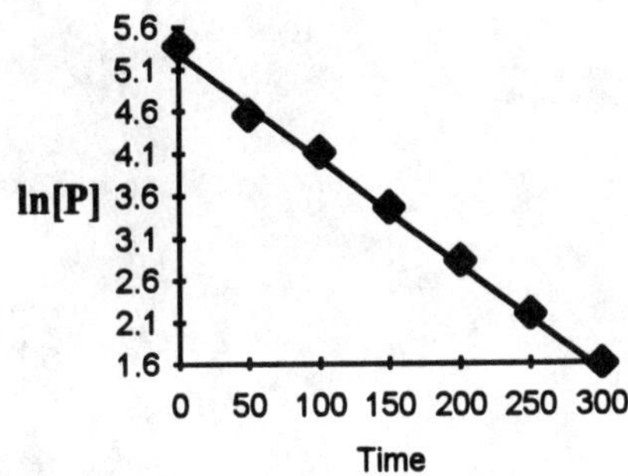

The rate constant for a first order reaciton is equal to the slope of the line when lnP is plotted versus time. A linear regression analysis gives a slope of -0.0124, and a correlation coefficient of 0.9988 indicates very good agreement between the data and first-order reaction.
Hence, **k = 0.0124 s⁻¹**.

Since $kt_{\frac{1}{2}} = 0.693$ for a first-order process,

$$t_{\frac{1}{2}} = \frac{0.693}{k} \qquad t_{\frac{1}{2}} = \frac{0.693}{0.0124\ s^{-1}} = \mathbf{55.9\ s}$$

Note: If your calculator does not have the ability to perform linear regression analyses or you do not have acess to a computer graphics program, you can caculate the slope of the line by using

$$\frac{(\ln[P]_2 - \ln[P]_1)}{(t_2 - t_1)}$$ with two points from the line which are far apart.

Chapter 17: Electrochemistry

17.2	***See Sections 8.3 and 17.1 and Examples 8.4, 5 and 6.***

(a) NO_3^-

Total valence electrons $= [1 \times 5(N) + 3 \times 6(O) + 1(charge)] = 24$.
Eighteen electrons remain after assigning three single bonds, and twenty unshared electrons are needed to give each atom a noble gas configuration (6 for each O and 2 for N). Hence, two electrons (20-18) must be used to form one additional bond. Twenty four electrons are used in writing the Lewis structure.

A total of eight valence shell electrons are assigned to each oxygen atom, two more than are present in a nonbonded oxygen atom, giving each oxygen atom an oxidation state of -2. A nitrogen atom having its five valence shell electrons assigned to the more electronegative oxygen atoms has an oxidation state of +5. Hence, oxidation state of **N = +5, O = -2**.

(b) NO_2^-

Total valence electrons $= [1 \times 5(N) + 2 \times 6(O) + 1(charge)] = 18$.
Fourteen electrons remain after assigning two single bonds, and sixteen unshared electrons are needed to give each atom a noble gas configuration (6 for each O and 4 for N). Hence, two electrons (16-14) must be used to form one additional bond. Eighteen electrons are used in writing the Lewis structure.

A total of eight valence shell electrons are assigned to each oxygen atom, two more than are present in a nonbonded oxygen atom, giving each oxygen atom an oxidation state of -2. A nitrogen atom having three of its five valence shell electrons assigned to the more electronegative oxygen atoms has an oxidation state of +3. Hence, oxidation state of **N = +3, O = -2**.

(c) NH_4^+

Total valence electrons $= [1 \times 5(N) + 4 \times 1(H) - 1(charge)] = 8$.
All eight electrons are used in writing four single bonds.

All eight shared electrons are assigned to the valence shell of the more electronegative nitrogen atom. Since there are five electrons in the valence shell of a nonbonded nitrogen atom, nitrogen has an oxidation state of -3 in NH_4^+. Each hydrogen atom having its only valence shell electron assigned to the more electronegative nitrogen atom has an oxidation state of +1. Hence, oxidation state of **N = -3, H = +1**.

(d) Br_2

Br—Br

:Br—Br:

Total valence electrons $= [2 \times 7(\text{Br})] = 14$.
Twelve electrons remain after assigning one single bond and twelve unshared electrons are needed to give each Br atom a noble gas configuration. Fourteen electrons are used in writing the Lewis structure.

The shared electrons are assigned equally to the bromine atoms giving the valence shell of each bromine atom seven valence electrons, the same number as a nonbonded bromine atom. Therefore, each bromine atom has an oxidation state of zero. Hence, oxidation state of **Br = 0**.

(e) H_2SO_4

```
        O
        |
H—O—S—O—H
        |
        O

       :O:
   ..  ||  ..
H—O—S—O—H
   ..  ||  ..
       :O:
```

Total valence electrons $= [2 \times 1(\text{H}) + 1 \times 6(\text{S}) + 4 \times 6(\text{O})] = 32$.
Twenty electrons remain after assigning six single bonds, and twenty electrons are needed to give each atom a noble gas configuration (6 for each lone O atom and 4 for each O in a O-H bond). Hence, no multiple bonds are needed to provide each atom with a noble gas configuration. However, sulfur is capable of using an expanded valence shell and double bonds are assigned to each lone O atom to reduce the formal charges. Thirty electrons are used to write the Lewis structure.

The shared electrons are assigned to the more electronegative oxygen atoms giving each oxygen atom eight electrons in its valence shell. This gives each oxygen atom two more valence electrons than a nonbonded oxygen atom and an oxidation state of -2. Each hydrogen atom having its only valence shell electron assigned to oxygen has an oxidation state of +1. The sulfur atom having its six valence shell electrons assigned to the more electronegative oxygen atoms has an oxidation state of +6. Hence, oxidation state of **H = +1, S = +6, O = -2.**

(f) CO_2

O—C—O

O=C=O

Total valence electrons $= [1 \times 4(\text{C}) + 2 \times 6(\text{O})] = 16$.
Twelve electrons remain after assigning two single bonds, and sixteen electrons are needed to give each atom a noble gas configuration (6 for each O and 4 for C). Hence, four electrons (16-12) must be used to form additional bonds. A carbon-oxygen double bond is formed with each O atom, and sixteen electrons are used to write the Lewis structure.

The shared electrons are assigned to the more electronegative oxygen atoms giving each oxygen atom eight electrons in its valence shell. This gives each oxygen atom two more valence electrons than a nonbonded oxygen atom and an oxidation state of -2. The carbon atom having its four valence shell electrons assigned to the more electronegative oxygen atoms has an oxidation state of +4. Hence, oxidation state of **C = +4, O = -2.**

17.4 ***See Section 17.1 and Example 17.2.***

(a) $KMnO_4$: Applying Rule 2, K^+ is assigned an oxidation state of +1. Applying Rule 5, O is assigned an oxidation state of -2. This gives $1 + y + 4(-2) = 0$ and $y = +7$ for the oxidation state of the manganese in $KMnO_4$. Hence, oxidation state of **K = +1, Mn = +7, O = -2.**

(b) H_2O: Applying Rule 4, hydrogen is assigned an oxidation state of +1. Applying Rule 5, oxygen is assigned an oxidation state of -2. This gives the expected sum of zero for H_2O. Hence, oxidation state of **H = +1, O = -2.**

(c) Cl_2: Applying Rule 1, each chlorine is assigned an oxidation state of zero. This gives the expected sum of zero for Cl_2. Hence, oxidation state of **Cl = 0.**

(d) NO_2: Applying Rule 5, each oxygen atom is assigned an oxidation state of -2. This gives $y + 2(-2) = 0$ and $y = +4$ for the oxidation state of the nitrogen in NO_2. Hence, oxidation state of **N = +4, O = -2.**

(e) CrO_2^-: Applying Rule 5, each oxygen atom is assigned an oxidation state of -2. This gives y + 2(-2) = -1 and y = +3 for the oxidation state of chromium in CrO_2^-. Hence, oxidation state of **Cr = +3, O = -2.**

(f) $Co(NO_3)_3$: Applying Rule 2, cobalt is assigned an oxidation state of +3. Applying Rule 5, each oxygen atom is assigned an oxidation state of -2. This gives y + 3(-2) = -1 and y = +5 for the oxidation state of nitrogen in NO_3^- and $Co(NO_3)_3$. Hence, oxidation state of **Co = +3, N = +5, O = -2.**

(g) $CaCO_3$: Applying Rule 2, calcium is assigned an oxidation state of +2. Applying Rule 5, oxygen is assigned an oxidation state of -2. This gives y + 3(-2) = -2 and y = +4 for the oxidation state of carbon in CO_3^{2-} and $CaCO_3$. Hence, oxidation state of **Ca = +2, C = +4, O = -2.**

(h) $HBrO_4$: Applying Rule 4, hydrogen is assigned an oxidation state of +1. Applying Rule 5, oxygen is assigned an oxidation state of -2. This gives 1 + y + 4(-2) = 0 and y = +7 for the oxidation state of bromine in $HBrO_4$. Hence, oxidation state of **H = +1, Br = +7, O = -2.**

(i) Fe^{3+}: Applying Rule 1, iron is assigned an oxidation state of +3. Hence, oxidation state of **Fe = +3.**

17.6 ***See Section 17.2.***

(a) $Cr^{3+}(aq) + 3e^- \rightarrow Cr(s)$, gain of electrons and therefore reduction.

(b) $2I^-(aq) \rightarrow I_2(aq) + 2e^-$, loss of electrons and therfore oxidation.

(c) $NO_2^-(aq) + H_2O(l) \rightarrow NO_3^-(aq) + 2H^+(aq) + 2e^-$, loss of electrons and therefore oxidation.

(d) $Fe^{2+}(aq) \rightarrow Fe^{3+}(aq) + e^-$, loss of electrons and therefore oxidation.

(e) $Cr_2O_7^{2-}(aq) + 14H^+(aq) + 6e^- \rightarrow 2Cr^{3+}(aq) + 7H_2O(l)$, gain of electrons and therefore reduction.

(f) $VO_2^+(aq) + 4H^+(aq) + 2e^- \rightarrow V^{3+}(aq) + 2H_2O(l)$, gain of electrons and therefore reduction.

17.8 ***See Section 17.2 and Example 17.3.***

(a) Oxidation half-reaction: $Sn(s) \rightarrow Sn^{2+}(aq) + 2e^-$

Reduction half-reaction: $2\left[Fe^{3+}(aq) + e^- \rightarrow Fe^{2+}(aq)\right]$

Balanced redox equation: $Sn(s) + 2Fe^{3+}(aq) \rightarrow Sn^{2+}(aq) + 2Fe^{2+}(aq)$

(b) Oxidation half-reaction: $HAsO_3^{2-}(aq) + H_2O(l) \rightarrow H_2AsO_4^-(aq) + H^+(aq) + 2e^-$

Reduction half-reaction: $I_2(aq) + 2e^- \rightarrow 2I^-(aq)$

Balanced redox equation: $HAsO_3^{2-}(aq) + H_2O(l) + I_2(aq) \rightarrow H_2AsO_4^-(aq) + H^+(aq) + 2I^-(aq)$

(c) Oxidation half-reaction: $Cu(s) \rightarrow Cu^{2+}(aq) + 2e^-$

Reduction half-reaction: $2[Ag^+(aq) + e^- \rightarrow Ag(s)]$

Balanced redox equation: $Cu(s) + 2Ag^+(aq) \rightarrow Cu^{2+}(aq) + 2Ag(s)$

(d) Oxidation half-reaction: $5[H_2C_2O_4(aq) \rightarrow 2CO_2(g) + 2H^+(aq) + 2e^-]$

Reduction half-reaction: $2[MnO_4^-(aq) + 8H^+(aq) + 5e^- \rightarrow Mn^{2+}(aq) + 4H_2O(l)]$

Balanced redox equation:

$$2MnO_4^-(aq) + 6H^+(aq) + 5H_2C_2O_4(aq) \rightarrow 2Mn^{2+}(aq) + 8H_2O(l) + 10CO_2(g)]$$

(e) Oxidation half-reaction: $2Br^-(aq) \rightarrow Br_2(l) + 2e^-$

Reduction half-reaction: $Cl_2(g) + 2e^- \rightarrow 2Cl^-(aq)$

Balanced redox equation: $Cl_2(g) + 2Br^-(aq) \rightarrow 2Cl^-(aq) + Br_2(l)$

(f) Oxidation half-reaction: $3[Cu(s) \rightarrow Cu^{2+}(aq) + 2e^-]$

Reduction half-reaction: $2[NO_3^-(aq) + 4H^+(aq) + 3e^- \rightarrow NO(g) + 2H_2O(l)]$

Balanced redox equation: $3Cu(s) + 2NO_3^-(aq) + 8H^+(aq) \rightarrow 3Cu^{2+}(aq) + 2NO(g) + 4H_2O(l)$

(g) Oxidation half-reaction: $6[VO^{2+}(aq) + H_2O(l) \rightarrow VO_2^+(aq) + 2H^+(aq) + e^-]$

Reduction half-reaction: $Cr_2O_7^{2-}(aq) + 14H^+(aq) + 6e^-(aq) \rightarrow 2Cr^{3+}(aq) + 7H_2O(l)$

Balanced redox equation: $6VO^{2+}(aq) + Cr_2O_7^{2-}(aq) + 2H^+(aq) \rightarrow 6VO_2^+(aq) + 2Cr^{3+}(aq) + H_2O(l)$

17.10 ***See Section 17.2 and Example 17.4.***

(a) Oxidation half-reaction: $2[Al(s) + 4H_2O(l) \rightarrow Al(OH)_4^-(aq) + 4H^+(aq) + 3e^-]$

Reduction half-reaction: $3[ClO^-(aq) + 2H^+(aq) + 2e^- \rightarrow Cl^-(aq) + H_2O(l)]$

Balanced redox equation: $2Al(s) + 5H_2O(l) + 3ClO^-(aq) \rightarrow 2Al(OH)_4^-(aq) + 2H^+(aq) + 3Cl^-(aq)$

Balanced redox equation: $2Al(s) + 5H_2O(l) + 3ClO^-(aq) \rightarrow 2Al(OH)_4^-(aq) + 2H^+(aq) + 3Cl^-(aq)$

Base conversion equation: $2H^+(aq) + 2OH^-(aq) \rightarrow 2H_2O(l)$

Final redox equation:

$$2Al(s) + 3H_2O(l) + 3ClO^-(aq) + 2OH^-(aq) \rightarrow 2Al(OH)_4^-(aq) + 3Cl^-(aq)$$

(b) Oxidation half-reaction: $3\left[SO_3^{2-}(aq) + H_2O(l) \rightarrow SO_4^{2-}(aq) + 2H^+(aq) + 2e^-\right]$

Reduction reaction: $2\left[MnO_4^-(aq) + 4H^+(aq) + 3e^- \rightarrow MnO_2(s) + 2H_2O(l)\right]$

Balanced redox equation:

$$3SO_3^{2-}(aq) + 2MnO_4^-(aq) + 2H^+(aq) \rightarrow 3SO_4^{2-}(aq) + 2MnO_2(s) + H_2O(l)$$

Base conversion equation: $2H_2O(l) \rightarrow 2H^+(aq) + 2OH^-(aq)$

Final redox equation:

$$3SO_3^{2-}(aq) + 2MnO_4^-(aq) + H_2O(l) \rightarrow 3SO_4^{2-}(aq) + 2MnO_2(s) + 2OH^-(aq)$$

(c) Oxidation half-reaction: $4\left[Zn(s) + 4H_2O(l) \rightarrow Zn(OH)_4^{2-}(aq) + 4H^+(aq) + 2e^-\right]$

Reduction half-reaction: $NO_3^-(aq) + 9H^+(aq) + 8e^- \rightarrow NH_3(aq) + 3H_2O(l)$

Balanced redox eqution: $4Zn(s) + 13H_2O(l) + NO_3^-(aq) \rightarrow 4Zn(OH)_4^{2-}(aq) + 7H^+(aq) + NH_3(aq)$

Base conversion equation: $7H^+(aq) + 7OH^-(aq) \rightarrow 7H_2O(l)$

Final redox equation:

$$4Zn(s) + 6H_2O(l) + NO_3^-(aq) + 7OH^-(aq) \rightarrow 4Zn(OH)_4^{2-}(aq) + NH_3(aq)$$

(d) Oxidation half-reaction: $2\left[CrO_2^-(aq) + 2H_2O(l) \rightarrow CrO_4^{2-}(aq) + 4H^+(aq) + 3e^-\right]$

Reduction half-reaction: $3\left[ClO^-(aq) + 2H^+(aq) + 2e^- \rightarrow Cl^-(aq) + H_2O(l)\right]$

Balanced redox equation:

$$2CrO_2^-(aq) + H_2O(l) + 3ClO^-(aq) \rightarrow 2CrO_4^{2-}(aq) + 2H^+(aq) + 3Cl^-(aq)$$

Base conversion equation: $2H^+(aq) + 2OH^-(aq) \rightarrow 2H_2O(l)$

Final redox equation:

$$2CrO_2^-(aq) + 3ClO^-(aq) + 2OH^-(aq) \rightarrow 2CrO_4^{2-}(aq) + 3Cl^-(aq) + H_2O(l)$$

(e) Oxidation half-reaction: $Br_2(aq) + 6H_2O(l) \rightarrow 2BrO_3^-(aq) + 12H^+(aq) + 10e^-$

Reduction half-reaction: $5\left[Br_2(aq) + 2e^- \rightarrow 2Br^-(aq)\right]$

Balanced redox equation: $6Br_2(aq) + 6H_2O(l) \rightarrow 2BrO_3^-(aq) + 12H^+(aq) + 10Br^-(aq)$

Smallest coefficients: $3Br_2(aq) + 3H_2O(l) \rightarrow BrO_3^-(aq) + 6H^+(aq) + 5Br^-(aq)$

Base conversion equation: $6H^+(aq) + 6OH^-(aq) \rightarrow 6H_2O(l)$

Final redox equation: $3Br_2(aq) + 6OH^-(aq) \rightarrow BrO_3^-(aq) + 5Br^-(aq) + 3H_2O(l)$

(f) Oxidation half-reaction: $N_2H_4(aq) \rightarrow N_2(g) + 4H^+(aq) + 4e^-$

Reduction half-reaction: $2\left[H_2O_2(aq) + 2H^+(aq) + 2e^- \rightarrow 2H_2O(l)\right]$

Balanced redox equation: $N_2H_4(aq) + 2H_2O_2(aq) \rightarrow N_2(g) + 4H_2O(l)$

17.12 ***See Section 17.2, Example 17.5 and Solution 17.8d.***

$$2MnO_4^-(aq) + 5C_2O_4^{2-}(aq) + 16H^+(aq) \rightarrow 2Mn^{2+}(aq) + 10CO_2(g) + 8H_2O(l)$$

Strategy: $g\ Na_2C_2O_4 \rightarrow mol\ Na_2C_2O_4 \rightarrow mol\ KMnO_4 \rightarrow M\ KMnO_4\ soln$

$$?\text{ mol KMnO}_4 = 0.103\text{ g Na}_2\text{C}_2\text{O}_4 \times \frac{1\text{ mol Na}_2\text{C}_2\text{O}_4}{134.0\text{ g Na}_2\text{C}_2\text{O}_4} \times \frac{2\text{ mol KMnO}_4}{5\text{ mol Na}_2\text{C}_2\text{O}_4} = 3.08 \times 10^{-4}\text{ mol KMnO}_4$$

$$?\ M\text{ KMnO}_4\text{ soln} = \frac{3.08 \times 10^{-4}\text{ mol KMnO}_4}{0.02430\text{ L KMnO}_4\text{ soln}} = \mathbf{1.27 \times 10^{-2}}\ \boldsymbol{M}\ \mathbf{KMnO_4}$$

17.16 ***See Section 17.3.***

(a) Oxidation: $2Hg(l) + 2Cl^-(aq) \rightarrow Hg_2Cl_2(s) + 2e^-$
Reduction: $Cu^{2+}(aq) + 2e^- \rightarrow Cu(s)$

(b) Cell: $2Hg(l) + 2Cl^-(aq) + Cu^{2+}(aq) \rightarrow Cu(s) + Hg_2Cl_2(s)$

(c) The electrode at which reduction occurs is the positive electrode in a voltaic cell. In this case, the copper electrode is the positive electrode.

(d) As $Cu^{2+}(aq)$ ions come into contact with Hg(l), a direct electron transfer will take place.

17.18 ***See Section 17.4.***

(a) Anode:	$Zn(s) \rightarrow Zn^{2+}(aq) + 2e^-$	$E° = +0.76$ V
Cathode:	$Ni^{2+}(aq) + 2e^- \rightarrow Ni(s)$	$E° = -0.25$ V
Cell:	$Zn(s) + Ni^{2+}(aq) \rightarrow Zn^{2+}(aq) + Ni(s)$	$E° =$ **0.51 V**
(b) Anode:	$Pb(s) \rightarrow Pb^{2+}(aq) + 2e^-$	$E° = 0.126$ V
Cathode:	$2Ag^+(aq) + 2e^- \rightarrow 2Ag(s)$	$E° = 0.80$ V
Cell:	$Pb(s) + 2Ag^+(aq) \rightarrow Pb^{2+}(aq) + 2Ag(s)$	$E° =$ **0.93 V**

17.20 ***See Section 17.4, Example 17.8 and Appendix H.***

(a)

$$H_2(g) \rightarrow 2H^+(aq) + 2e^- \qquad E° = 0.00 \text{ V}$$
$$Cl_2(g) + 2e^- \rightarrow 2Cl^-(aq) \qquad E° = 1.36 \text{ V}$$

$$H_2(g) + Cl_2(g) \rightarrow 2H^+(aq) + 2Cl^-(aq) \qquad E° = \mathbf{1.36 \text{ V}}$$

Since the calculated voltage is positive, the reaction is spontaneous in the direction shown under standard conditions.

(b)

$$3Cr^{2+}(aq) \rightarrow 3Cr^{3+}(aq) + 3e^- \qquad E° = 0.41 \text{ V}$$
$$Al^{3+}(aq) + 3e^- \rightarrow Al(s) \qquad E° = -1.66 \text{ V}$$

$$3Cr^{2+}(aq) + Al^{3+}(aq) \rightarrow 3Cr^{3+}(aq) + Al(s) \qquad E° = \mathbf{-1.25 \text{ V}}$$

Since the calculated voltage is negative, the reaction is spontaneous in the reverse direction (nonspontaneous in the forward direction) under standard conditions.

(c)

$$Sn^{2+}(aq) \rightarrow Sn^{4+}(aq) + 2e^- \qquad E° = -0.15 \text{ V}$$
$$2Fe^{3+}(aq) + 2e^- \rightarrow 2Fe^{2+}(aq) \qquad E° = 0.771 \text{ V}$$

$$Sn^{2+}(aq) + 2Fe^{3+}(aq) \rightarrow Sn^{4+}(aq) + 2Fe^{2+}(aq) \qquad E° = \mathbf{0.62 \text{ V}}$$

Since the calculated voltage is positive, the reaction is spontaneous in the direction shown under standard conditions.

17.22 ***See Sections 17.9 and 12.***

(a)

$$Cu(s) \rightarrow Cu^{2+}(aq) + 2e^- \qquad E° = -0.34 \text{ V}$$
$$2H_2O(l) + 2e^- \rightarrow H_2(g) + 2OH^-(aq) \qquad E° = -0.83 \text{ V}$$

$$Cu(s) + 2H_2O(l) \rightarrow Cu^{2+}(aq) + H_2(g) + 2OH^-(aq) \qquad E° = \mathbf{-1.17 \text{ V}}$$

Since the calculated voltage is negative, the reaction of Cu(s) with $H_2O(l)$ is nonspontaneous under standard conditions. Hence, Cu(s) will not dissolve in water.

(b)

$$2[Na(s) \rightarrow Na^{+}(aq) + e^{-}] \qquad E^\circ = 2.71\ V$$

$$2H_2O(l) + 2e^{-} \rightarrow H_2(g) + 2OH^{-}(aq) \qquad E^\circ = -0.83\ V$$

$$2Na(s) + 2H_2O(l) \rightarrow 2Na^{+}(aq) + H_2(g) + 2OH^{-}(aq) \qquad E^\circ = \mathbf{1.88\ V}$$

Since the calculated voltage is positive, the reaction of Na(s) with H_2O(l) is spontaneous under standard conditions. Indeed, Na(s) dissolves in water in a very vigorous reaction.

(c)

$$2[Al(s) \rightarrow Al^{3+}(aq) + 3e^{-}] \qquad E^\circ = 1.66\ V$$

$$3[2H_2O(l) + 2e^{-} \rightarrow H_2(g) + 2OH^{-}(aq)] \qquad E^\circ = -0.83\ V$$

$$2Al(s) + 6H_2O(l) \rightarrow 2Al^{3+}(aq) + 3H_2(g) + 6OH^{-}(aq) \qquad E^\circ = \mathbf{0.83\ V}$$

Since the calculated voltage is positive, the reaction of Al(s) with H_2O(l) is spontaneous under standard conditions. Hence, it is predicted Al(s) will dissolve in water. However, it doesn't commonly do so; see answer d below.

(d) Aluminum forms an impervious coating of aluminum oxide, Al_2O_3.

17.24 ***See Section 17.4 and Table 17.1.***

$$Pb(s) \rightarrow Pb^{+}(aq) + 2e^{-} \qquad E^\circ = 0.126\ V$$

$$UO_2^{2+}(aq) + 4H^{+}(aq) + 2e^{-} \rightarrow U^{4+}(aq) + 2H_2O(l) \qquad E^\circ = y$$

$$Pb(s) + UO_2^{2+}(aq) + 4H^{+}(aq) \rightarrow Pb^{2+}(aq) + U^{4+}(aq) + 2H_2O(l) \qquad E^\circ = 0.460\ V$$

Solving 0.126 V + y = 0.460 V gives y = **0.334 V**.

Oxidation of Pb(s) occurs at the negative electrode in a voltaic cell. Reduction of UO_2^{2+}(aq) occurs at the positive electrode in a voltaic cell.

17.26 ***See Section 17.4 and Table 17.1.***

Oxidation occurs at the negative electrode in a voltaic cell. Hence, Cu(s) must be oxidized and Pu^{4+}(aq) must be reduced. This gives

$$Cu(s) \rightarrow Cu^{2+}(aq) + 2e^{-} \qquad E^\circ = -0.34\ V$$

$$2Pu^{4+}(aq) + 2e^{-} \rightarrow 2Pu^{3+}(aq) \qquad E^\circ = y$$

$$Cu(s) + 2Pu^{4+}(aq) \rightarrow Cu^{2+}(aq) + 2Pu^{3+}(aq) \qquad E^\circ = 0.642$$

Solving -0.34 V + y = 0.642 V gives y = **0.98 V**.

17.28 ***See Section 17.5, Examples 17.9 and 10 and Appendix H.***

(a)

$$Ni(s) \rightarrow Ni^{2+}(aq) + 2e^- \qquad E^\circ = 0.25\ V$$

$$Cu^{2+}(aq) + 2e^- \rightarrow Cu(s) \qquad E^\circ = 0.34\ V$$

$$Ni(s) + Cu^{2+}(aq) \rightarrow Ni^{2+}(aq) + Cu(s) \qquad E^\circ = \mathbf{0.59\ V}$$

$$\Delta G^\circ = -nFE^\circ \qquad \Delta G^\circ = -(2\ \text{mol e}^-)\left(\frac{9.65 \times 10^4\ C}{1\ \text{mol e}^-}\right)\left(\frac{0.59\ J}{1\ C}\right) = \mathbf{-1.1 \times 10^5\ J}$$

$$\log K_{eq} = \frac{nE^\circ}{0.0591} \qquad \log K_{eq} = \frac{(2)(0.59)}{0.0591} = 20.0 \qquad K_{eq} = 10^{20.0} = \text{inv}\log 20.0 = \mathbf{1 \times 10^{20}}$$

Since the calculated voltage is positive, the reaction is **spontaneous** in the direction shown under standard conditions. This is consistent with a negative vlaue for ΔG° and $K_{eq} > 1$.

(b)

$$2Ag(s) + 2Cl^-(aq) \rightarrow 2AgCl(s) + 2e^- \qquad E^\circ = -0.222\ V$$

$$Cl_2(g) + 2e^- \rightarrow 2Cl^-(aq) \qquad E^\circ = 1.36\ V$$

$$2Ag(s) + Cl_2(g) \rightarrow 2AgCl(s) \qquad E^\circ = \mathbf{1.14\ V}$$

$$\Delta G^\circ = -nFE^\circ \qquad \Delta G^\circ = -(2\ \text{mol e}^-)\left(\frac{9.65 \times 10^4\ C}{1\ \text{mol e}^-}\right)\left(\frac{1.14\ J}{1\ C}\right) = \mathbf{-2.20 \times 10^5\ J}$$

$$\log K_{eq} = \frac{nE^\circ}{0.0591} \qquad \log K_{eq} = \frac{(2)(1.14)}{0.0591} = 38.6 \qquad K_{eq} = 10^{38.6} = \text{inv}\log 38.6 = \mathbf{4 \times 10^{38}}$$

Since the calculated voltage is positive, the reaction is **spontaneous** in the direction shown under standard conditions. This is consistent with a negative vlaue for ΔG° and $K_{eq} > 1$.

(c)

$$2I^-(aq) \rightarrow I_2(s) + 2e^- \qquad E^\circ = -0.54\ V$$

$$Cl_2(g) + 2e^- \rightarrow 2Cl^-(aq) \qquad E^\circ = 1.36\ V$$

$$2I^-(aq) + Cl_2(g) \rightarrow I_2(s) + 2Cl^-(aq) \qquad E^\circ = \mathbf{0.82\ V}$$

$$\Delta G^\circ = -nFE^\circ \qquad \Delta G^\circ = -(2\ \text{mol e}^-)\left(\frac{9.65 \times 10^4\ C}{1\ \text{mol e}^-}\right)\left(\frac{0.82\ J}{1\ C}\right) = \mathbf{-1.6 \times 10^5\ J}$$

$$\log K_{eq} = \frac{nE^\circ}{0.0591} \qquad \log K_{eq} = \frac{(2)(0.82)}{0.0591} = 28 \qquad K_{eq} = 10^{28} = \text{inv}\log 28 = \mathbf{1 \times 10^{28}}$$

Since the calculated voltage is positive, the reaction is **spontaneous** in the direction shown under standard conditions. This is consistent with a negative vlaue for ΔG° and $K_{eq} > 1$.

17.30 ***See Section 17.5 and Examples 17.9 and 10.***

$$\Delta G^\circ = -RT\ln K_{eq} \qquad \Delta G^\circ = -8.314\ \text{J}\cdot\text{mol}^{-1}\cdot\text{K}^{-1} \times 298\ \text{K} \times \ln\left(1.58 \text{ x } 10^2\right) = \mathbf{-1.25 \text{ x } 10^4\ J}$$

$$E^\circ = \frac{0.0591\ \text{V}\cdot\text{mol e}^-}{n}\log K_{eq} \qquad E^\circ = \frac{0.0591\ \text{V}\cdot\text{mol e}^-}{2\ \text{mol e}^-}\log\left(1.58 \text{ x } 10^2\right) = \mathbf{0.0650\ V}$$

17.32 ***See Section 17.6, Examples 17.11 and 12 and Solution for 17.28.***

(a) For $Ni(s) + Cu^{2+}(aq) \rightarrow Ni^{2+}(aq) + Cu(s)$,

$$E = E^\circ - \frac{0.0591}{n}\log\left(\frac{[Ni^{2+}]}{[Cu^{2+}]}\right) \qquad E = 0.59\ \text{V} - \frac{0.0591\ \text{V}\cdot\text{mol e}^-}{2\ \text{mol e}^-}\log\left(\frac{1.40}{0.050}\right) = \mathbf{0.55\ V}$$

(b) For $2Ag(s) + Cl_2(g) \rightarrow 2AgCl(s)$

$$E = E^\circ - \frac{0.0591}{n}\log\left(\frac{1}{P_{Cl_2}}\right) \qquad E = 1.14\ \text{V} - \frac{0.0591\ \text{V}\cdot\text{mol e}^-}{2\ \text{mol e}^-}\log\left(\frac{1}{320\ \text{torr} \times \dfrac{1\ \text{atm}}{760\ \text{torr}}}\right) = \mathbf{1.13\ V}$$

(c) For $Cl_2(g) + 2I^-(aq) \rightarrow 2Cl^-(aq) + I_2(s)$,

$$E = E^\circ - \frac{0.0591}{n}\log\left(\frac{[Cl^-]^2}{P_{Cl_2}[I^-]^2}\right) \qquad E = 0.82\ \text{V} - \frac{0.0591\ \text{V}\cdot\text{mol e}^-}{2\ \text{mol e}^-}\log\left(\frac{(0.60)^2}{(0.300)(0.0010)^2}\right) = \mathbf{0.64\ V}$$

17.34 ***See Section 17.6 and Example 17.12.***

For $2H^+(aq) + 2e^- \rightarrow H_2(g)$ with pH = 6.00 and $P_{H_2} = 2.5$ atm,

$$E = E^\circ - \frac{0.0591}{n}\log\left(\frac{P_{H_2}}{[H^+]^2}\right) \qquad E = 0.00\ \text{V} - \frac{0.0591\ \text{V}\cdot\text{mol e}^-}{2\ \text{mol e}^-}\log\left(\frac{2.5}{\left(1.00 \text{ x } 10^{-6}\right)^2}\right) = \mathbf{-0.37\ V}$$

17.36 *See Section 17.6.*

The electrode at which reduction occurs is the positive electrode in a voltaic cell. Hence, $Pb^{2+}(aq)$ ions are reduced in this cell: $Pb^{2+}(aq) + 2e^- \rightarrow Pb(s)$. The Nernst equation expression for this cell is

$$E = E^\circ - \frac{0.0591\ V\cdot mol\ e^-}{n}\log\left(\frac{1}{\left[Pb^{2+}\right]}\right),$$

since a reference electrode is an electrode that has an accurately known potential that is independent of the composition of the test solution.

For the conditions given, $0.053\ V = E^\circ - \frac{0.0591\ V\cdot mol\ e^-}{2\ mol\ e^-}\log\left(\frac{1}{0.100}\right)$

This gives, $0.053\ V = E^\circ - 0.030\ V$ and $E^\circ = 0.083\ V$.

For $\left[Pb^{2+}\right] = 1.6 \times 10^{-2}\ M$, $E = 0.083\ V - \frac{0.0591\ V\cdot mol\ e^-}{2\ mol\ e^-}\log\left(\frac{1}{1.6 \times 10^{-2}}\right) = \mathbf{0.030\ V}$.

17.38 *See Section 17.7, Example 17.13, and Solution for 17.36.*

Substituting in $E = E^\circ - \frac{0.0591\ V\cdot mol\ e^-}{n}\log\left(\frac{1}{\left[Pb^{2+}\right]}\right)$ yields

$$0.010\ V = 0.083\ V - \frac{0.0591\ V\cdot mol\ e^-}{2\ mol\ e^-}\log\left(\frac{1}{\left[Pb^{2+}\right]}\right)$$

So, $\log\left(\frac{1}{\left[Pb^{2+}\right]}\right) = \frac{-2\ mol\ e^-(0.010\ V - 0.083\ V)}{0.0591\ V\cdot mol\ e^-} = 2.47$ and $\frac{1}{\left[Pb^{2+}\right]} = 10^{2.47} = \text{inv}\log 2.47 = 295$

Hence, $\left[Pb^{2+}\right] = \frac{1}{295} = \mathbf{3.4 \times 10^{-3}}\ \boldsymbol{M}\ \mathbf{Pb^{2+}}$

17.40 *See Sections 17.5 and 7.*

$$Ba(IO_3)_2(s) \rightleftarrows Ba^{2+}(aq) + 2IO_3^-(aq) \qquad \Delta G_1^\circ$$

$$Ba^{2+}(aq) + 2e^- \rightarrow Ba(s) \qquad \Delta G_2^\circ$$

$$Ba(IO_3)_2(s) + 2e^- \rightarrow Ba(s) + 2IO_3^-(aq) \qquad \Delta G_3^\circ$$

$\Delta G_1^\circ = -RT\ln K_{sp}$ $\quad \Delta G_1^\circ = -8.314\ J\cdot mol^{-1}\cdot K^{-1} \times 298\ K \times \ln(1.5 \times 10^{-9}) = 5.0 \times 10^4\ J$

$\Delta G_2^\circ = -n\mathcal{F}E^\circ$ $\quad \Delta G_2^\circ = -(2\ mol\ e^-)\left(\frac{9.65 \times 10^4\ C}{1\ mol\ e^-}\right)\left(\frac{-2.90\ J}{1\ C}\right) = 5.60 \times 10^5\ J$

$\Delta G_3^\circ = \Delta G_1^\circ + \Delta G_2^\circ$ $\quad 5.0 \times 10^4\ J + 5.60 \times 10^5\ J = 6.10 \times 10^5\ J$

Solving $\Delta G^\circ = -n\mathcal{F}E^\circ$ for E° gives $E^\circ = \frac{-\Delta G^\circ}{n\mathcal{F}}$ $\quad E^\circ = \frac{-6.10 \times 10^5\ J}{(2\ mol\ e^-)\left(\frac{9.65 \times 10^4\ C}{1\ mol\ e^-}\right)} = -3.16\ J\cdot C^{-1} = \mathbf{-3.16\ V}$

17.44 ***See Sections 17.2, 4 and 6 and Appendix H.***

$$Zn(s) + 2OH^-(aq) \rightarrow Zn(OH)_2(s) + 2e^- \qquad E^\circ = 1.25\ V$$

$$Ag_2O(s) + H_2O(l) + 2e^- \rightarrow 2Ag(s) + 2OH^-(aq) \qquad E^\circ = 0.342\ V$$

$$Zn(s) + Ag_2O(s) + H_2O(l) \rightarrow Zn(OH)_2(s) + 2Ag(s) \qquad E^\circ = \mathbf{1.59\ V}$$

(b) Since the reactants and products are all solids and pure liquids, the value of the reaction quotient in the Nernst equation does not change during the course of the reaction, and the voltage therefore remains the same.

17.46 ***See Sections 15.4, 17.5 and 8 and Examples 15.4 and 17.9.***

(a) Oxidation half-reaction: $\quad C_3H_8(g) + 20OH^-(aq) \rightarrow 3CO_2(g) + 14H_2O(l) + 20e^-$

Reduction half-reaction: $\quad O_2(g) + 2H_2O(l) + 4e^- \rightarrow 4OH^-(aq)$

(b) For $C_3H_8(g) + 5O_2(g) \rightarrow 3CO_2(g) + 4H_2O(l)$,

$$\Delta G^\circ_{rxn} = [3\Delta G_f^\circ CO_2(g) + 4\Delta G_f^\circ H_2O(l)] - [\Delta G_f^\circ C_3H_8(g) + 5\Delta G_f^\circ O_2(g)]$$

$$\Delta G^\circ_{rxn} = [(3\ mol \times -394.36\ kJ\cdot mol^{-1}) + (4\ mol \times -237.18\ kJ\cdot mol^{-1})]$$

$$-[(1\ mol \times -23.49\ kJ\cdot mol^{-1}) + (5\ mol \times 0.\ kJ\cdot mol^{-1})] = \mathbf{-2108.31\ kJ}$$

Solving $\Delta G^\circ = -n\mathcal{F}E^\circ$ for E° gives $E^\circ = \frac{-\Delta G^\circ}{n\mathcal{F}}$ $\quad E^\circ = \frac{-(-2108.31 \times 10^5\ J)}{(20\ mol\ e^-)\left(\frac{9.65 \times 10^4\ C}{1\ mol\ e^-}\right)} = \mathbf{1.09\ V}$

(c) Recognizing that ΔG° represents the maximum amount of work that can be obtained from a system gives

$$? \text{ kJ for consumption of } 1.00 \text{ g } C_3H_8 = 1.00 \text{ g } C_3H_8 \times \frac{1 \text{ mol } C_3H_8}{44.0 \text{ g } C_3H_8} \times \frac{2108.31 \text{ kJ}}{1 \text{ mol } C_3H_8} = \mathbf{47.9 \text{ kJ}}$$

17.48 ***See Section 17.9 and Example 17.14.***

The most easily oxidized species in an electrolysis is the one with the highest (most positive) oxidation potential. The most easily reduced species in an electrolysis is the one with the highest (most positive) reduction potential.

(a) At the anode the possible reaction are:

$Pb(s) \rightarrow Pb^{2+}(aq) + 2e^-$ $\quad E^\circ = 0.126$ V

$2H_2O(l) \rightarrow O_2(g) + 4H^+(aq) + 4e^-$ $\quad E^\circ = -1.23$ V

Since Pb has the more positive potential for oxidation, it will be oxidized at the anode.

At the cathode the possible reactions are:

$2H^+(aq) + 2e^- \rightarrow H_2(g)$ $\quad E^\circ = 0.00$ V

$Pb^{2+}(aq) + 2e^- \rightarrow Pb(s)$ $\quad E^\circ = -0.126$ V

However, the overvoltage of approximately 0.4 V associated with producing $H_2(g)$ means it is actually more difficult to reduce H^+ than Pb^{2+}. Hence, Pb^{2+} will be reduced at the cathode.

Since Pb(s) is oxidized and Pb^{2+} is reduced, no net reaction occurs.

(b) At the anode the possible reactions are:

$2I^-(aq) \rightarrow I_2(aq) + 2e^-$ $\quad E^\circ = -0.54$ V

$2H_2O(l) \rightarrow 4H^+(aq) + O_2(g) + 4e^-$ $\quad E^\circ = -1.23$ V

Since I^- has a much more positive potential for oxidation, it will be oxidized at the anode.

At the cathode the possible reactions are:

$2H^+(aq) + 2e^- \rightarrow H_2(g)$ $\quad E^\circ = 0.00$ V

$Na^+(aq) + e^- \rightarrow Na(s)$ $\quad E^\circ = -2.71$ V

Since H^+ has a potential for reduction that is far more positive than that for Na^+, the hydrogen overvoltage of 0.4 V is not a factor. Hence, H^+ will be reduced at the cathode.

The net reaction will be:

$2H^+(aq) + 2I^-(aq) \rightarrow H_2(g) + I_2(s)$

(c) At the anode the possible reactions are:

$2Br^-(aq) \rightarrow Br_2(l) + 2e^-$ $\quad E^\circ = -1.06$ V

$2H_2O(l) \rightarrow 4H^+(aq) + O_2(g) + 4e^-$ $\quad E^\circ = -1.23$ V

Since $Br^-(aq)$ has a more positive potential for oxidation and there is an additional oxygen overvoltage of 0.4 V, $Br^-(aq)$ will be oxidized at the anode.

At the cathode the possible reactions are:

$Cu^{2+}(aq) + 2e^- \rightarrow Cu(s)$ $E^\circ = +0.34$ V

$2H^+(aq) + 2e^- \rightarrow H_2(g)$ $E^\circ = 0.00$ V

Since Cu^{2+} has a more positive potential for reduction and there is an additional hydrogen overvoltage of 0.4 V, Cu^{2+} will be reduced at the cathode.

The net reaction will be:

$Cu^{2+}(aq) + 2Br^-(aq) \rightarrow Cu(s) + Br_2(l)$

17.50 ***See Section 17.9 and Table 17.1.***

The most easily oxidized species in an electrolysis is the one with the highest (most positive) oxidation potential. The most easily reduced species in an electrolysis is the one with the highest (most positive) reduction potential.

The pertinent reduction potentials are:

$Ba^{2+}(aq) + 2e^- \rightarrow Ba(s)$ $E^\circ = -2.90$ V

$Pb^{2+}(aq) + 2e^- \rightarrow Pb(s)$ $E^\circ = -0.126$ V

$2H^+(aq) + 2e^- \rightarrow H_2(g)$ $E^\circ = 0.00$ V

$Ag^+(aq) + e^- \rightarrow Ag(s)$ $E^\circ = 0.80$ V

However, the overvoltage for producting hydrogen gas makes the potential for the H^+/H_2 half-reaction approximately 0.4 V less favorable. Hence,

(a) $Ag^+(aq)$ will be reduced first.
(b) Pb^{2+} (aq) will be reduced second.
(c) $Ba^{2+}(aq)$ cannot be reduced by electrolysis of an aqueous solution, since H_2O is more easily reduced than

$Ba^{2+}(aq)$. $\left[2H_2O(l) + 2e^- \rightarrow H_2(g) + 2OH^-(aq) \quad \text{has } E^\circ = -0.83 \text{ V}\right]$

17.52 ***See Section 17.11.***

(a) HF is a molecular substance whereas KF is an ionic substance. Molten KF provides the ions necessary to conduct electricity through the electrolyte and thus completes the electrical circuit.

(b) Potassium would be produced by reduction of K^+ ions at the cathode, and fluorine would be produced by oxidation of F^- ions at the anode.

17.54 ***See Section 17.10.***

(a) Three moles of electrons are required to produce one mole of Al via $Al^{3+}(aq) + 3e^- \rightarrow Al(s)$. Hence,

$$? \text{ C} = 0.50 \text{ mol Al} \times \frac{3 \text{ mol e}^-}{1 \text{ mol Al}} \times \frac{9.65 \times 10^4 \text{ C}}{1 \text{ mol e}^-} = \mathbf{1.4 \times 10^5 \text{ C}}$$

(b) Two moles of electrons are required to produce one mole of Cu via $Cu^{2+}(aq) + 2e^- \rightarrow Cu(s)$. Hence,

$$? \text{ mol Cu}^{2+} \text{ present} = 0.100 \text{ L Cu(NO}_3)_2 \text{ soln} \times \frac{0.20 \text{ mol Cu(NO}_3)_2}{1 \text{ L Cu(NO}_3)_2 \text{ soln}} \times \frac{1 \text{ mol Cu}^{2+}}{1 \text{ mol Cu(NO}_3)_2}$$

$$= 0.020 \text{ mol Cu}^{2+}$$

$$? \text{ C} = 0.020 \text{ mol Cu}^{2+} \times \frac{2 \text{ mol e}^-}{1 \text{ mol Cu}} \times \frac{9.65 \times 10^4 \text{ C}}{1 \text{ mol e}^-} = \mathbf{3.9 \times 10^3 \text{ C}}$$

(c) Two moles of electrons are involved in producting one mole of Cl_2 via $2Cl^-(aq) \rightarrow Cl_2(g) + 2e^-$. Hence,

$$? \text{ C} = 10.0 \text{ g Cl}_2 \times \frac{1 \text{ mol Cl}_2}{70.9 \text{ g Cl}_2} \times \frac{2 \text{ mol e}^-}{1 \text{ mol Cl}_2} \times \frac{9.65 \times 10^4 \text{ C}}{1 \text{ mol e}^-} = \mathbf{2.72 \times 10^4 \text{ C}}$$

(d) One mole of electrons is required to produce one mole of Ag via $Ag^+(aq) + e^- \rightarrow Ag(s)$. Hence,

$$? \text{ C} = 0.32 \text{ g Ag} \times \frac{1 \text{ mol Ag}}{107.9 \text{ g Ag}} \times \frac{1 \text{ mol e}^-}{1 \text{ mol Ag}} \times \frac{9.65 \times 10^4 \text{ C}}{1 \text{ mol e}^-} = \mathbf{2.9 \times 10^2 \text{ C}}$$

17.56 ***See Section 17.10.***

The equation for producing H_2 via electrolysis of an aqueous hydrochloric acid solution is: $2H^+(aq) + 2e^- \rightarrow H_2(g)$. Hence, we can calculate mass of hydrogen produced by calculating moles of electrons involved.

Strategy: time and current $\rightarrow C \rightarrow mole\ e^- \rightarrow mol\ H_2 \rightarrow g\ H_2$

$$? \text{ g H}_2 = 59.0 \text{ min} \times \frac{60 \text{ s}}{1 \text{ min}} \times \frac{0.500 \text{ C}}{\text{s}} \times \frac{1 \text{ mol e}^-}{9.65 \times 10^4 \text{ C}} \times \frac{1 \text{ mol H}_2}{2 \text{ mol e}^-} \times \frac{2.02 \text{ g H}_2}{1 \text{ mol H}_2} = \mathbf{0.0185 \text{ g H}_2}$$

17.58 ***See Section 17.10.***

The key to solving this problem involves recognizing that the same number of coulombs is passing through both cells. In the one cell, Ag is being produced via $Ag^+(aq) + e^- \rightarrow Ag(s)$, and in the other, Sn^{2+} is being produced via $Sn^{4+}(aq) + 2e^- \rightarrow Sn^{2+}(aq)$. Hence, calculating the number of coulombs that are being used to produce 0.158 g Ag will tell us the number of coulombs that are being used to reduce $Sn^{4+}(aq)$ to $Sn^{2+}(aq)$.

Strategy: $g\ Ag \rightarrow mol\ Ag \rightarrow mol\ e^- \rightarrow C$

$$? \text{ C to produce } 0.158 \text{ g Ag} = 0.158 \text{ g Ag} \times \frac{1 \text{ mol Ag}}{107.9 \text{ g Ag}} \times \frac{1 \text{ mol e}^-}{1 \text{ mol Ag}} \times \frac{9.65 \times 10^4 \text{ C}}{1 \text{ mol e}^-} = 1.41 \times 10^2 \text{ C}$$

Strategy: $C \rightarrow mol\ e^- \rightarrow mol\ Sn^{4+}(aq)$ *reduced*

$$? \text{ mol Sn}^{4+}(aq) \text{ reduced} = 1.41 \times 10^2 \text{ C} \times \frac{1 \text{ mol e}^-}{9.65 \times 10^4 \text{ C}} \times \frac{1 \text{ mol Sn}^{4+}(aq) \text{ reduced}}{2 \text{ mol e}^-}$$

$$= \mathbf{7.32 \times 10^{-4} \text{ mol Sn}^{4+}(aq)}$$

17.60 *See Section 17.10.*

The equation for producing Au via electrolysis of $Au(CN)_4^-(aq)$ is: $Au(CN)_4^-(aq) + 3e^- \rightarrow Au(s) + 4CN^-(aq)$. The key to solving this problem involves determining the number of moles of Au to be plated.

Strategy: area and thickness → volume Au → g Au → mol Au

$$? \text{ mol Au plated} = 100 \text{ cm}^2 \times \left(0.0020 \text{ mm} \times \frac{1 \text{ cm}}{10 \text{ mm}}\right) \times \frac{19.3 \text{ g Au}}{\text{cm}^3} \times \frac{1 \text{ mol Au}}{197.0 \text{ g Au}} = 2.0 \times 10^{-3} \text{ mol Au}$$

Strategy: mol Au → mol e⁻ → C → time

$$? \text{ min for plating} = 2.0 \times 10^{-3} \text{ mol Au} \times \frac{3 \text{ mol e}^-}{1 \text{ mol Au}} \times \frac{9.65 \times 10^4 \text{ C}}{1 \text{ mol e}^-} \times \frac{1 \text{ s}}{0.500 \text{ C}} \times \frac{1 \text{ min}}{60 \text{ s}} = \mathbf{19 \text{ min}}$$

17.62 *See Section 17.10 and Examples 17.16 and 17.*

(a) The equation for producing pure Cu^{2+} by the electrorefining process is $Cu^{2+}(aq) + 2e^- \rightarrow Cu(s)$.

Strategy: g Cu/hr → mol Cu/hr → mole e⁻/hr → C/hr → C/s

$$? \text{ current in C/s} = \frac{500 \text{ g Cu}}{1 \text{ hr}} \times \frac{1 \text{ mol Cu}}{63.5 \text{ g Cu}} \times \frac{2 \text{ mol e}^-}{1 \text{ mol Cu}} \times \frac{9.65 \times 10^4 \text{ C}}{1 \text{ mol e}^-} \times \frac{1 \text{ hr}}{60 \text{ min}} \times \frac{1 \text{ min}}{60 \text{ s}}$$
$$= 4.22 \times 10^2 \text{ C/s} = \mathbf{4.22 \times 10^2 \text{ A}}$$

(b) The total charge required to produce 500 g of refined Cu is:

$$Q = 500 \text{ g Cu} \times \frac{1 \text{ mol Cu}}{63.5 \text{ g Cu}} \times \frac{2 \text{ mol e}^-}{1 \text{ mol Cu}} \times \frac{9.65 \times 10^4 \text{ C}}{1 \text{ mol e}^-} = 1.52 \times 10^6 \text{ C}$$

The electrical energy consumed is :

$$E = Q \times V \times \frac{1 \text{ kw} \cdot \text{hr}}{3.60 \times 10^6 \text{ J}} \qquad E = 1.52 \times 10^6 \text{ C} \times \frac{0.100 \text{ J}}{1 \text{ C}} \times \frac{1 \text{ kw} \cdot \text{hr}}{3.60 \times 10^6 \text{ J}} = \mathbf{4.22 \times 10^{-2} \text{ kw} \cdot \text{hr}}$$

17.68 *See Sections 17.5 and 10 and Example 17.9.*

(a) For $Zn(s) + HgO(s) \rightarrow Hg(l) + ZnO(s)$, $E^\circ = 1.35$ V.

$$\Delta G^\circ = -nFE^\circ \qquad \Delta G^\circ = -(2 \text{ mol e}^-)\left(\frac{9.65 \times 10^4 \text{ C}}{1 \text{ mol e}^-}\right)\left(\frac{1.35 \text{ J}}{1 \text{ C}}\right) = \mathbf{-2.61 \times 10^5 \text{ J}}$$

Note: The 2 mole e⁻ can be deduced from the changes in the oxidation states. Zn goes from 0 to +2 and Hg from +2 to 0.

$$\text{(b) ? max work for } 1.00 \text{ g HgO} = 1.00 \text{ g HgO} \times \frac{1 \text{ mol HgO}}{216.6 \text{ g HgO}} \times \frac{2.61 \times 10^5 \text{ J}}{1 \text{ mol HgO}} = \mathbf{1.20 \times 10^3 \text{ J}}$$

(c) *Strategy: g HgO → mol HgO → mol e⁻ → C → s → hr*

$$? \text{ time in hr} = 3.50 \text{ g HgO} \times \frac{1 \text{ mol HgO}}{216.6 \text{ g HgO}} \times \frac{2 \text{ mol e}^-}{1 \text{ mol HgO}} \times \frac{9.65 \times 10^4 \text{ C}}{1 \text{ mol e}^-} \times \frac{1 \text{ s}}{10^{-2} \text{ C}} \times \frac{1 \text{ min}}{60 \text{ s}} \times \frac{1 \text{ hr}}{60 \text{ min}} = \mathbf{86.6 \text{ hr}}$$

Note: $10 \text{mA} \times \frac{1 \text{ A}}{10^3 \text{ mA}} = 10^{-2} \text{ A} = 10^{-2} \text{ C/s}$

17.70 ***See Sections 17.5, 6 and 7.***

(a)

$$PbC_2O_4(s) \rightleftarrows Pb^{2+}(aq) + C_2O_4^{2-}(aq) \qquad \Delta G_1^\circ$$

$$Pb^{2+}(aq) + 2e^- \rightarrow Pb(s) \qquad \Delta G_2^\circ$$

$$PbC_2O_4(s) + 2e^- \rightarrow Pb(s) + C_2O_4^{2-}(aq) \qquad \Delta G_3^\circ$$

$\Delta G_1^\circ = -RT\ln K_{sp}$ $\qquad \Delta G_1^\circ = -8.314 \text{ J}\cdot\text{mol}^{-1}\cdot\text{K}^{-1} \times 298 \text{ K} \times \ln(8.5 \times 10^{-10}) = 5.17 \times 10^4 \text{ J}$

$\Delta G_2^\circ = -nFE^\circ$ $\qquad \Delta G_2^\circ = -(2 \text{ mol e}^-)\left(\frac{9.65 \times 10^4 \text{ C}}{1 \text{ mol e}^-}\right)\left(\frac{-0.126 \text{ J}}{1 \text{ C}}\right) = 2.43 \times 10^4 \text{ J}$

$\Delta G_3^\circ = \Delta G_1^\circ + \Delta G_2^\circ$ $\qquad \Delta G_3^\circ = 5.17 \times 10^4 \text{ J} + 2.43 \times 10^4 \text{ J} = 7.60 \times 10^4 \text{ J}$

Solving $\Delta G^\circ = -nFE^\circ$ for E° gives $E^\circ = \frac{-\Delta G^\circ}{nF}$ $\qquad E^\circ = \frac{-7.60 \times 10^4 \text{ J}}{(2 \text{ mol e}^-)\left(\frac{9.65 \times 10^4 \text{ C}}{1 \text{ mol e}^-}\right)} = -0.394 \text{ J}\cdot\text{C}^{-1} = \mathbf{-0.394 \text{ V}}$

(b) For $PbC_2O_4(s) + 2e^- \rightarrow Pb(s) + C_2O_4^{2-}(aq)$,

$E = E^\circ - \frac{0.0591 \text{ V}\cdot\text{mol e}^-}{n}\log[C_2O_4^{2-}]$ $\qquad E = -0.394 \text{ V} - \frac{0.0591 \text{ V}\cdot\text{mol e}^-}{2 \text{ mol e}^-}\log(.025) = \mathbf{-0.347 \text{ V}}$

17.72 ***See Section 14.1.***

$[H_3O^+] = C_{HA}$ for a monoprotic strong acid. Hence, $[H_3O^+] = 0.032\,M$, and pH = -log(0.032) = 1.49.
Using this pH to solve for k gives

$k = E_{cell} + 0.0591\text{pH}$ $\qquad k = 0.135 \text{ V} + 0.0591(1.49) = 0.223 \text{ V}$

and $E_{cell} = 0.223 \text{ V} - 0.0591\text{pH}$ for this particular pH meter.

The pH at the equivalence point for titrating a strong acid with a strong base is 7.00. Hence, at the equivalence point,

$E_{cell} = 0.223 \text{ V} - 0.0591\text{pH}$ $\qquad E_{cell} = 0.223 \text{ V} - 0.0591(7.00) = \mathbf{-0.191 \text{ V}}$

17.74 ***See Section 17.4, Example 17.8 and Appendix H.***

(a)

$$2\left[Fe^{3+}(aq) + e^- \rightarrow Fe^{2+}(aq)\right] \qquad E^\circ = 0.771\ V$$

$$H_2SO_3(aq) + H_2O(l) \rightarrow HSO_4^-(aq) + 3H^+(aq) + 2e^- \qquad E^\circ = -0.17\ V$$

$$2Fe^{3+}(aq) + H_2SO_3(aq) + H_2O(l) \rightarrow 2Fe^{2+}(aq) + HSO_4^-(aq) + 3H^+(aq) \qquad E^\circ = \mathbf{0.60\ V}$$

(b)

$$6\left[Fe^{2+}(aq) \rightarrow Fe^{3+}(aq) + e^-\right] \qquad E^\circ = -0.771\ V$$

$$Cr_2O_7^{2-}(aq) + 14H^+(aq) + 6e^- \rightarrow 2Cr^{3+}(aq) + 7H_2O(l) \qquad E^\circ = 1.33\ V$$

$$6Fe^{2+}(aq) + Cr_2O_7^{2-}(aq) + 14H^+(aq) \rightarrow 6Fe^{3+}(aq) + 2Cr^{3+}(aq) + 7H_2O(l) \qquad E^\circ = \mathbf{0.56\ V}$$

(c)

$$Fe^{2+}(aq) \rightarrow Fe^{3+}(aq) + e^- \qquad E^\circ = -0.771\ V$$

$$HNO_2(aq) + H^+(aq) + e^- \rightarrow NO(g) + H_2O(l) \qquad E^\circ = 0.983\ V$$

$$Fe^{2+}(aq) + HNO_2(aq) + H^+(aq) \rightarrow Fe^{3+}(aq) + NO(g) + H_2O(l) \qquad E^\circ = \mathbf{.212\ V}$$

Chapter 18: Metallurgy

18.6 ***See Section 7.1, Table 7.1 and Section 18.2.***

(a) Sc has $[Ar]4s^2 3d^1$ as its electron configuration and form as Sc^{3+} with [Ar] as its electron configuration in its only common oxidation state.

(b) The elements in Group IIB(12), Zn, Cd, and Hg, are found in the "d-block" of the periodic table but do not have partially filled d orbitals in either the metal atoms or in one of their oxidation states. Hence, these elements do not meet the definition of a transition element.

(c) Ac has $[Rn]7s^2 6d^1$ as its electron configuration and therefore meets the definition of a transition element. (See discussion of La and Lu in text.)

18.10 ***See Section 18.2 and Table 18.3.***

(a) Cr has a higher melting point than Cu. The melting points of the transition elements generally reach a maximum in Group VB, VIB or VIIB.

(b) Os has a higher melting point than Fe. The melting points of the transition elements generally increase going down a group in the periodic table.

(c) Cr has a higher melting point than V. The melting points of the trasition elements generally reach a maximum in Group VB, VIB or VIIB.

(d) W has a higher melting point than La. The melting points of the transition elements generally reach a maximum in Groups VB, VIB or VIIB.

18.12 ***See Section 18.2.***

The atomic radii of the transition elements within a period decrease more rapidly from Group IIIB through Group VIB than through the rest of the transition elements in that period because increases in effective nuclear charge are outweighed by repulsions among d electrons toward the end of the d-block. This is partially due to pairing in the d orbitals.

18.14 ***See Section 18.2 and Table 18.4.***

The order of decreasing atomic radii is: **Nb > W > V > Co**. The atomic radii of the transition elements generally decrease going across a period and increase going down groups. However, the atomic radii of transition elements in the sixth period are almost identical to the radii of transtion elements in the fifth period as a result of the lanthanide contraction. This accounts for the relative sizes of Nb and W. The radius of W would be approximately equal to the radius of Nb if it were directly below Nb but is less because it is to the right of that position.

18.16 ***See Section 18.2.***

The maximum positive oxidation state is equal to the group number for elements in Groups IIIB through VIIB. This gives: (a) **Ti**, IVB, **+4** (b) **W**, VIB, **+6** (c) **Ta**, VB, **+5** (d) **Re**, VIIB, **+7**

18.18 *See Section 7.1 and 18.2 and Table 7.1.*

The atoms of elements in Group IB have just one electron in the outermost s subshell whereas the atoms of most other transition elements have two electrons in their outermost s subshell.

18.20 *See Section 18.2 and Figure 18.6.*

(a) Mn has a higher first ionization energy than Ti. Ionization energies generally increase with increasing effective nuclear charge and decreasing size proceeding from left to right across a period in the periodic table.

(b) Ta has a higher first ionization energy than V. The increase in effective nuclear charge that occurs during filling the poor shielding 4f orbitals outweighs the increase in size from V to Ta causing the ionization energy of Ta to be higher than that of V.

(c) Rh has a higher first ionization energy than Ru; see explanation given in answer for part a.

(d) Os has a higher first ionization energy than Mo. Transition metals of the sixth period generally have higher first ionization energies than transition metals in the fifth period. This is due to the increase in effective nuclear charge that occurs during filling the poor shielding 4f orbitals. In addition, Os is further to the right in the periodic table. This leads to an additional increase in effective nuclear charge and a decrease in size, causing Os to be smaller than Mo. This also contributes to the higher ionization energy of Os.

18.22 *See Section 18.3 and Appendix H.*

(a) $FeCr_2O_4 + 4C \rightarrow Fe + 2Cr + 4CO$

(b) $Cr_2O_3 + 2Al \rightarrow 2Cr + Al_2O_3$

(c) $4Cr^{2+}(aq) + O_2(g) + 4H^+(aq) \rightarrow 4Cr^{3+}(aq) + 2H_2O(l)$

(d) $2Cr(OH)_4^- + 3ClO^-(aq) + 2OH^-(aq) \rightarrow 2CrO_4^{2-}(aq) + 3Cl^-(aq) + 5H_2O(l)$

(e) $6Fe^{2+}(aq) + Cr_2O_7^{2-}(aq) + 14H^+(aq) \rightarrow 6Fe^{3+}(aq) + 2Cr^{3+}(aq) + 7H_2O(l)$

18.24 *See Sections 7.1 and 18.2.*

Fe: $[Ar]4s^23d^6$ Fe^{2+}: $[Ar]3d^6$ Fe^{3+}: $[Ar]3d^5$

18.26 *See Section 18.3.*

$Cu(s) + 2H_2SO_4(aq) \rightarrow CuSO_4(aq) + SO_2(g) + 2H_2O(l)$

The oxidizing agent is H_2SO_4; the S is reduced from the +6 oxidation state to +4 oxidation state in SO_2.

18.28 *See Section 18.4, Table 18.7 and Example 18.1.*

(a) $[Co(en)_3][Co(CN)_6]$ (b) $[Pt(NH_3)_4](NO_3)_2$

(c) $Na_2[RhCl_5(H_2O)]$, since a complex composed of Rh(III), five Cl^- ions and one water molecule has a 2- charge.

18.30 See Section 18.4, Tables 18.7 and 8 and Example 18.2.

(a) $[Fe(CO)_5]$, pentacarbonyliron (0)

(b) $K_2[Cr(CN)_5NO]$, potassium pentacyanonitrosylchromate(III)

(c) $[Ru(NH_3)_5Cl]Cl_2$, pentaamminechlororuthenium(III) chloride

(d) $[Co(dien)Br_3]$, tribromo(diethylenetriamine)cobalt(III)

(e) $[Cr(NH_3)_6][Cr(CN)_6]$, hexaamminechromium(III) hexacyanochromate(III)

18.32 See Section 18.4, Tables 18.7 and 8 and Example 18.3.

(a) hexaaquachromium(III) hexacyanoferrate(III), $[Cr(H_2O)_6][Fe(CN)_6]$

(b) bromochlorobis(ethylenediamine)cobalt(III), $[Co(en)_2BrCl]^+$

(c) carbonylpentacyanocolbaltate(III), $[CoCO(CN)_5]^{2-}$

(d) trinitro(diethylenetriamine)chromium(III), $[Cr(dien)(NO_2)_3]$

(e) pentaaquathiocyanotoiron(III), $[Fe(H_2O)_5SCN]^{2+}$

18.34 See Section 18.5 and Figure 18.17.

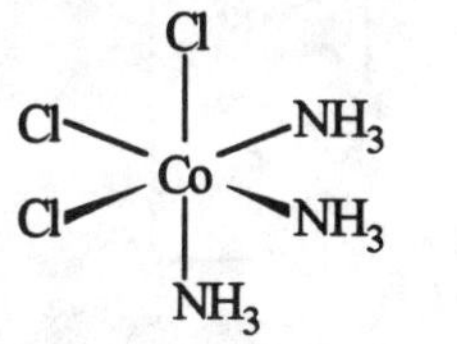

Cl
Cl — Co — NH_3
Cl — Co — NH_3
NH_3

fac-triamminetrichlorocobalt(III)

Cl
Cl — Co — NH_3
H_3N — Co — Cl
NH_3

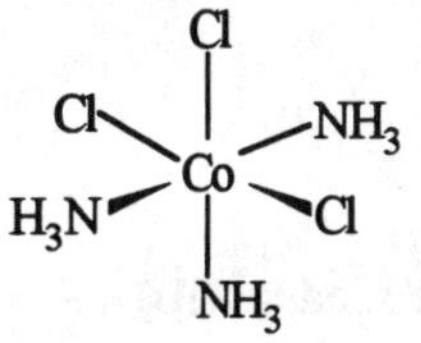

mer-triamminetrichlorocobalt(III)

18.36 See Section 18.5.

(a) Linkage isomerism of NCS^-; see Figure 18.15.

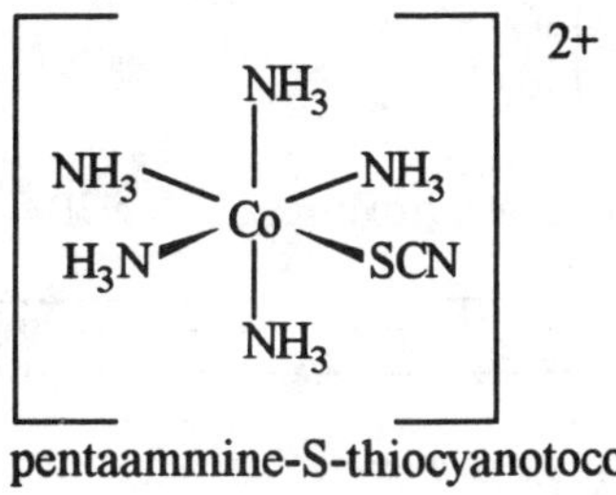

NH_3
NH_3 — Co — NH_3
H_3N — Co — SCN
NH_3
2+

pentaammine-S-thiocyanotocobalt(III)

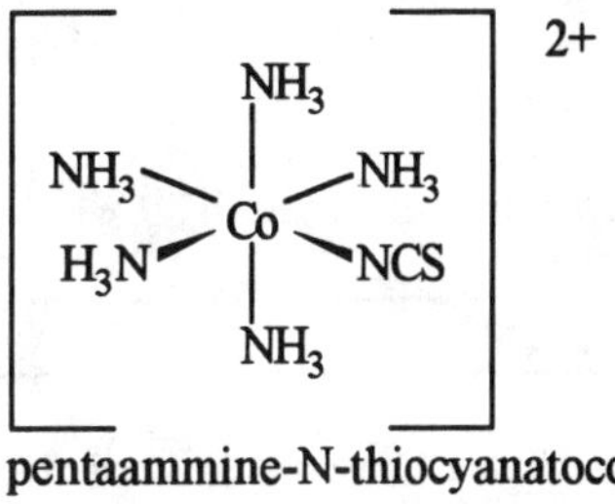

NH_3
NH_3 — Co — NH_3
H_3N — Co — NCS
NH_3
2+

pentaammine-N-thiocyanatocobalt(III)

(b) Geometric isomerism; see Figure 18.16.

Br, Br, Pd, NH_3, NH_3

cis-diamminedibromopalladium(II)

Br, H_3N, Pd, NH_3, Br

trans-diamminedibromopalladium(II)

(c) The oxalate anion, $C_2O_4^{2-}$, is a symmetrical bidentate ligand (Table 18.6) which can be abbreviated as O—O. Optical isomersism; see Figure 18.21.

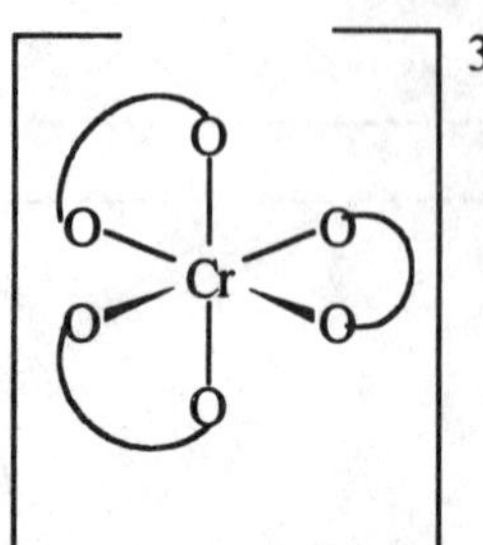

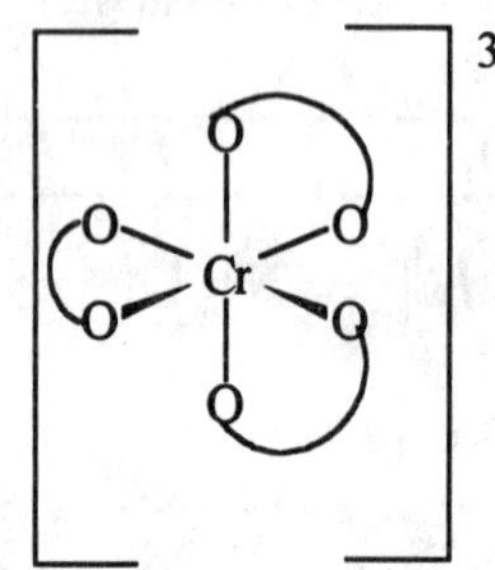

tris(oxalato)chromate(III)

(d) Ethylenediammine (en) is a symmetrical bidenate ligand (Table 18.6) which can be abbreviated and N-N. Geometric and optical isomerism; see Example 18.5.

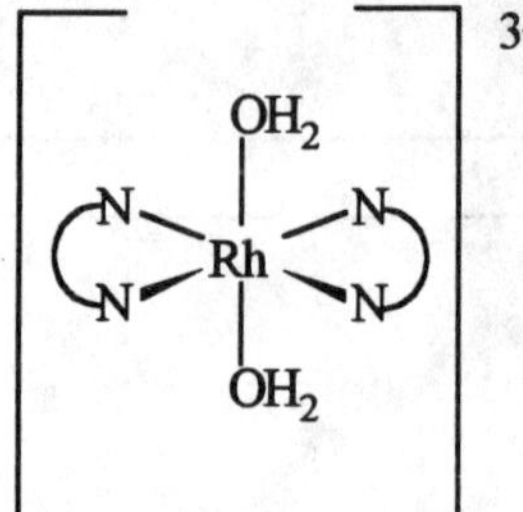

trans-diaquabis(ethylenediamine)rhodium(III)

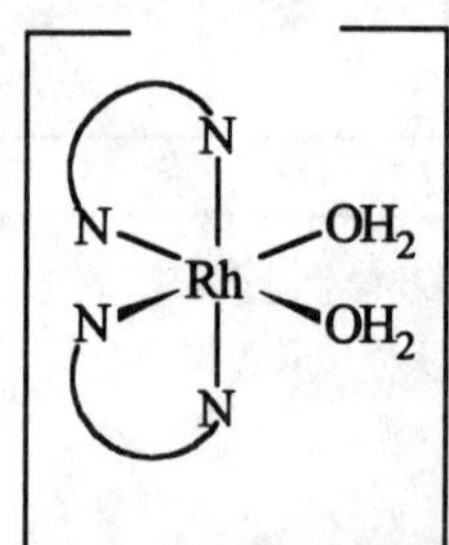

3+ H_2O, H_2O, Rh, N, N, N, N

cis-diaquabis(ethylenediamine)rhodium(III)

(e) Geometric isomerism; see Figure 18.17.

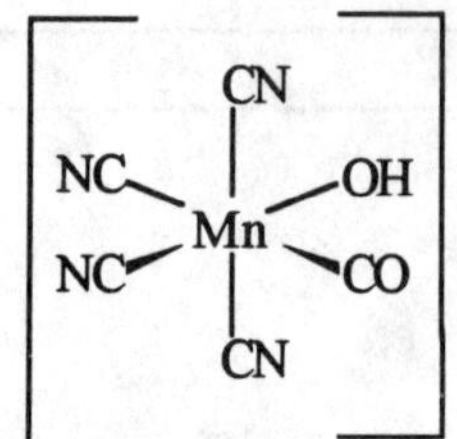

cis-carbonyltetracyanohydroxomanganate(II)

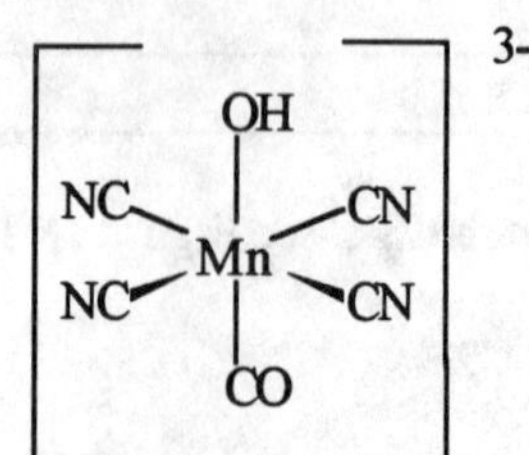

trans-carbonyltetracyanohydroxomanganate(II)

18.38 ***See Section 18.5.***

Linkage isomerism is possible when lone pairs of electrons on two or more atoms of different elements in a ligand can be used to form monodentate linkages to metals. Hence, linkage isomerism is possible with :

[:S—C≡N:]⁻ , [:O—N=O]⁻ and [:N≡C—O:]⁻

but not with

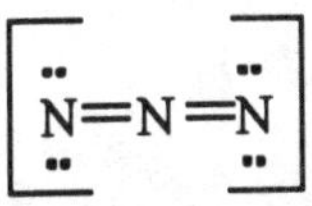

$$\left[\ddot{N}=N=\ddot{N}\right]^{-} \quad \text{or} \quad H_2\ddot{N}-CH_2-CH_2-\ddot{N}H_2$$

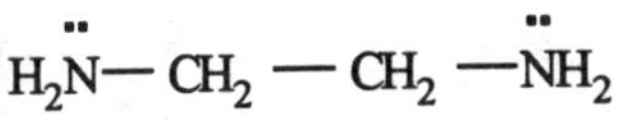

18.42 ***See Section 18.6 and Example 18.6.***

For one photon, $\Delta = \dfrac{hc}{\lambda}$

$$\Delta = \frac{(6.63 \times 10^{-34}\ \text{J}\cdot\text{s})(3.00 \times 10^{8}\ \text{m}\cdot\text{s}^{-1})}{450\ \text{nm} \times \dfrac{1\ \text{m}}{10^{9}\ \text{nm}}} = 4.42 \times 10^{-19}\ \text{J}$$

For one mole of photons,

$$\Delta = \frac{4.42 \times 10^{-19}\ \text{J}}{\text{photon}} \times \frac{6.02 \times 10^{23}\ \text{photons}}{\text{mol}} \times \frac{1\ \text{kJ}}{10^{3}\ \text{J}} = \mathbf{266\frac{kJ}{mol}}$$

18.44 ***See Section 18.6 and Example 18.6.***

The complex having the higher Δ absorbs light at a shorter wavelength. Hence,

(a) $[Rh(CN)_6]^{3-}$ will absorb at a shorter wavelength than $[Rh(NH_3)_6]^{3+}$; see spectrochemical series and Table 18.10.

(b) $[Fe(H_2O)_6]^{3+}$ will absorb at a shorter wavelength than $[Fe(H_2O)_6]^{2+}$; see text discussion of influence of metal charge on Δ.

(c) $[Rh(H_2O)_6]^{3+}$ will absorb at a shorter wavelength than $[Co(H_2O)_6]^{3+}$; see text discusion of influence of period of metal on Δ and Table 18.10.

(d) $[Ti(H_2O)_6]^{3+}$ will absorb at a shorter wavelength than $[TiF_6]^{3-}$; see spectrochemical series.

18.46 ***See Section 18.6 and Figure 18.25.***

(a)

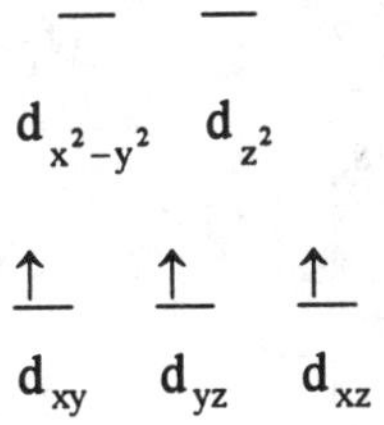

__ __

$d_{x^2-y^2}$ d_{z^2} $[Cr(H_2O)_6]^{3+}$

↑ ↑ ↑

d_{xy} d_{yz} d_{xz}

(b)

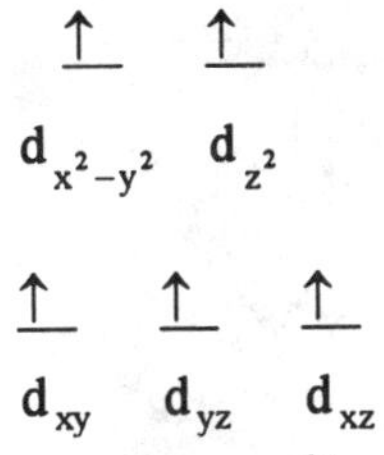

↑ ↑ __ __

$d_{x^2-y^2}$ d_{z^2} $d_{x^2-y^2}$ d_{z^2}

or

↑ ↑ ↑ ↑↓ ↑↓ ↑

d_{xy} d_{yz} d_{xz} d_{xy} d_{yz} d_{xz}

$[Fe(H_2O)_6]^{3+}$ $[Fe(CN)_6]^{3+}$

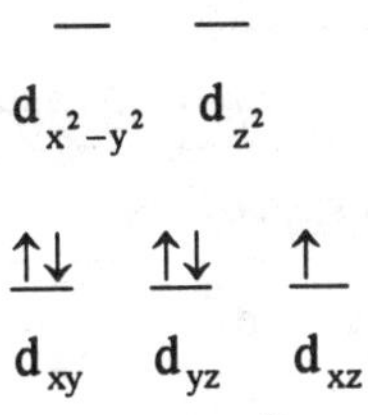

(c)

↑	↑			↑	—	
$d_{x^2-y^2}$	d_{z^2}		or	$d_{x^2-y^2}$	d_{z^2}	
↑↓	↑↓	↑		↑↓	↑↓	↑↓
d_{xy}	d_{yz}	d_{xz}		d_{xy}	d_{yz}	d_{xz}
$[CoF_6]^{4-}$				$[Co(CN)_6]^{4-}$		

(d)

↑↓	↑		
$d_{x^2-y^2}$	d_{z^2}		$[Cu(H_2O)_6]^{2+}$
↑↓	↑↓	↑↓	
d_{xy}	d_{yz}	d_{xz}	

18.48 ***See Section 18.6, Figure 18.30 and Example 18.7.***

Complex	Composition	Classification	Number of unpaired electrons
(a) $[MnF_6]^{3-}$	Mn^{3+} and F^-	high spin d^4	four unpaired electrons
(b) $[Fe(CN)_6]^{3-}$	Fe^{3+} and CN^-	low spin d^5	one unpaired electron
(c) $[Re(H_2O)_6]^{2+}$	Re^{2+} and H_2O	low spin d^5	one unpaired electron
(d) $[Fe(H_2O)_6]^{2+}$	Fe^{2+} and H_2O	high spin d^6	four unpaired electrons
(e) $[Ni(H_2O)_6]^{2+}$	Ni^{2+} and H_2O	d^8	two unpaired electrons

18.50 ***See Section 18.6, Figure 18.30, and Example 18.7.***

(a)

Complex	Composition	Classification	Number of unpaired electrons
$[Cr(H_2O)_6]^{2+}$	Cr^{2+} and H_2O	high spin d^4	four unpaired electrons
$[Mn(CN)_6]^{3-}$	Mn^{3+} and CN^-	low spin d^4	two unpaired electrons

(b)

Complex	Composition	Classification	Number of unpaired electrons
$[Fe(H_2O)_6]^{2+}$	Fe^{2+} and H_2O	high spin d^6	four unpaired electrons
$[Ru(H_2O)_6]^{2+}$	Ru^{2+} and H_2O	low spin d^6	zero unpaired electrons

	Complex	Composition	Classification	Number of unpaired electrons
(c)	$[Co(H_2O)_6]^{2+}$	Co^{2+} and H_2O	high spin d^7	three unpaired electrons
	$[Co(CN)_5(H_2O)]^{3-}$	Co^{2+}, CN^- and H_2O	low spin d^7	one unpaired electron

18.52 ***See Section 18.6 and Figure 18.31.***

↑ ↑ —

d_{xy} d_{yz} d_{xz}

↑↓ ↑↓

$d_{x^2-y^2}$ d_{z^2}

Co^{3+} is d^6. There are two unpaired electrons in a low spin tetrahedral complex of a d^6 metal ion.

18.54 ***See Section 18.6, Figures 18.31 and 32 and Example 18.8.***

Complex	Composition	Classification	Number of unpaired electrons
(a) $[Ni(CN)_4]^{2-}$	Ni^{2+} and CN^-	low spin d^8 square planar	zero unpaired electrons
(b) $[FeCl_4]^{2-}$	Fe^{2+} and Cl^-	high spin d^6 tetrahedral	four unpaired electrons
(c) $[Pd(NH_3)_4]^{2+}$	Pd^{2+} and NH_3	low spin d^8 square planar	zero unpaired electrons

Chapter 19: Hydrogen, Groups IIIA-VIA

19.4 ***See Section 19.1.***

	Element	Classification
(a)	carbon	nonmetal
(b)	tin	metal
(c)	chlorine	nonmetal
(d)	silicon	metalloid

19.14 ***See Section 19.1.***

$NaH(s) + H_2O(l) \rightarrow NaOH(aq) + H_2(g)$

The Ideal Gas Equation, PV = nRT, can be used to calculate the number of moles of H_2.

Known Quantities: $P = 1.00$ atm $V = 1.00$ L $T = 25 + 273 = 298$ K

Solving PV = nRT for n gives $n = \frac{PV}{RT}$ $\quad n = \frac{(1.00 \text{ atm})(1.00 \text{ L})}{\left(0.0821 \frac{\text{L} \cdot \text{atm}}{\text{mol} \cdot \text{K}}\right)(298 \text{ K})} = 0.0409 \text{ mol } H_2$

Strategy: mol H_2 → mol NaH → g NaH

$$? \text{ g NaH} = 0.0409 \text{ mol } H_2 \times \frac{1 \text{ mol NaH}}{1 \text{ mol } H_2} \times \frac{24.0 \text{ g NaH}}{1 \text{ mol NaH}} = \mathbf{0.981 \text{ g NaH}}$$

19.28 ***See Section 19.3.***

$BCl_3 + :NH_3 \rightarrow Cl_3B \leftarrow NH_3$

The BCl_3 acts as a Lewis acid and therefore provides and empty orbital to accept the unshared pair of electrons on the nitrogen atom in NH_3. The geometry about the B atom in BCl_3 is trigonal planar and that about the B atom in Cl_3BNH_3 is tetrahedral. The hybridization of the B atom changes from sp^2 to sp^3 and thereby provides an empty orbital that can overlap with the sp^3 orbital of N containing the unshared pair of electrons in NH_3.

19.40 ***See Sections 19.1 and 4.***

C uses sp hybridization in CO_2 and forms pπ-pπ bonds with each oxygen atom giving two carbon-oxygen double bonds. Si does not form simple molecules analogous to CO_2 because it is in the third period and is larger. Its sigma bond lengths are therefore longer, and it does not form strong pπ-pπ bonds. Instead it forms four sigma bonds to four different oxygen atoms as shown in Figure 19.1. In this arrangement, Si uses sp^3 hybridization.

19.46 ***See Section 8.3, 9.1 and 3 and 19.4.***

$SiCl_4$

Total valence electrons $= [1 \times 4(\text{Si}) + 4 \times 7(\text{Cl})] = 32$.

Twenty four electrons remain after assigning four single bonds, and twenty four unshared electrons are needed to give each atom a noble gas configuration (6 for each Cl). Twenty four electrons are used in writing the Lewis structure.

The Leiws structure predicts a tetrahedral electron-pair geometry and a tetrahedral molecular geometry. The hybridization of silicon in $SiCl_4$ is therefore predicted to be sp^3. Since $SiCl_4$ is totally symmetrical, it is nonpolar.

19.50 ***See Sections 8.3 and 19.5 and Figure 19.26.***

P_4

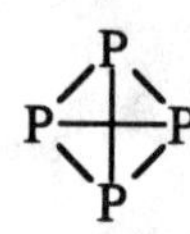

Total valence electrons $= [4 \times 5(\text{P})] = 20$.

Eight electrons remain after assigning six single bonds, and eight unshared electrons are needed to give each atom a noble gas configuration (2 for each P). Twenty electrons are used in writing the Lewis structure.

19.54 ***See Sections 8.3, 9.1 and 3 and 19.5 and Table 19.1.***

NO_2

Total valence electrons $= [1 \times 5(\text{N}) + 2 \times 6(\text{O})] = 17$.

Thirteen electrons remain after assigning two single bonds, and sixteen unshared electrons are needed to give each atom a noble gas configuration (6 for each O and 4 for N). Two electrons are used to form a nitrogen-oxygen double bond, and one unpaired electron is assigned to nitrogen to account for the difference of 3 electrons (16-13). Seventeen electrons are used in writing the Lewis structure.

The Lewis structure predicts a trigonal planar electron-pair geometry for NO_2 and sp^2 hybridization for N in NO_2.

N_2O_3

Total valence electrons $= [2 \times 5(\text{N}) + 3 \times 6(\text{O})] = 28$.

Twenty electrons remain after assigning four single bonds, and twenty four unshared electrons are needed to give each atom a noble gas configuration (6 for each O, 4 for left N and 2 for right N). Hence, a double bond is formed between each nitrogen atom and an oxygen atom. Twenty eight electrons are used in writing the Lewis structure.

The Lewis structure predicts a trigonal planar electron-pair geometry about each nitrogen atom in N_2O_3 and sp^2 hybridization for each N atom in N_2O_3.

19.56 *See Section 19.6.*

(a) $3\,Mg(s) + N_2(g) \rightarrow Mg_3N_2(s)$

(b) $P_4(s) + 5O_2(g) \rightarrow P_4O_{10}(s)$

(c) $2NO_2(g) + H_2O(l) \rightarrow HNO_2(aq) + HNO_3(aq)$

19.60 *See Section 19.5.*

$$4NH_3(g) + 5O_2(g) \xrightarrow{Pt} 4NO(g) + 6H_2O(g)$$

Strategy: $g\ NO \rightarrow mol\ NO \rightarrow mol\ NH_3 \rightarrow g\ NH_3$

$$?\ g\ NH_3 = 25 \times 10^3\ g\ NO \times \frac{1\ mol\ NO}{30.0\ g\ NO} \times \frac{4\ mol\ NH_3}{4\ mol\ NO} \times \frac{17.0\ g\ NH_3}{1\ mol\ NH_3} = \mathbf{1.4 \times 10^4\ g\ NH_3} \text{ or } \mathbf{14\ kg\ NH_3}$$

19.68 *See Section 19.6.*

$$2KClO_3(s) \xrightarrow[150^\circ]{MnO_2} 2KCl(s) + 3O_2(g)$$

Strategy: $PV = nRT \rightarrow mol\ O_2 \rightarrow mol\ KClO_3 \rightarrow g\ KClO_3$

Known Quantities: $P = 755\ torr \times \frac{1\ atm}{760\ torr} = 0.993\ atm$ $\quad V = 0.50\ L \quad T = 27 + 273 = 300\ K$

Solving PV = nRT for n gives $n = \frac{PV}{RT}$

$$n = \frac{(0.993\ atm)(0.50\ L)}{\left(0.0821 \frac{L \cdot atm}{mol \cdot K}\right)(300\ K)} = 0.020\ mol\ O_2$$

$$?\ g\ KClO_3 = 0.020\ mol\ O_2 \times \frac{2\ mol\ KClO_3}{3\ mol\ O_2} \times \frac{122.6\ g\ KClO_3}{1\ mol\ KClO_3} = \mathbf{1.6\ g\ KClO_3}$$

19.76 *See Sections 8.3, 9.1 and 19.7.*

XeF_2

F—Xe—F

:F—Xe—F: (with lone pairs: three on each F, three on Xe)

Total valence electrons $= [1 \times 8(Xe) + 2 \times 7(F)] = 22$. Eighteen electrons remain after assigning two single bonds, and sixteen unshared electrons are needed to give each atom a noble gas configuration (6 for each F and 4 for Xe). The remaining two electrons (18-16) are assigned to valence shell of Xe giving Xe an expanded valence shell. Twenty two electrons are used in writing the Lewis structure.

The Lewis structure predicts a trigonal bipyramidal electron-pair geometry and a symmetrical linear molecular geometry for XeF_2. Hence, XeF_2 is nonpolar.

XeO_3

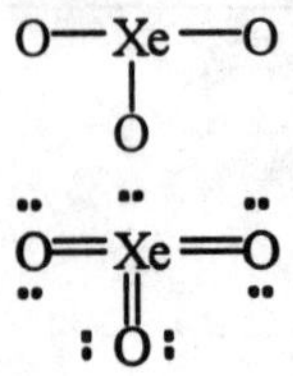

O—Xe—O
 |
 O

:Ö=Xe=Ö:
 ‖
 :O:

Total valence electrons $= [1 \times 8(\mathrm{Xe}) + 3 \times 6(\mathrm{O})] = 26$.
Twenty electrons remain after assigning three single bonds, and twenty unshared electrons are needed to give each atom a noble gas configuration (6 for each O and 2 for Xe). However, twenty six electrons also can be used to draw a Lewis structure which shows an expanded valence shell for Xe and has formal charges of zero for all atoms. Twenty six electrons are used in writing this Lewis structure.

The Lewis structure predicts a tetrahedral electron-pair geometry and an trigonal pyramidal molecular geometry for XeO_3. Hence, XeO_3 is polar.

Chapter 20: Nuclear Chemistry

20.2 ***See Introduction to Chapter 20.***

Symbol	Z	*A*	number of protons	number of neutrons
$^{23}_{11}Na$	**11**	**23**	**11**	**12**
$^{103}_{45}Rh$	45	103	**45**	**58**
$^{70}_{32}Ge$	**32**	70	32	38
$^{234}_{90}Th$	**90**	234	90	**144**

20.4 ***See Introduction to Chapter 20.***

Symbol	number of protons	number of neutrons	n/p ratio
(a) $^{12}_{6}C$	6	6	1.00
(b) $^{40}_{20}Ca$	20	20	1.00
(c) $^{90}_{40}Zr$	40	50	1.25
(d) $^{138}_{56}Ba$	56	82	1.46
(e) $^{208}_{82}Pb$	82	126	1.54

20.6 ***See Section 20.1 and Figure 20.1.***

	Symbol	number of protons	number of neutrons	n/p ratio	Comment
(a)	$^{17}_{8}O$	8	9	1.13	Within band of stability.
(b)	$^{93}_{38}Sr$	38	55	1.45	Above band of stability.
(c)	$^{67}_{30}Zn$	30	37	1.23	Within band of stability.
(d)	$^{233}_{92}U$	92	141	1.53	Beyond band of stability.
(e)	$^{28}_{12}Mg$	12	16	1.33	Above band of stability.

20.8 ***See Section 20.1.***

Beta emission has the net effect of converting a neutron in an original nucleus into a proton in the product nucleus via $^{1}_{0}n \rightarrow ^{1}_{1}p + ^{0}_{-1}\beta$, and thereby lowers the neutron-to-proton ratio in the nucleus. This produces a nuclide that is closer to the band of stability.

20.10 ***See Section 20.1.***

(a) Beta emission is characteristic of nuclei which are above the band of stability.

Hence, $^{32}_{14}Si$ is above the band of stability.

(b) Electron capture is characteristic of nuclei which are below the band of stability.

Hence, ${}^{44}_{22}Ti$ is below the band of stability.

(c) Positron emission is characteristic of nuclei which are below the band of stability.

Hence, ${}^{52}_{25}Mn$ is below the band of stability.

20.12 ***See Section 20.1.***

When a radioactive nuclide decomposes by alpha or beta particle emission, some of the nucleons are left in an excited nuclear state. Gamma rays carry away the energy that is released when the nucleons return to ground-state nuclear configuration.

20.14 ***See Section 20.1 and Example 20.1.***

(a) ${}^{201}_{83}Bi \rightarrow {}^{197}_{81}Tl + {}^{4}_{2}He$ (b) ${}^{184}_{77}Ir \rightarrow {}^{184}_{76}Os + {}^{0}_{1}\beta$ (c) ${}^{135}_{57}La + {}^{0}_{-1}e \rightarrow {}^{135}_{56}Ba$ (d) ${}^{80}_{35}Br \rightarrow {}^{80}_{36}Kr + {}^{0}_{-1}\beta$

20.16 ***See Section 20.1 and Example 20.1.***

(a) ${}^{227}_{90}Th \rightarrow {}^{223}_{88}Ra + {}^{4}_{2}He$ (b) ${}^{22}_{11}Na + {}^{0}_{-1}e \rightarrow {}^{22}_{10}Ne$ (c) ${}^{232}_{90}Th \rightarrow {}^{228}_{88}Ra + {}^{4}_{2}He$ (d) ${}^{72}_{31}Ga \rightarrow {}^{72}_{32}Ge + {}^{0}_{-1}\beta$

(e) ${}^{60}_{29}Cu \rightarrow {}^{60}_{28}Ni + {}^{0}_{1}\beta$

20.18 ***See Section 20.1 and Example 20.2.***

Solving $\ln\left(\frac{N}{N_0}\right) = \frac{-0.693t}{t_{1/2}}$ for $t_{1/2}$ gives $t_{1/2} = \frac{-0.693t}{\ln\left(\frac{N}{N_0}\right)}$. Since $\left(\frac{N}{N_0}\right)$ can be replaced by $\left(\frac{R}{R_0}\right)$,

$$t_{1/2} = \frac{-0.693t}{\ln\left(\frac{R}{R_0}\right)} \qquad t_{1/2} = \frac{-0.693(22.5\text{ hr})}{\ln\left(\frac{350}{1245}\right)} = \mathbf{12.3\ hr}$$

20.20 ***See Section 20.1 and Example 20.3.***

Solving rate = kN for k gives $k = \frac{\text{rate}}{N}$ $k = \frac{383\text{ atoms}\cdot\text{min}^{-1}}{3.75 \times 10^{13}\text{ atoms}} = 1.02 \times 10^{-11}\text{ min}^{-1}$

Substituting for k in $t_{1/2} = \frac{0.693}{k}$ gives $t_{1/2} = \frac{0.693}{1.02 \times 10^{-11}\text{ min}^{-1}} = 6.79 \times 10^{10}\text{ min}$

or $6.79 \times 10^{10}\text{ min} \times \frac{1\text{ hr}}{60\text{ min}} \times \frac{1\text{ day}}{24\text{ hours}} \times \frac{1\text{ yr}}{365\text{ days}} = \mathbf{1.29 \times 10^{5}\ yr}$

20.22 ***See Section 20.1 and Example 20.4.***

For each alpha particle emitted, the mass number decreases by 4, while a beta decay does not change the mass number. In the ${}^{239}_{94}Pu$ to ${}^{207}_{82}Pb$ series, the change in mass number is from 239 to 208 = 32. Hence, a total of eight (32/4) alpha particles must be emitted to balance the change in mass number in this series.

The change in atomic number in this series is from 94 to 82 = 12. This change can be represented as $[(\#\text{ alpha particles})(2)+(\#\text{ beta particles})(-1)] = 12$. Numerically, this yields $[(8)(2)+(x)(-1)] = 12$, where x represents the number of beta particles emitted. Hence, the number of beta particles emitted is 4.

20.24 ***See Section 20.1 and Example 20.6.***

$\left(\dfrac{\text{Current mass }^{238}U}{\text{Original mass }^{238}U}\right)$ can be used in place of $\left(\dfrac{N}{N_0}\right)$ in $\ln\left(\dfrac{N}{N_0}\right) = \dfrac{-0.693t}{t_{1/2}}$ which yields

$$t = \frac{-t_{1/2}}{0.693}\ln\left(\frac{\text{Current mass }^{238}U}{\text{Original mass }^{238}U}\right)$$

$$\text{Original mass }^{238}U = 6.73\text{ mg }^{238}U + \left(3.22\text{ mg }^{206}Pb\right)\left(\frac{238\text{ mg U}}{206\text{ mg Pb}}\right) = 10.45\text{ mg }^{238}U$$

Hence, $t = \dfrac{-4.51 \times 10^9\text{ yr}}{0.693}\ln\left(\dfrac{6.73}{10.45}\right) = \mathbf{2.86 \times 10^9\ yr}$

20.26 ***See Section 20.1.***

One can assume the mass of ${}^{87}Sr$ formed is equal to the mass of ${}^{87}Rb$ reacted, since these species both conatain 87 nuclear particles. Hence, original mass of ${}^{87}Rb$ is given as

$$\text{original mass }^{87}Rb = \left[\text{current mass }^{87}Rb + \text{ current mass }^{87}Sr\right]$$

$$= \left[\text{current mass }^{87}Rb + (0.51)\text{current mass }^{87}Rb\right]$$

$$= 1.051\left(\text{current mass }^{87}Rb\right).$$

Solving $\ln\left(\dfrac{N}{N_0}\right) = \dfrac{-0.693t}{t_{1/2}}$ for t and substituting gives

$$t = \frac{-t_{1/2}}{0.693}\ln\left(\frac{N}{N_0}\right) \qquad t = \frac{-5.0 \times 10^{10}\text{ yr}}{0.693}\ln\left(\frac{1.000}{1.051}\right) = \mathbf{3.6 \times 10^9\ yr}$$

20.28 *See Section 20.1.*

$$\text{percentage } {}^{14}_{6}\text{C atoms} = \frac{\text{number of } {}^{14}_{6}\text{C atoms in 1.00 g C}}{\text{total number of carbon atoms in 1.00 g C}} \times 100\%$$

The number of ${}^{14}_{6}$C atoms in 1.00 g C can be determined from the rate of decay.

Solving rate = kN for N gives $N = \frac{\text{rate}}{k}$, and the value of k can be determined from the half-life for ${}^{14}_{6}$C.

Solving $t_{1/2} = \frac{0.693}{k}$ for k gives $k = \frac{0.693}{t_{1/2}}$ $k = \frac{0.693}{\left(5730 \text{ yr} \times \frac{365 \text{ days}}{\text{yr}} \times \frac{24 \text{ hr}}{\text{day}} \times \frac{60 \text{ min}}{\text{hr}}\right)} = 2.30 \times 10^{-10} \text{ min}^{-1}$

and $N = \frac{15.3 \ {}^{14}_{6}\text{C atoms disintegrating/ min}}{2.30 \times 10^{-10} \text{ min}^{-1}} = 6.65 \times 10^{10} \ {}^{14}_{6}\text{C atoms}$

$$? \text{ total C atoms in 1.00 g C} = 1.00 \text{ g C} \times \frac{1 \text{ mol C}}{12.0 \text{ g C}} \times \frac{6.02 \times 10^{23} \text{ C atoms}}{1 \text{ mol C}} = 5.02 \times 10^{22} \text{ C atoms}$$

$$\text{Hence, percentage } {}^{14}_{6}\text{C atoms} = \frac{6.65 \times 10^{10} \ {}^{14}_{6}\text{C atoms}}{5.02 \times 10^{22} \text{ C atoms}} \times 100\% = \mathbf{1.30 \times 10^{-10}\%}$$

20.30 *See Section 20.2 and Example 20.7.*

(a) ${}^{54}_{26}\text{Fe} + {}^{4}_{2}\text{He} \rightarrow 2\,{}^{1}_{1}\text{H} + \mathbf{{}^{56}_{26}Fe}$ (b) $\mathbf{{}^{27}_{13}Al} + {}^{1}_{0}\text{n} \rightarrow {}^{24}_{11}\text{Na} + {}^{4}_{2}\text{He}$ (c) ${}^{238}_{92}\text{U} + {}^{16}_{8}\text{O} \rightarrow \mathbf{{}^{249}_{100}Fm} + 5\,{}^{1}_{0}\text{n}$

(d) ${}^{96}_{42}\text{Mo} + \mathbf{{}^{2}_{1}H} \rightarrow {}^{97}_{43}\text{Tc} + {}^{1}_{0}\text{n}$ (e) ${}^{250}_{98}\text{Cf} + {}^{11}_{5}\text{B} \rightarrow 5\,{}^{1}_{0}\text{n} + \mathbf{{}^{256}_{103}Lr}$

20.32 *See Section 20.2.*

A bombarding charge particle must be accelerated to a very high energy to overcome the electrostatic repulsion of the protons in the nucleus of the atom that will be hit.

20.34 *See Section 20.3, Table 20.3 and Example 20.8.*

(a) Mass defect of one atom of ${}^{19}_{9}$F:

mass of 9 ${}^{1}_{1}$H: = 9×1.007825 u = 9.070425 u

mass of 10 ${}^{1}_{0}$n: = 10×1.008665 u = <u>10.008665 u</u>

total mass of nucleons of ${}^{19}_{9}$F = 26.214370 u

mass of one atom of ${}^{19}_{9}$F = <u>18.9984 u</u>

Mass defect of one atom of ${}^{19}_{9}$F = 0.1587 u

Mass defect for one mole of ${}^{19}_{9}$F atoms = 0.1587 g

(b) ? total binding energy for 1 atom of ${}^{19}_{9}F$ in MeV $= \frac{0.1587\ u}{1\ atom\ {}^{19}_{9}F} \times \frac{931.5\ MeV}{1\ u} = \mathbf{147.8} \frac{\mathbf{MeV}}{\mathbf{atom\ {}^{19}_{9}F}}$

(c) ? binding energy per nucleon= $\frac{147.8\ MeV}{1\ atom\ {}^{19}_{9}F} \times \frac{1\ atom\ {}^{19}_{9}F}{19\ nucleons} = \mathbf{7.779} \frac{\mathbf{MeV}}{\mathbf{nucleon}}$

20.36 ***See Section 20.3, Table 20.3 and Example 20.8.***

(a) Mass defect of ${}^{26}Al$:

$$
\begin{aligned}
\text{mass of } 13\ {}^{1}_{1}H: &= 13 \times 1.007825\ u &= 13.101725\ u \\
\text{mass of } 13\ {}^{1}_{0}n: &= 13 \times 1.008665\ u &= \underline{13.112645\ u} \\
\text{total mass of nucleons of } {}^{26}Al & &= 26.214370\ u \\
\text{mass of } {}^{26}Al & &= \underline{25.9869\ u} \\
\text{Mass defect of } {}^{26}Al & &= 0.2275\ u
\end{aligned}
$$

Binding energy per nucleon for ${}^{26}Al$:

? binding energy per nucleon= $\frac{0.2775\ u}{26\ nucleons} \times \frac{931.5\ MeV}{1\ u} = 8.151 \frac{MeV}{nucleon}$

Mass defect for ${}^{27}Al$:

$$
\begin{aligned}
\text{mass of } 13\ {}^{1}_{1}H: &= 13 \times 1.007825\ u &= 13.101725\ u \\
\text{mass of } 14\ {}^{1}_{0}n: &= 14 \times 1.008665\ u &= \underline{14.12131\ u} \\
\text{total mass of nucleons of } {}^{27}Al & &= 27.223035\ u \\
\text{mass of } {}^{27}Al & &= \underline{26.9815\ u} \\
\text{Mass defect of } {}^{27}Al & &= 0.2415\ u
\end{aligned}
$$

Binding energy per nucleon for ${}^{27}Al$:

? binding energy per nucleon= $\frac{0.2415\ u}{27\ nucleons} \times \frac{931.5\ MeV}{1\ u} = 8.332 \frac{MeV}{nucleon}$

Mass defect for ${}^{28}Al$:

$$
\begin{aligned}
\text{mass of } 13\ {}^{1}_{1}H: &= 13 \times 1.007825\ u &= 13.101725\ u \\
\text{mass of } 15\ {}^{1}_{0}n: &= 15 \times 1.008665\ u &= \underline{15.129975\ u} \\
\text{total mass of nucleons of } {}^{28}Al & &= 28.231700\ u \\
\text{mass of } {}^{28}Al & &= \underline{27.9819\ u} \\
\text{Mass defect of } {}^{28}Al & &= 0.2498\ u
\end{aligned}
$$

Binding energy per nucleon for ^{28}Al:

$$? \text{ binding energy per nucleon} = \frac{0.2498 \text{ u}}{28 \text{ nucleons}} \times \frac{931.5 \text{ MeV}}{1 \text{ u}} = 8.310 \frac{\text{MeV}}{\text{nucleon}}$$

(b) ^{27}Al has the highest binding energy per nucleon and is therefore likely to be the stable isotope of Al. ^{26}Al and ^{28}Al have lower binding energies per nucleon than ^{27}Al and are therefore likely to be the radioactive nuclides of Al.

20.38 ***See Section 20.4.***

The neutron-to-proton ratios of fission products are too high, so nearly all of them are radioactive, decaying by a series of beta emissions to approach the band of stability shown in Figure 20.1.

20.40 ***See Section 20.3 and 4.***

$$^{1}_{0}n \quad + \quad ^{239}_{94}Pu \quad \rightarrow \quad ^{98}_{40}Zr \quad + \quad ^{139}_{54}Xe \quad + \quad 3\,^{1}_{0}n$$

1.008665 u   239.052 u   97.913 u   138.919 u   3(1.008665 u)

(a) Δm = mass defect for the reaction = mass of reactants - mass of products

$$\Delta m = [(1 \times 1.008665 \text{ u}) + (1 \times 239.052 \text{ u})] - [(1 \times 97.913 \text{ u}) + (1 \times 138.919 \text{ u}) + (3 \times 1.008665 \text{ u})] = \mathbf{0.203 \text{ u}}$$

(b) $$? \text{ J for one fission} = \frac{0.203 \text{ u}}{1 \text{ fission}} \times \frac{931.5 \text{ MeV}}{1 \text{ u}} \times \frac{1.602 \times 10^{-13} \text{ J}}{1 \text{ MeV}} = \mathbf{3.03 \times 10^{-11} \frac{J}{fission}}$$

(c) $$? \text{ J for } 1.00 \text{ g Pu} = 1.00 \text{ g Pu} \times \frac{1 \text{ mol Pu}}{239.05 \text{ g Pu}} \times \frac{6.02 \times 10^{23} \text{ atoms Pu}}{1 \text{ mol Pu}} \times \frac{3.03 \times 10^{-11} \text{ J}}{1 \text{ Pu atom}} = \mathbf{7.63 \times 10^{10} \text{ J}}$$

20.42 ***See Section 20.4.***

(a) $$? \text{ J for 1 kg } ^{235}U = 1.0 \text{ kg } ^{235}U \times \frac{10^3 \text{ g } ^{235}U}{\text{kg } ^{235}U} \times \frac{1 \text{ mol } ^{235}U}{235.0493 \text{ g } ^{235}U} \times \frac{1.9 \times 10^{10} \text{ kJ}}{1 \text{ mol } ^{235}U} = 8.1 \times 10^{10} \text{ kJ}$$

$$? \text{ metric tons coal supplying equal energy} = 8.1 \times 10^{10} \text{ kJ} \times \frac{1.0 \text{ kg coal}}{2.8 \times 10^4 \text{ kJ}} \times \frac{10^3 \text{ g coal}}{1 \text{ kg coal}} \times \frac{1 \text{ metric ton coal}}{10^6 \text{ g coal}}$$

$$= \mathbf{2.9 \times 10^3 \text{ metric tons coal}}$$

(b) $$? \text{ metric tons } SO_2 = 2.9 \times 10^3 \text{ metric tons coal} \times \frac{0.90 \text{ metric ton S}}{100.00 \text{ metric tons coal}} \times \frac{64.0 \text{ metric tons } SO_2}{32.0 \text{ metric tons S}}$$

$$= \mathbf{52 \text{ metric tons } SO_2}$$

20.44 ***See Sections 20.3 and 4.***

$$ {}^{2}_{1}H \quad + \quad {}^{3}_{1}H \quad \rightarrow \quad {}^{4}_{2}He \quad + \quad {}^{1}_{0}n $$

2.0140 u   3.016 u   4.0026 u   1.008665 u

Δm = mass defect for the reaction = mass of reactants - mass of products

$$\Delta m = \left[(1\times 2.0140\ u)+(1\times 3.016\ u)\right]-\left[(1\times 4.0026u\ )+(1\times 1.008665\ u)\right] = \mathbf{0.0188\ u}$$

$$?\ \text{J per gram } {}^{4}\text{He formed} = 1.00\ \text{g}\ {}^{4}\text{He} \times \frac{1\ \text{mol}\ {}^{4}\text{He}}{4.0026\ \text{g}\ {}^{4}\text{He}} \times \frac{6.02\ \text{x}\ 10^{23}\ {}^{4}\text{He atoms}}{1\ \text{mol}\ {}^{4}\text{He}} \times \frac{0.0188\ \text{u}}{1\ \text{atom}\ {}^{4}\text{He}}$$

$$\times \frac{931.5\ \text{MeV}}{1\ \text{u}} \times \frac{1.602\ \text{x}\ 10^{-13}\ \text{J}}{1\ \text{MeV}} = \mathbf{4.22\ x\ 10^{11}\ J}$$

The energy released with the formation of one gram of ${}^{4}He$ via fusion of ${}^{2}_{1}H$ and ${}^{3}_{1}H$ is 5.3 times greater than the energy released by fission of one gram of ${}^{235}U$ (4.22 x 10^{8} kJ vs. 8 x 10^{7} kJ).

20.48 ***See Sections 6.1 and 20.3.***

$$ {}^{1}_{0}n \quad + \quad {}^{113}_{48}Cd \quad \rightarrow \quad {}^{114}_{48}Cd \quad + \quad \gamma $$

1.008665 u   112.9044 u   113.9034 u

(a) Δm = mass defect for the reaction = mass of reactants - mass of products

$$\Delta m = \left[(1\times 1.008665\ u)+(1\times 112.9044\ u)\right]-\left[(1\times 113.9034\ u)\right] = \mathbf{9.7\ x\ 10^{-3}\ u}$$

$$?\ \text{energy of}\ \gamma\ \text{ray in MeV} = 9.7\ \text{x}\ 10^{-3}\ \text{u} \times \frac{931.5\ \text{MeV}}{1\ \text{u}} = \mathbf{9.0\ MeV}$$

(b) Solving $\Delta E = \frac{hc}{\lambda}$ for λ gives $\lambda = \frac{hc}{\Delta E}$. Substituting in the energy of the gamma ray yields

$$\lambda = \frac{(6.63\ \text{x}\ 10^{-34}\ \text{J}\cdot\text{s})(3.00\ \text{x}\ 10^{8}\ \text{m}\cdot\text{s}^{-1})}{\left(9.0\ \text{MeV} \times \frac{1.602\ \text{x}\ 10^{-13}\ \text{J}}{1\ \text{MeV}}\right)} = \mathbf{1.4\ x\ 10^{-13}\ m}$$

Chapter 21: Organic and Biochemistry

21.4 *See Section 21.1.*

Noncyclic alkanes have the general formula C_nH_{2n+2}. Hence, only (b) C_6H_{14} and (d) C_9H_{20} are alkanes.

21.6 *See Section 21.1, Example 21.1 and Table 21.1.*

The structural formula for the straight chain isomer of C_5H_{12} is:

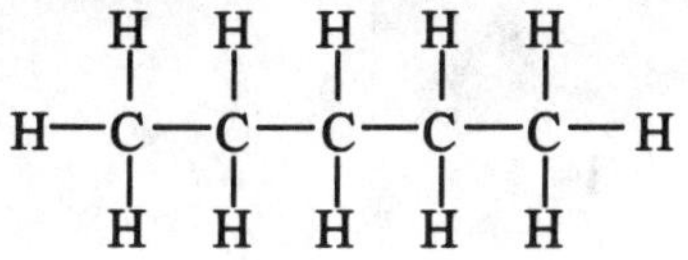

```
    H   H   H   H   H
    |   |   |   |   |
H---C---C---C---C---C---H
    |   |   |   |   |
    H   H   H   H   H
```

The name of the alkyl group that is formed by removing a hydrogen atom from one of the terminal carbon atoms is **n-pentyl**.

21.8 *See Section 21.1 and Example 21.2.*

Any three of the following are acceptable:

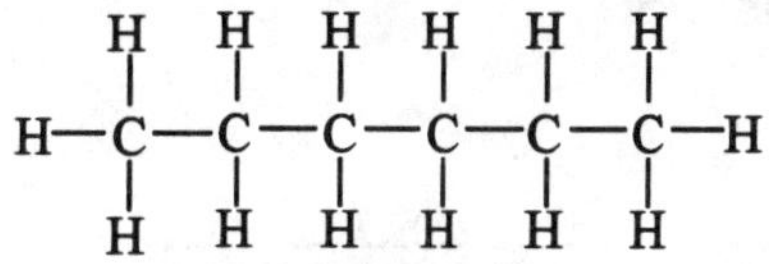

```
    H   H   H   H   H   H
    |   |   |   |   |   |
H---C---C---C---C---C---C---H
    |   |   |   |   |   |
    H   H   H   H   H   H
```

n-hexane

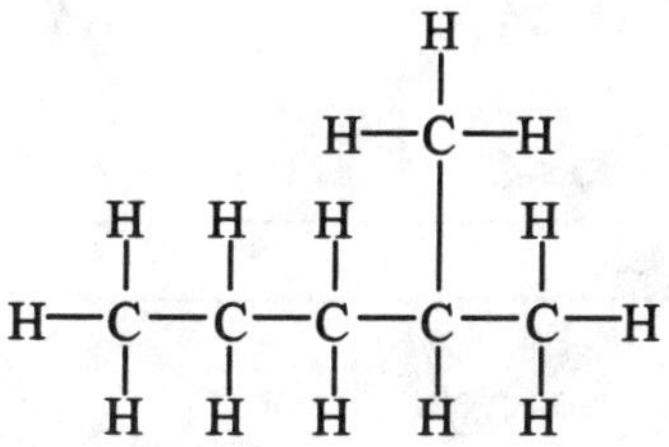

```
                H
                |
            H---C---H
    H   H   H   |   H
    |   |   |   |   |
H---C---C---C---C---C---H
    |   |   |   |   |
    H   H   H   H   H
```

2-methylpentane

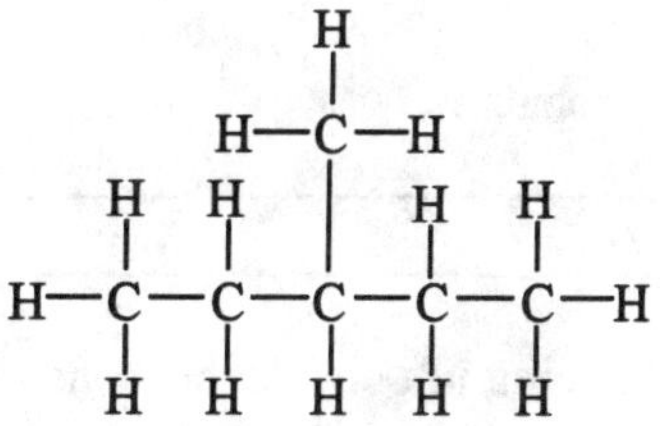

```
            H
            |
        H---C---H
    H   H   |   H   H
    |   |   |   |   |
H---C---C---C---C---C---H
    |   |   |   |   |
    H   H   H   H   H
```

3-methylpentane

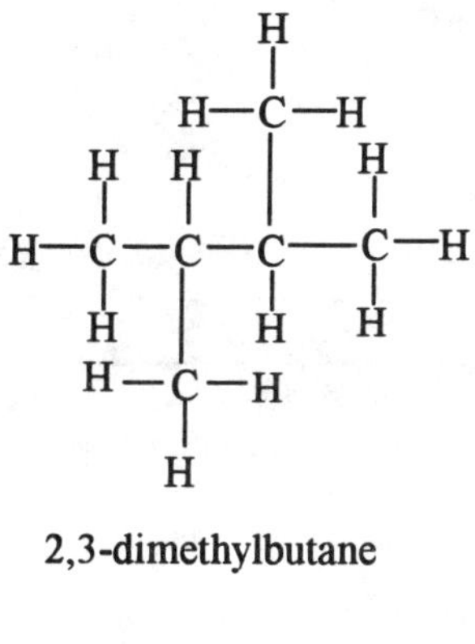

```
            H
            |
        H---C---H
    H   H   |       H
    |   |   |       |
H---C---C---C-------C---H
    |   |   |       |
    H   |   H       H
    H---C---H
        |
        H
```

2,3-dimethylbutane

```
            H
            |
        H---C---H
    H   H   |   H
    |   |   |   |
H---C---C---C---C---H
    |   |   |   |
    H   H   |   H
        H---C---H
            |
            H
```

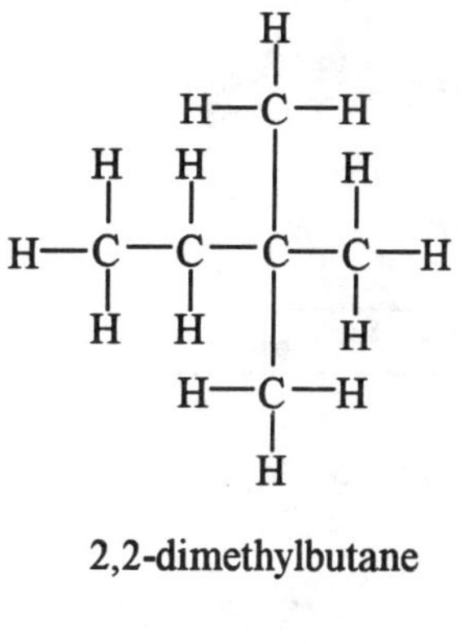

2,2-dimethylbutane

21.10 *See Section 21.1 and Example 21.2.*

(a) 2-methylhexane

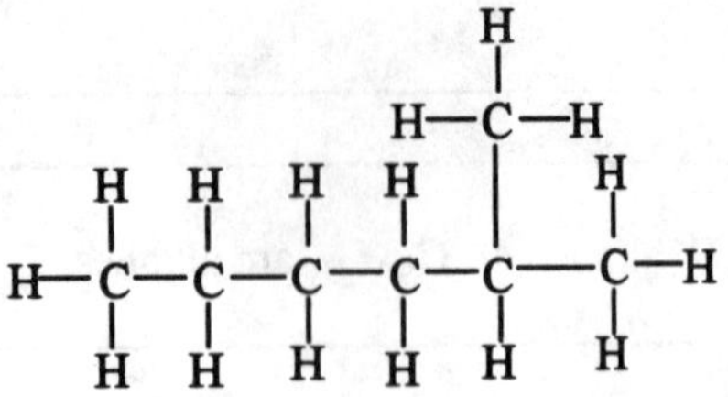

(b) 3,3-dichloroheptane

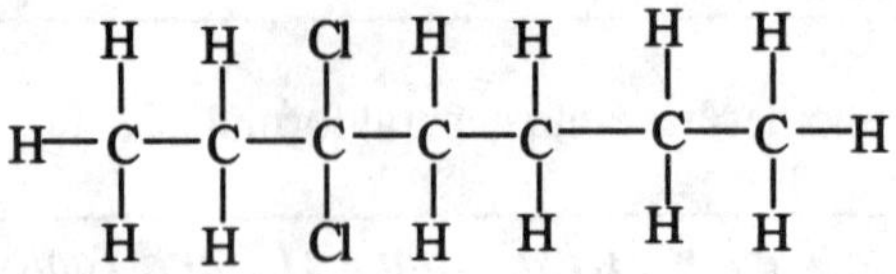

(c) 2-methyl-3-phenyloctane

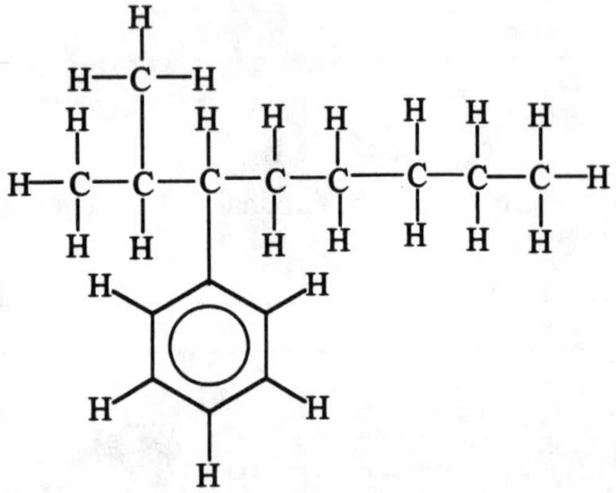

(d) 1,1-diethylcyclohexane

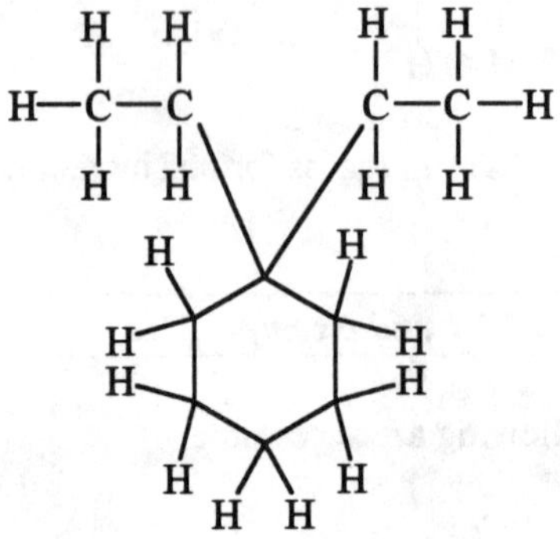

21.12 *See Section 21.1 and Examples 21.2 and 3.*

(a) 1-fluoro-2-methylpentane
(b) 3-methylhexane
(c) 2,2-dimethylbutane
(d) 1-chloro-2-ethylcyclobutane

21.16 *See Section 21.2.*

Noncyclic alkenes containing just one double bond have the general formula $\mathbf{C_nH_{2n}}$. Noncyclic alkynes containing just one triple bond have the general formaula $\mathbf{C_nH_{2n-2}}$.

21.18 *See Section 21.2 and Example 21.4.*

(a) *trans*-1-bromopropene
(b) *cis*-3-heptene
(c) 5-fluoro-2-pentyne
(d) *cis*-1-chloro-3-hexene

21.20 *See Section 21.2 and Example 21.4.*

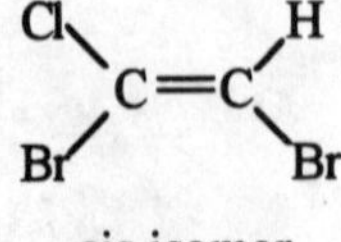

cis isomer

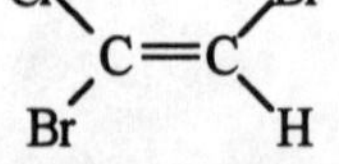

trans isomer

21.22 *See Section 21.2.*

The structural isomers for the substituted alkene C_4H_7Cl with the chlorine atom attached to a double-bonded carbon atom are:

$(Cl)(H_3C)C{=}C(CH_3)(H)$ $(Cl)(H_3C)C{=}C(H)(CH_3)$ $(Cl)(H)C{=}C(CH_3)(CH_3)$

$(Cl)(H)C{=}C(CH_2CH_3)(H)$ $(Cl)(H)C{=}C(H)(CH_2CH_3)$ $(Cl)(H_3CH_2C)C{=}C(H)(H)$

21.24 *See Section 21.3.*

(a) 2,3-dimethyl-1-pentene

$(H)(H)C{=}C(CH_3)(CHCH_2CH_3)$, with CH_3 on the CH

(b) 1-methylcyclohexene

Cyclohexene ring with CH_3 on a double-bonded carbon

(c) *cis*-1-chloro-2-butene

$(H)(H_3C)C{=}C(H)(CH_2Cl)$

(d) 4-methyl-1-hexene

$(H)(H)C{=}C(H)(CH_2CHCH_2CH_3)$, with CH_3 on the CH

21.26 *See Section 21.2.*

(a) $CH_2{=}CHCH_3 + H_2 \rightarrow CH_3CH_2CH_3$

(b) $CH{\equiv}CCH_2CH_3 + 2Cl_2 \rightarrow CHCl_2CCl_2CH_2CH_3$

(c) $C_6H_6 + Cl_2 \xrightarrow{FeCl_3} C_6H_5Cl + HCl$

21.28 *See Section 21.3 and Table 21.3.*

(a) **carboxylic acid**

$-C(=O)-OH$

(b) **aldehyde**

$-C(=O)-H$

(c) **ether**

C—O—C

21.30 ***See Section 21.3.***

(a) 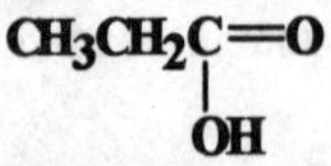 $CH_3CH_2C(=O)OH$ has O-H bonds and hydrogen bonding betweeen molecules. Hence, it has a higher boiling point than $CH_3CH_2OCH_2CH_3$.

(b) $\mathbf{CH_3CH_2CH_2CH_2NH_2}$ has N-H bonds and hydrogen bonding between molecules. Hence, it has a higher boiling point than $CH_3CH_2C(=O)—CH_3$.

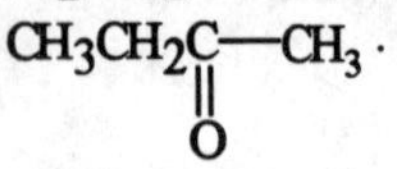

(c) $\mathbf{CH_3CH_2C{\equiv}CF}$ has a triple bond and is larger than $CH_2{=}CH_2$. It therefore has a more polarizable electron cloud than $CH_2{=}CH_2$ and has higher London dispersion forces between molecules. Hence, it has a higher boiling point than $CH_2{=}CH_2$.

21.32 ***See Section 21.3.***

$$CH_3OH + HOCH_3 \xrightarrow{H+} H_2O + CH_3OCH_3$$ dimethylether

21.34 ***See Section 21.3.***

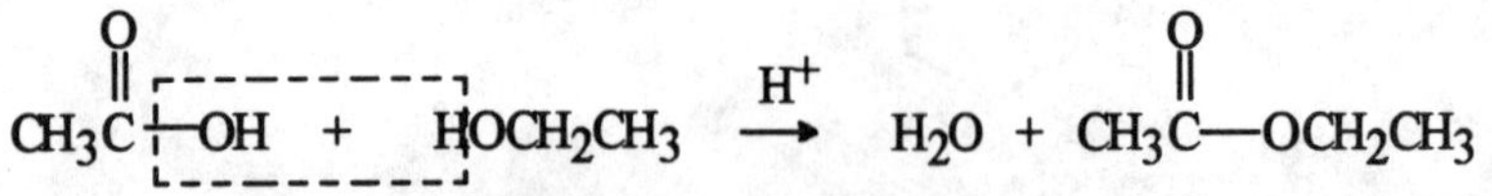

$$CH_3C(=O)—OH + HOCH_2CH_3 \xrightarrow{H^+} H_2O + CH_3C(=O)—OCH_2CH_3$$ ethyl acetate (ethyl ethanoate)

21.36 ***See Section 21.3 and Example 21.7.***

(a) butanol

$CH_3CH_2CH_2CH_2OH$

(b) 3-methyl-2-pentanone

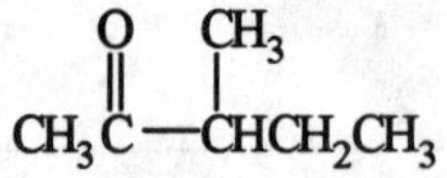

$CH_3C(=O)—CH(CH_3)CH_2CH_3$

(c) methyl acetate

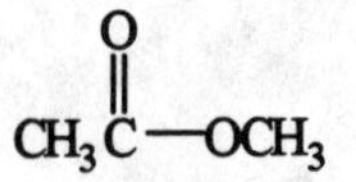

$CH_3C(=O)—OCH_3$

(d) ethyl phenyl amine

$C_6H_5—N(H)CH_2CH_3$

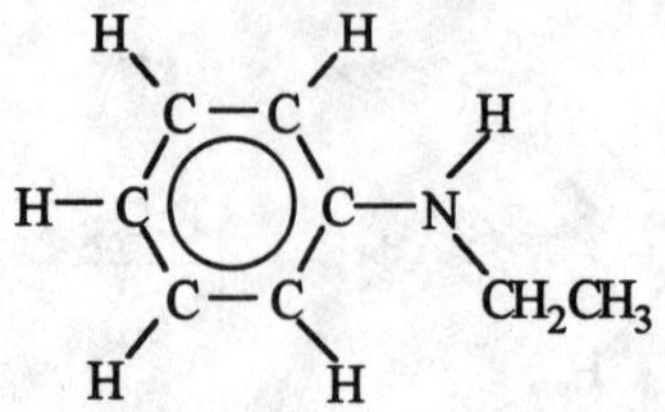

21.38 ***See Section 21.3 Example 21.7.***

(a) 3-fluoro-1-propanol
(b) propanoic acid
(c) isopropyl acetate
(d) propyl amine

21.40 ***See Section 21.3.***

A carbon atom that is bonded to four different substituents is chiral. Only (c) contains a chiral carbon atom, the central carbon atom.

21.44 *See Section 21.4.*

Phenylethylene (styrene) is $C(C_6H_5)H{=}CH_2$. The repeating unit of polystyrene is:

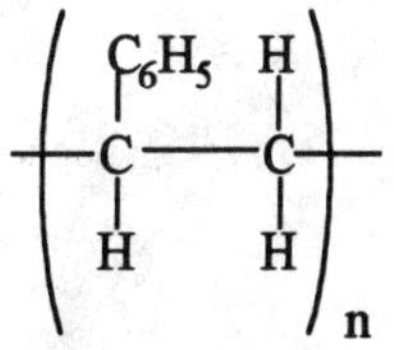

$$\left(\begin{array}{cc} C_6H_5 & H \\ | & | \\ -C- & -C- \\ | & | \\ H & H \end{array}\right)_n$$

21.46 *See Section 21.4.*

1,1-difluorethylene is $CF_2{=}CH_2$, and hexafluoropropene is $CF_2{=}CFCF_3$. The repeating unit of viton is:

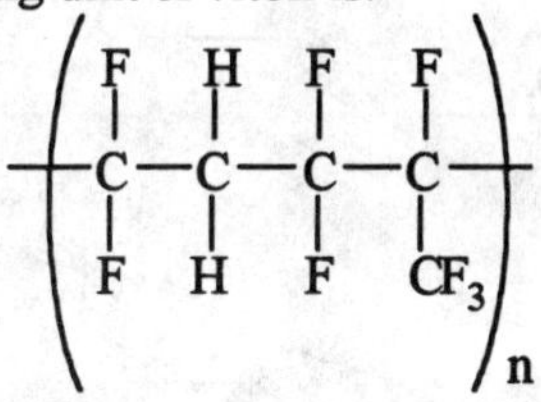

$$\left(\begin{array}{cccc} F & H & F & F \\ | & | & | & | \\ -C- & -C- & -C- & -C- \\ | & | & | & | \\ F & H & F & CF_3 \end{array}\right)_n$$

21.50 *See Section 21.4.*

$$\left(-O-C_6H_4-C(CH_3)_2-C_6H_4-O-\overset{O}{\overset{\|}{C}}-\right)_n$$

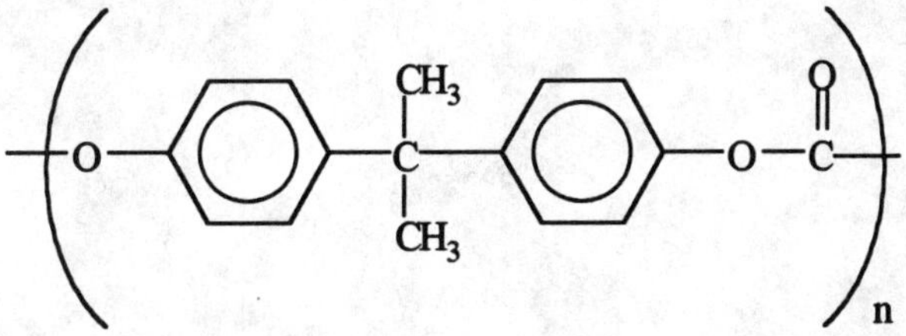

21.66 *See Section 21.3.*

(a) cis-2-bromo-3-hexene

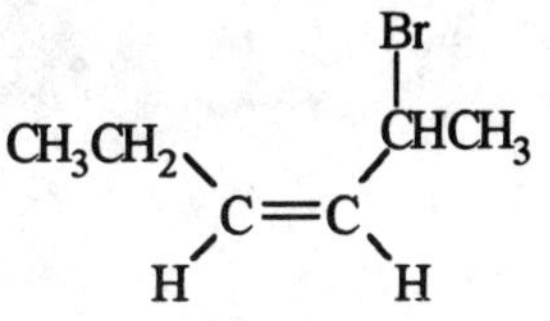

$CH_3CH_2(H)C{=}C(H)CH(Br)CH_3$

(b) 2-nitrophenol

$C_6H_4(OH)(NO_2)$

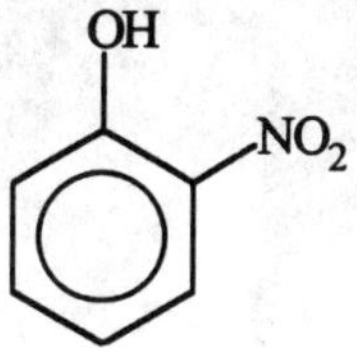